COMPUTER-AIDED DESIGN
AND MANUFACTURE

COMPUTER-AIDED DESIGN AND MANUFACTURE

3rd Edition

C. B. BESANT, BsC(Eng), PhD, DIC
Reader in Mechanical Engineering,
Imperial College of Science & Technology
University of London

C. W. K. LUI, BsC(Eng), PhD, DIC
Senior Assistant in CAD/CAM Group
Imperial College of Science & Technology
University of London

ELLIS HORWOOD LIMITED
Publishers · Chichester

Halsted Press: a division of
JOHN WILEY & SONS
New York · Chichester · Brisbane · Toronto

First published in 1986 by
ELLIS HORWOOD LIMITED
Market Cross House, Cooper Street, Chichester, West Sussex, PO19 1EB,
England

The publisher's colophon is reproduced from James Gillison's drawing of the ancient Market Cross, Chichester.

Distributors:
Australia and New Zealand:
Jacaranda-Wiley Ltd., Jacaranda Press,
JOHN WILEY & SONS INC.
GPO Box 859, Brisbane, Queensland 4001, Australia

Canada:
JOHN WILEY & SONS CANADA LIMITED
22 Worcester Road, Rexdale, Ontario, Canada

Europe and Africa:
JOHN WILEY & SONS LIMITED
Baffins Lane, Chichester, West Sussex, England

North and South America and the rest of the world:
Halsted Press: a division of
JOHN WILEY & SONS
605 Third Avenue, New York, NY 10158, USA

©1986 C.B. Besant and C.W.K. Lui/Ellis Horwood Limited

British Library Cataloguing in Publication Data
Besant, C.B.
Computer-aided design and manufacture. — 3rd ed. —
(Ellis Horwood Series in mechanical engineering)
1. Engineering design — Data processing
I. Title II. Lui, C.W.K.
620'.00425'02854   TA174

ISBN 0–85312–909–6 (Ellis Horwood Limited – Library Edn.)
ISBN 0–85312–952–5 (Ellis Horwood Limited — Student Edn.)
ISBN 0–470–20180–0 (Halsted Press — Library Edn.)
ISBN 0–470–20242–4 (Halsted Press — Student Edn.)

Typeset by Ellis Horwood Limited
Printed in Great Britain by R.J. Acford, Chichester

Table of Contents

Preface 11

Chapter 1 — Introduction
    1.1   Historical background 13
    1.2   Concepts of integrated CAD/CAM 15
    1.3   Benefits of CAD/CAM 21
    1.4   Applications of CAD/CAM 22

Chapter 2 — Computer Systems
    2.1   Introduction 25
    2.2   Central processing unit 26
        2.2.1   Multiple CPUs 28
        2.2.2   Categories of computer 30
    2.3   Mass storage devices 31
        2.3.1   Magnetic tape 32
        2.3.2   Magnetic disk 33
        2.3.3   Magnetic drum 35
        2.3.4   Other storage devices 36
    2.4   Input/output devices 36
        2.4.1   Card reader/punch 36
        2.4.2   Paper reader/punch 37
        2.4.3   Terminals 38
        2.4.4   Printers 39
        2.4.5   Input/output devices 40

2.5   Data representation. 42
2.6   Programming languages. 44
2.7   Operating system 46
2.8   System configurations. 48

Chapter 3 — Computer-aided Design System Hardware
3.1   Introduction 53
3.2   Graphics input devices 53
   3.2.1   Light-pens 54
   3.2.2   Analogue devices. 54
   3.2.3   Keyboard devices. 57
3.3   Graphics display devices 59
   3.3.1   CRT displays 59
   3.3.2   Plasma panel displays. 66
3.4   Graphics output devices 67
   3.4.1   Pen plotters. 68
   3.4.2   Electrostatic plotters 70
   3.4.3   Other graphics output devices 72
   3.4.4   Modes of operation. 72
3.5   CAD system configuration 74

Chapter 4 — Computer-aided design system software
4.1   Introduction 79
4.2   Operating system 79
4.3   Graphics system. 80
   4.3.1   The overlay system 81
   4.3.2   Graphics database structure and handling 86
   4.3.3   Operating features 92
   4.3.4   Symbols 106
   4.3.5   Macros 108
   4.3.6   Editing facility 110
   4.3.7   Data selection 112
   4.3.8   Graphic transformation. 114
   4.3.9   Plotting 116
4.4   Graphics standards 117
   4.4.1   GKS and CORE .118
   4.4.2   GKS-3D and PHIGS 119
   4.4.3   IGES 119
   4.4.4   Other graphics standards 121

Chapter 5 — Transformation systems
5.1   Display 125
5.2   Windowing and Clipping 126
5.3   Two-dimensional transformations 128
5.4   Three-dimensional transformations. 130
5.5   Linear transformations .132
5.6   Display files for three-dimensional data. 134
5.7   Visualisation of three-dimensional data. 135
5.8   Eye coordinates system .139
5.9   Joystick function 144
   5.9.1   Distortion 146

Chapter 6 — Geometric modelling
6.1   Introduction 152
6.2   Dimensions of models 152
6.3   Types of models. 154
6.4   Construction of solid models 157

Chapter 7 — Draughting
7.1   Introduction 166
7.2   Annotation 167
   7.2.1   Arrows and pointers 167
   7.2.2   Dimensioning .168
   7.2.3   Text 168
7.3   Cross-hatching 171
7.4   Draughting examples 172

**Chapter 8 — Application of CAD techniques to finite element data
         Preparation**
8.1   Automatic mesh generation. 181
8.2   The finite element method 181
8.3   A general finite element mesh generating system GFEMGS 183
8.4   An application of GFEMGS 185
8.5   Results presentation 189
8.6   Three-dimensional shape description and mesh generation. 196
8.7   Input methods and the macroblocks 196
8.8   Formulation of the mesh 202
8.9   Using the 3-D system 205

Chapter 9 — Computer-aided manufacture
9.1   Existing CAM systems — the APT system 215
   9.1.1   APT symbols and words 217
   9.1.2   Geometric definitions in APT 218
   9.1.3   Motion statements 220
   9.1.4   Additional APT statements 224
   9.1.5   Conclusions 225
9.2   CAD/CAM systems. 228
9.3   The use of microcomoputers in CAD/CAM systems 232
9.4   A CNC system with integrated interactive graphics. 233
   9.4.1   The hardware. .233
   9.4.2   The software 236
   9.4.3   The front panel .236
   9.4.4   NC machining 236
   9.4.5   NC display 237
   9.4.6   Diagnostics 237
   9.4.7   CAPPS programming. 237
   9.4.8   Geometry definition 237
   9.4.9   Toolpath generation 237
   9.4.10 Tool library. 238
   9.4.11 Material library. 238
   9.4.12 Files 238

Chapter 10 — The use of microcomputers in CAD/CAM systems
10.1   Microcomputer systems. 240
10.2   CAD/CAM systems based on microcomputers 242
10.3   How to choose the microcomputer system 245
    10.3.1   The microprocessor. 245
    10.3.2   Primary memory 246
    10.3.3   Back-up storage .247
    10.3.4   Communications system 249
    10.3.5   Software for CAD/CAM systems. 250

Chapter 11 — A microcomputer-based CAD/CAM system
11.1   The hardware for the CAD/CAM work station. 252
11.2   Software development aids .253
11.3   The software for the work station. 254
    11.3.1   The database 254
    11.3.2   The creator/editor program. 257
    11.3.3   The display programs. 260
    11.3.4   Cutter path derivation and simulation 260
11.4   The machine tool control system 263
11.5   Using the CAD/CAM system 266

**Chapter 12 — The application of CAD/CAM techniques to the
                design and manufacture of complex shapes**
12.1   Introduction 272
12.2   The CAD/CAM system. 274
12.3   System software. 276
    12.3.1   CAD software 276
    12.3.2   CAM software 278
    12.3.3   Summary of CAD/CAM system operation 285

Chapter 13 — Industrial robots
13.1   Basic robot elements 297
13.2   Types of industrial robot .298
13.3   The robot controller 300
13.4   Assembly robots 304
    13.4.1   Structure of assembly robot systems 304
    13.4.2   Mechanical structure of robots 306
    13.4.3   Axis control 306
    13.4.4   Control of the end effector (tool). 306
    13.4.5   Sensor systems and processing of data 308
13.5   The robot controller 308
    13.5.1   Trajectory planner .309
    13.5.2   Coordinate transformation 309
    13.5.3   Dynamic world processing 310
13.6   The structure of the robot controller 310

Chapter 14 — CAD and the programming of robots
14.1   Problems associated with robot programming 314
14.2   Robot-programming systems 314

14.3   Automatic robot-programming 318
   14.3.1   Computing safe positions 319
   14.3.2   Dealing with uncertainty 320
14.4   Conclusions 321

Chapter 15 — Automated guided vehicles
15.1   Vehicles guided by off-board fixed paths 324
   15.1.1   Wire-guided AGVs 324
   15.1.2   Painted-line guided AGVs 328
15.2   Vehicles guided by on-board, software-programmable paths 328
   15.2.1   Various guidance techniques for 'free-ranging' vehicles 328
15.3   The free-ranging AGV 330
   15.3.1   Free-range AGV concept .330
   15.3.2   The vehicle 338
   15.3.3   Position determination via fixed beacons in the
             surroundings 338
15.4   The supervisory system 338
   15.4.1   Supervisory fleet control 339

Chapter 16 — Flexible manufacturing systems
16.1   FMS applications 343
16.2   Equipment used in FMS 344
16.3   Materials handling 345
16.4   Economies of FMS technology .345

Chapter 17 — Process planning
17.1   Part families, classification and coding 348
17.2   Common classification and coding 348
17.3   Planning functions 350
17.4   Retrieval-type planning system 350
17.5   Generative process planning systems 352
17.6   Machinability information 354
17.7   Process optimization 354
17.8   Machinability data system 356
17.9   Computerised machinability data systems 357
   17.9.1   Database systems 357
   17.9.2   Mathematical modelling systems 358
17.10   Machinability data collection system 358
   17.10.1   Adaptive optimization 359

Chapter 18 — Factory management
18.1   Introduction 364
18.2   Computer-assisted factory management 365
18.3   Material requirements planning (MRP) 368
18.4   Manufacturing resource planning (MRP II) 374

Chapter 19 — Implementation of CAD/CAM System
19.1   Introduction 376
19.2   Setting up a committee 377

19.3   Justifications 378
19.4   Specification and selection 382
19.5   Installation 390
19.6   CAD/CAM personnel 392

Chapter 20 — Implications of CAD/CAM for industry
20.1   CAD/CAM and productivity 400
20.2   NC and computer graphics 401
20.3   The planning of manufacturing processes 402
20.4   Materials handling and control 403
20.5   Inspection and quality control 403
20.6   Selective assembly 404
20.7   CAD/CAM and the future 404
20.8   The trade union view 405
20.9   The employers view 406

Index 408

Preface

Computers are now beginning to be used extensively in the engineering industry for both design (CAD) and manufacture (CAM). Much of the industrial design is performed within the drawing office by design and production draughtspersons and engineers. Many organisations have introduced CAD techniques into the drawing office in the form of automatic draughting systems or computer graphics systems. Also computer graphics is also playing a significant role in aiding manufacture. Furthermore, CAD and CAM techniques are now being taught in a large number of universities as part of engineering or computer science courses. At present there are few books on CAD/CAM which meet the requirements of designer draughtspersons or engineering undergraduates who have had little computing experience.

This book aims to introduce the subject of computing as an aid to design and manufacture, and to take the reader through from the basics of computers to their application in real engineering design and manufacture. It provides a description of both the hardware and software of CAD/CAM systems, together with a practical discussion of their use in engineering. A chapter is devoted to the specialised application of computer graphics to draughting and this is illustrated with practical examples. Design is very much the linking of graphics with engineering analysis and this is demonstrated in a chapter on the application of CAD techniques to stressing engineering designs.

Manufacture and the subject of using computers in manufacture is receiving particular prominence as companies seek to improve product quality, increase productivity and flexibility and to reduce inventory costs. Therefore emphasis has been attributed to the subject of CAD and its link with CAM. Chapters are devoted to the use of computers in machine tools, robots and automated guided vehicles. Each of these topics are considered in some depth with the aim of showing the reader how such systems are programmed and used together with the advantages that they bring to manufacturing industry. A chapter is also devoted to complete systems of computer control machines consisting of machine tools, robots and automated guided vehicles which make up manufacturing systems known as flexible manufacturing systems (FMS).

Particular emphasis is given to the reader who has very little knowledge of computing and for those who become sufficiently interested to write their own software whether it be for graphics, robots or other systems used in manufacture. The aim has been to give the reader some ideas from which new work might also grow.

The book is in part based on the work of the CAD/CAM Section at Imperial College and we are indebted to all who have worked with us in this field over the past five years and in particular to the hard work of research students who have contributed so much to this field. Thanks are also due to the many organisations who have sponsored CAD/CAM research and whose staff contributed so many ideas.

Finally thanks are also due to Miss Nesta Coxeter for patiently typing the manuscript.

1

Introduction

1.1 HISTORICAL BACKGROUND

In the nineteenth century, the industrial revolution considerably enhanced man's physical power. In the present century, a second industrial revolution is taking place, with computers offering an enhancement of man's mental capabilities. It is quite unthinkable these days to undertake a major engineering project without the use of a computer. Since the late 1950s, the applications of computers and computing techniques in engineering disciplines of all types have increased dramatically because computers have become larger in memory capacity and faster in processing speed which mean that more complex problems can be tackled and that more calculations can be performed in a given time. More importantly, with the advent of microelectronics such as Very Large Scale Integration (VLSI) Technology, computer hardware is gradually becoming cheaper and cheaper every day and it is now within the financial reach of most industrial companies that wish to take advantage of its capabilities. Also due to VLSI, computer hardware is getting smaller in size. Because of this size reduction, its applications are increasingly being spread to other areas of industry which formerly could not use the traditional computers owing to their cumbersome size. As a result of these developments in computer science, computer-aided design and computer-aided manufacture (CAD/CAM) was conceived and is rapidly gaining acceptance in engineering industries for their ability to create major increases in productivity.

CAD/CAM can be very simply defined, as its name implies, as the use of computers to aid the design and manufacture process. To explain more clearly, CAD/CAM is concerned with the application of computers to the manufacture

of engineering components starting from the drawing office, to the production
department, to the machine and assembly shops, to the quality control department, right through to the finished parts store. The technology of CAD/CAM
represents an efficient, accurate and consistent method to design and manufacture high-quality products. CAD/CAM involves two separate disciplines called
Computer-Aided Design (CAD) and Computer-Aided Manufacture (CAM) that
were originally developed independently over the past thirty years. They are now
being combined into integrated CAD/CAM systems, with which a design can be
developed and the manufacturing process can be monitored and controlled from
start to finish with a single system.

Computer-aided design is essentially based on a versatile and powerful
technique called computer graphics which basically means the creation and
manipulation of pictures on a display device with the aid of a computer.
Computer graphics originated at the Massachusetts Institute of Technology
(MIT) in 1950 when the first computer-driven display, linked to a Whirlwind 1
computer, was used to generate some simple pictures. The first important step
forward in computer graphics came in 1963 when a system called SKETCHPAD
was demonstrated at the Lincoln Laboratory of MIT. This system consists of
a cathode ray tube (CRT) driven by a TX2 computer. The CRT had a keyboard
and a light-pen. Pictures cauld be drawn on the screen and then manipulated
interactively by the user *via* the light-pen. This demonstration clearly showed
that the CRT could potentially be used as a designer's electronic drawing board
with common graphic operations such as scaling, translation, rotation, animation
and simulation automatically performed at the 'push of a button'. At that time,
these systems were very expensive, therefore, they were adopted only in such
major industries as the aircraft and automotive industries where their use in
design justified the high capital costs. Another crucial factor preventing
computer graphics from being generally applied to engineering industries was
that there was a lack of appropriate graphics and application software to run on
these systems. However, a computer-based design system was clearly emerging.
Since these pioneering developments in computer graphics, which had captured
the imagination of the engineering industry all over the world, new and improved
hardware, which is faster in processing speed, larger in memory, cheaper in
cost and smaller in size, have become widely available. Sophisticated software
techniques and packages have also gradually been developed. Consequently, the
application of CAD in industry has been growing rapidly. Initially, CAD systems
primarily were automated draughting stations in which computer-controlled
plotters produced engineering drawings. The systems were later linked to graphic
display terminals where geometric models describing part dimensions were
created, and the resulting database in the computer was then used to produce
drawings. Nowadays, CAD systems can do much more than mere draughting.
Some systems have analytical capabilities that allow parts to be evaluated with
techniques such as the finite element method. There are also kinematic analysis
programs that enable the motion of mechanisms to be studied. In addition, CAD
systems include testing techniques to perform modal analysis on structures,
and to evaluate their response to pinpoint any possible defects.

Computer-aided manufacture started originally from numerical control (NC) which was pioneered by a man called John T. Parsons who invented a way of controlling a machine tool in the late 1940s. His method involved using punched cards to contain coordinate data that directed the machine to move in small steps to various positions that defined the desired surface to be machined. Positional movements of the machine tool were found to be improved by specifying numerical values for the movements of each axis of the machine tool. Subsequently, the US Air Force became seriously interested in this new method and sponsored further research on this concept at MIT. In 1952, the first NC prototype was demonstrated to show the potential benefit of NC. Then, machine tool makers and manufacturing industry put in their efforts and resources to develop different NC machines to meet their own particular requirements.

During the late 1950s, computers had become available and it had been recognised that they could be used to generate numerical values for NC machines. The US Air Force once again continued its interest in NC development by providing more funds for additional research at MIT to design a computer part programming language that could be used to describe geometric tool movements for NC machines. As a result of this work, APT, which stands for Automatically Programmed Tools, has become a universally known standard NC part programming language. APT provides a means by which the part programmer can communicate the machining instructions to the machine tool. There were three versions of APT developed over the years. APT I was never completed and APT II was developed at MIT. APT III was developed for use in the US aerospace industry by the Illinois Institute of Technology Research Institute (IITRI). APT is a composite language with a vocabulary of over 300 'English-like' words that can be used for the machining of three-dimensional curved components such as a sphere, cylinder, paraboloid and lofted surfaces. With APT, the part programmer can define tool shapes, tolerances, geometric definitions, tool motions and auxiliary machine commands. Since its beginning APT has been enhanced and it is still commonly used in industry today. However, owing to new developments in NC machines, there are now several custom-built part programming languages based on APT concepts such as NELPT in the UK, EXAPT and AUTOPIT in Germany, ADAPT, AUTOSPOT, FMILL, APTLFT, PMT2 and FACUT's FAPT.

As described above, the initial developments of CAM were being made mainly in numerical control. Until recently, NC instructions had still to be produced and verified by hand. Now, CAM systems can automatically generate NC programs, and simulate tool paths quickly on graphics display for verification. In addition, most systems also have full process planning features for determining sequences of fabrication steps, and they may even have factory management capabilities for directing the flow of work and materials through the factory. The latest feature incorporated into CAM is robotics in which automated manipulator arms handle workpieces and tools.

1.2 CONCEPTS OF INTEGRATED CAD/CAM

Computer-aided design (CAD) is a technique in which man and machine are

blended into a problem-solving team, initmately coupling the best characteristics of each. The result of this combination works better than either man
or machine would work alone, and by using a multi-disciplinary approach it
offers the advantage of integrated team-work.

CAD implies by definition that the computer is not used when the designer
is most effective, and vice versa. This being so, it is therefore useful to examine
some individual characteristics of man and computer in order to identify which
processes can best be separately perfomed by each, and where one can aid the
other. Table 1.1 compares the capabilities of man and computer for a range of
tasks. It can be seen that in most cases the two are complementary, that for
some tasks man is far superior to the computer, and that in others the computer

Table 1.1
Characteristics of man and computer

	Man	*Computer*
1. Method of logic and reasoning	Intuitive by experience, imagination, and judgement	Systematic and stylised
2. Level of intelligence	Learns rapidly but sequential. Unreliable intelligence	Little learning capability but reliable level of intelligence
3. Method of information input	Large amounts of input at one time by sight or hearing	Sequential stylised input
4. Method of information input	Slow sequential output by speech or manual actions	Rapid stylised sequential output by the equivalent of manual actions
5. Organisation of information	Informal and intuitive	Formal and detailed
6. Effort involved in organising information	Small	Large
7. Storage of detailed information	Small capacity, highly time dependent	Large capacity, time independent
8. Tolerance for repetitious and mundane work	Poor	Excellent
9. Ability to extract significant information	Good	Poor

Table 1.1 (*Contd.*)

	Man	*Computer*
10. Production of errors	Frequent	Rare
11. Tolerance for erroneous information	Good intuitive correction of errors	Highly intolerant
12. Method of error detection	Intuitive	Systematic
13. Method of editing information	Easy and instantaneous	Difficult and involved
14. Analysis capabilities	Good intuitive analysis, poor numerical analysis ability	No intuitive analysis, good numerical analysis ability

excels. It is, therefore, the marriage of the characteristics of each which is so important in CAD. The characteristics affect the design of a CAD system in the following areas:

(a) DESIGN CONSTRUCTION LOGIC – the method of constructing the design.
(b) INFORMATION HANDLING – the storing and communication of design information.
(c) MODIFICATION – the handling of errors and design changes.
(d) ANALYSIS – the examination of the design and factors influencing it.

Let us now consider these four areas which are of such significance in CAD.

(a) Design construction logic

The use of experience combined with judgement is a necessary ingredient of the design process. The design construction must therefore be controlled by the designer. This means that the designer must have the flexibility to work on various parts of the design at any time and in any sequence, and be able to follow his own intuitive design logic rather than a stylised computer logic.

The computer cannot cope with any significant learning. This must be left to the designer, who can learn from past designs. The computer can, however, provide rapid recall of old designs for reference. Thus, in some ways the designer can pass on his experience to the computer, and other designers can then have access to it.

(b) Information handling

Information is required from the specification before the design solution stage can proceed. Similarly, when the design solution is complete, information must in turn be output to enable the design to be manufactured. Figure 1.1 shows the

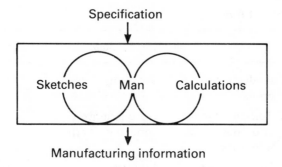

Fig. 1.1 – The conventional design process.

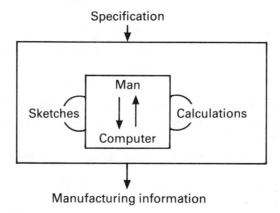

Fig. 1.2 – The design process using CAD techniques.

application of this process to manual design. Information is assimilated by the designer from the input specification. The design solution process then takes place, whereby information is passed from the designer to paper and back again in the form of sketches and calculations. When this process is completed, manufacturing information in the form of drawings and instructions is produced.

Figure 1.2 shows the process extended to the combination of designer and computer. The design solution stage now includes a flow of information between the designer and computer in the form of graphics and alphanumeric characters.

The initial specification must be input to the designer in order that selected parts can be communicated to the computer in a form it can 'understand' and use. The first role of the computer is to check the information for human errors, which must then be corrected by the intervention of the designer.

The human brain is able to store information in an intuitively ordered manner, but its storage capacity is limited, and the information is not all retained as time passes. By contrast, computers have no ability to organise data intuitively but have large permanent storage capacities. Information storage should therefore be carried out by the computer under the direction of the designer.

The output of manufacturing information, from the solution stage of the design process, usually involves the production of drawings. This is a slow and mundane process when carried out manually but is quite suitable for execution by the computer. It is therefore desirable to allow the computer to generate as much production information as possible, so freeing the designer from repetitious work at all stages of the design process.

(c) Modification

Design descriptive information must frequently be modified to make correction of errors, to make design changes, and to produce new designs from previous ones. The computer has the ability to detect those design errors which are systematically definable; whereas man can exercise an intuitive approach to error detection. For example, the computer can calculate the torque capacity of a shaft, whilst a designer can tell from experience and judgement that the shaft is too small.

The automatic correction of errors is generally difficult for a computer. It should therefore be left to the designer to monitor corrections of errors and any other design changes which may be needed.

(d) Analysis

A computer is very good at performing those analytical calculations of a 'numerical analysis nature' which man finds time-consuming and tedious. As much as possible of the numerical analysis involved in the design should be done by the computer, leaving the designer free to make decisions based on the results of this and his own inutitive analysis.

It can be seen from the discussion so far that there exists a clear division between the functions of man and computer in CAD.

The computer has three main functions:

(1) To serve as an extension to the memory of the designer.
(2) To enhance the analytical and logical power of the designer.
(3) To relieve the designer from routine repetitious tasks.

The designer is left to perform the following activities:

(1) Control of the design process in information distribution.
(2) Application of creativity, ingenuity, and experience.
(3) Organisation of design information.

The main purpose of CAD is to produce a definition of the part or system to be manufactured in the form of a geometric database, or a drawing derived from this database, which establishes the physical configuration of the part or system. On the other hand, the purpose of CAM is to translate this definition into tangible hardware based on that database. In computer-aided manufacture, the computer plays many diverse roles in manufacturing functions such as numerical control, process planning, robotics and factory management. In numerical control, a human programmer develops manually an NC tool path required to generate the part shape. The calculations involved are often tedious and time-consuming, not to mention error-prone. Computer-assisted NC part programming uses a computer to perform automatically these laborious calculations, thus eliminating may of the problems involved. In processing planning applications, the function of the computer is to capture the years of manufacturing experience that normally resides with a few individuals, and make this information available to less-experienced planners. Computer-assisted process planning systems provide management control of the process planning function, enable fast process plan updating, and reduce proliferation of methods. Robots can now perform complex tasks under computer control in many industrial applications. They are usually programmable, and can carry out various different tasks. Robots are especially suited to tasks that require a great deal of repetition, and many were designed to work to close tolerances. In many industries, robots are generally employed to replace human operators in jobs that are dirty, dangerous, monotonous, and routine in nature such as tool and material handling, welding and painting which robots can easily perform with a consistency and reliability beyond the physical capability of humans. In factory management applications, computer systems can use inventory and production information in the master schedule to monitor and control the manufacturing process. Thus, manufacturing inventories are optimised relative to the master schedule, lot-sizing policies, and limiting factors of production lead time. As the above four manufacturing functions have evolved and developed separately, a great potential may be exploited if these discrete elements can be linked into an integrated manufacturing system which is the ultimate goal that the technology of CAM attempts to achieve.

In summary, the basic comcept of CAD/CAM is that individual functions in design and manufacturing are computerised, and that these functions are tied together through a central computer database shared among them. As a result, a CAD system allows a user to interact with a computer through a graphics terminal to define a design configuration, analyse the structure and its mechanical behaviour, perform kinematic study and modal testing, and automatically produce engineering drawings. Then the production people can make use of the geometric description provided by CAD as a starting point in CAM to create NC programs for machine tools, determine process plans for fabricating the complete assembly, instruct robots to handle tools and workpieces, and schedule plant operation with a factory management system.

The above overall concept of CAD/CAM is illustrated as a block diagram in Fig. 1.3, which shows an idealised structure of a CAD/CAM system. Many existing CAD/CAM systems available in the market already have all the CAD functions integrated and even have full NC capabilities. However, the other CAM functions are still in different stages of development and are usually performed independently of the CAD/CAM system. Although Fig. 1.3 represents only an idealised model of a CAD/CAM system, it does serve to outline the ultimate principles of CAD/CAM. Major efforts from universities, private industry and government are now put together to combine these individual areas into fully integrated CAD/CAM systems, so a practical, rather than idealised, CAD/CAM system may become a reality before too long. In fact, some experts confidently predict that integrated CAD/CAM systems will be achieved, and it will eventually lead to the completely automated factory.

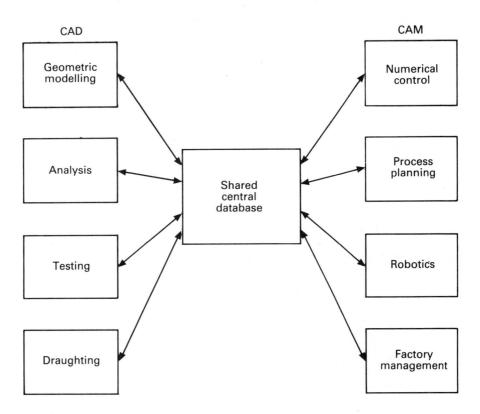

Fig. 1.3 – The concept of integrated computer-aided design and computer-aided manufacture (CAD/CAM).

From Fig. 1.3, it can be seen that CAD features consist of geometric modelling, analysis, testing and draughting, whereas CAM functions include numerical control, process planning, robotics and factory management. The advantage of organising the various CAD/CAM functions in the kind of structure shown in Fig. 1.3 is that it is very easy to introduce new functions into the system if the need arises without affecting the other functions already in the system. As each function is modular in nature within the structure, new enhancements can be made to an existing function with little difficulty or effect on the other functions.

1.3 BENEFITS OF CAD/CAM

It is rather difficult to outline all the possible benefits and advantages of using CAD/CAM, but it can be easily imagined from Fig. 1.3 that they are considerable in terms of working efficiency and productivity alone. CAD can drastically reduce the number of steps involved in the design process for a particular product and can also make each design step much easier and less tedious for the designer to perform. As a result, an immense increase in the work output of a designer can be made possible and an enormous amount of time-saving can also be achieved between the initial conception of an idea and its final implementation as a product design. CAD enables an accurate representation of a design and provides the designer with a versatile tool to graphically manipulate it. With this flexibility of manipulating the drawing of a design, the designer can easily obtain a deep insight into complex problems arising from the design. This will help the designer make better decisions and will reduce the possibility of having errors which may be difficult to spot by any other method. Hence the designer will be more likely to arrive at an optimal solution which will eventually lead to an improvement in the quality of the final product and will therefore make it more competitive in the market.

CAM can greatly increase productivity on the shop floor in various ways. It can automatically generate NC programs and generates them with fewer errors as they can be checked on a graphics display. With CAM, scheduling of components and tools through manufacturing workstations is made much easier. As a result of these, delivery times of products are remarkably reduced. As for as management is concerned, CAD/CAM can improve capacity planning and production scheduling. Thus better inventory control can be achieved, and the working capital required by the company is reduced. It is also possible to simplify product forecasting, estimating and price-fixing, so that the process of preparing tenders is made much easier and is greatly speeded up.

Finally, one other benefit of CAD/CAM, which I think is crucial to its success, is that as all information is stored in the computer instead of on paper; the transfer of data from department to department is quicker, more reliable and less redundant.

1.4 APPLICATIONS OF CAD/CAM

Interactive computer graphics has been widely applied in many different areas of science and technology where drawings of any kind are a vital element for illustrating new scientific or design ideas to another person. Its applications include study of molecular structures in chemistry, medical research, animation, aircraft flight simulation, structural design in aircraft, shipbuilding and automobile industries, integrated circuits and printed circuit board design in the electronics industry, town planning and architectural design, pipe routing and layout in chemical plant design, mesh data preparation for finite element analysis and draughting, etc.

Nowadays, aircraft pilots are trained not in a real aircraft in the air but on the ground at the controls of a flight simulator which very closely resembles an aircraft flight deck in both its functions and its appearance. A flight simulator contains all the usual control instrumentation together with television screens displaying computer-generated views of the terrain visible on takeoff and touch-down. These views change instantaneously according to the actions taken by the trainee pilot to control his 'aircraft' so as to continuously keep a precise impression of the aircraft's motion. The major advantages of using flight simulators for training pilots are the cost savings on fuel, better safety and the ability to prepare the trainee pilot to fly an aircraft to numerous airports all over the world and not just the one where he obtained his training.

The electronics industry also extensively uses interactive computer graphics in the design of integrated electronic circuits, which can become so complicated that it may take a human being weeks to draw by hand and a similar length of time, if not longer, to redraw if a major modification is required because of a change in the present design. If an interactive computer graphics system is used, the circuit diagram can be drawn in a much shorter time and the computer can be utilised to check for any ambiguity and inconsistency in the design. Corrections and modifications to the design can then be made in a matter of minutes. This latest development in integrated circuit design has contributed enormously to the trend towards low-cost electronic equipment.

 Finite element analysis has rapidly established itself as an important engineering technique in the last decade or so for carrying out stress analysis on many types of structures. This engineering stress analysis technique was initially applied in the aircraft industry where more accurate methods of analysis were needed for the increasingly complex airframes being developed. Then interest in this powerful technique soon spread to nearly all fields of engineering, as well as to reactor physics, because it offered the designer a potent and effective computational tool for analysing complex engineering structures. However, the finite element method is often not easy to use, despite its increasing popularity among industrial designers over the years. The most difficult problem lies in choosing suitable elements of regular shape to represent the geometry of the structure under consideration. This process is called the idealisation. Then the required input data has to be formulated for a finite element analysis problem. This input data consists of the geometric idealisation, the material properties, and the loading and boundary conditions. A significant portion of the input data is the geometric representation of the structure by a suitable mesh. If this task of data preparation is to be performed manually, it is obviously time-consuming, tedious and no doubt prone to considerable errors as enormous numbers of punched cards are usually needed to describe an average-size problem. Incorrect input data due to human errors will lead to the runs of the finite element analysis program being aborted and thus it can be a very costly exercise. Using a CAD system in the data preparation process can at least reduce, if not eliminate, most of the above problems. The user is able to see the element connections and position of each element directly on a display as the element generation is in progress. The user can also change a mesh instantaneously to arrive at the best mesh arrangement to suit a particular problem by adding or deleting elements. The CAD system can then be used after the finite element analysis to present data graphically so that the results can be quickly assessed.

In draughting, computer-aided design is now very commonly used to ease the task of the draughtsman as well as to help increase his efficiency. Computer-aided design systems provide many automatic features to speed up draughting. Essentially, the draughtsman need not manually draw each line in an engineering drawing. Rather, the computer-aided design system constructs the lines based on user-specified points and identification codes indicating their nature, e.g., the start and end of a line. Circles may be produced by specifying a centre point and a radius, or three points on the circumference. Other automatic features provide additional draughting aids. For example, points and lines created in one view may be automatically projected into other views. Similarly, any subsequent drawing modification made on one view will be automatically added to the others. In addition, most CAD systems include automatic scaling and dimensioning features as well. When the drawing is completed, one or more copies may be produced automatically on a plotter in a much shorter time than it takes if done manually by a draughtsman. This shows that CAD systems can greatly improve the production of drawings in terms of both quality and speed.

Apart from the previous examples, CAD/CAM is used in various degrees in many other branches of science and technology which are just too numerous to mention here. However, CAD/CAM has been applied extensively to mechanical engineering and manufacturing industries, and it is these particular areas that the rest of this text is intended to cover.

REFERENCES

[1] Besant, C. B., *Computer-aided design and manufacture,* 2nd edn, Ellis Horwood, 1982.

[2] Lui, C. W. K., *The use of computer-aided techniques for hypoid gear design*, Ph.D. thesis, Imperial College of Science and Technology, 1983.

[3] Khurmi, S., *The microprocessor and computer-aided design and manufacture*, Ph.D. thesis, Imperial College of Science and Technology, 1982.

[4] Smith, W.A., *A guide to CADCAM,* The Institution of Production Engineers, 1983.

[5] Groover, M.P. and Zimmers, Jr, E.W., *Computer-aided design and computer-aided manufacturing,* Prentice-Hall, 1984.

[6] *CAE systems and software annual,* Penton/IPC Publication, 1982.

2

Computer systems

2.1 INTRODUCTION

Computers are now in common use in both scientific and commercial fields. New advances in microelectronics have led to the development of computers that have considerable computing power and which are physically small, reliable, and sufficiently low in cost to make their use acceptable for many applications. They have made it possible to perform functions that are just too complicated and time-consuming to do manually. The digital computer is a major and central component of a CAD/CAM system, so it is essential to be familiar with the technology of the digital computer and the principles on which it works.

The digital computer is an electronic computing machine that can perform mathematical and logical operations and data processing functions according to a predefined series of instructions known as a program. The physical component units making up a computer system are often called the 'hardware', whereas the programs are commonly referred to as the 'software'.

The three main categories of hardware components are as follows and they will be described individually in subsequent sections of this chapter.

(a) Central processing unit (CPU).
(b) Mass storage devices.
(c) Input/output (I/O) devices.

Figure 2.1 illustrates schematically the relationship between the above three main categories of computer hardware. The way in which these hardware components are linked together defines the system architecture and thus influences its performance. The simple computer system as shown in Fig. 2.1 can

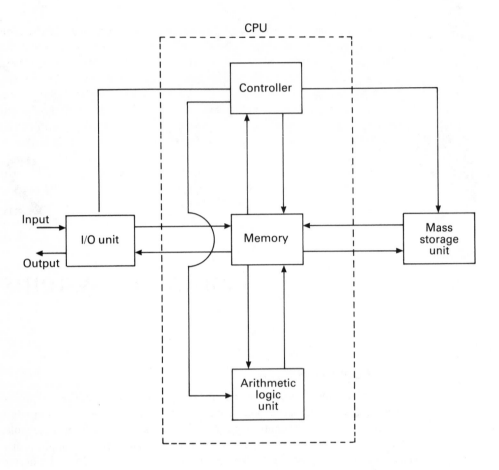

Fig. 2.1 — General organisation of a simple computer system.

be expanded if necessary into a complex system with many input/output and mass storage devices. Indeed, it sometimes can even have more than one CPU to do processing in parallel, as in some multiple processor systems.

2.2 CENTRAL PROCESSING UNIT

The central processing unit is the nerve centre of any digital computer system, since it coordinates and controls the activities of all the other units and performs all the arithmetic and logical processes to be applied to data as specified by program instructions. The CPU can be considered as comprising three separate operating units:

(a) controller
(b) arithmetic and logical unit (ALU)
(c) memory.

The controller basically acts as an administrator in a computer. It examines sequentially each individual instruction in the user's program, interprets it and decides how and when to perform the operation. It then informs the ALU what to do and where to obtain the necessary information. It knows when the ALU has completed a calculation, and it tells the ALU what to do with the results and what to do next. The arithmetic and logical unit, as its name implies, deals with arithmetic and logical operations. It performs the actual work of computation and calculation such as additions and subtractions which are as fundamental to the functions of computers.

Both the controller and the ALU make use of registers to perform their functions. Computer registers are small memory units within the CPU that can receive, hold and transfer data. Each register contains binary cells to represent a binary digit, called a bit. The number of bits in the register constitutes the computer's word length which can indicate to a certain degree the processing capability of the computer. Different computers have a different number of registers depending on the complexity of the architecture of their CPU. In general, these registers are arranged in such a way that they provide support for the CPU to carry out many of its processing operations. These registers can be classified according to their functions as follows.

Program counter

The program counter, sometimes referred to as the control register, contains the address of the next instruction to be performed by the CPU. During program execution, the CPU fetches each instruction word from a memory location whose address is indicated by the contents of the program counter. In effect, the program counter acts as a pointer to instruction words of a program. After CPU has taken the current instruction word, the program counter is automatically incremented to point to the next one.

Instruction register

This register stores the current instruction so that it can be translated into the corresponding machine code for the CPU to perform the required operation.

Memory address register

The memory address register is used to hold the address of data stored in memory so that the CPU knows from this register where to fetch the required data for each instruction as almost every instruction involves some kind of operation on data.

Accumulator

An accumulator is a register used to temporarily store intermediate results from arithmetic or logical operations. Using the addition of two numbers as an example, the accumulator first stores one number, and on receipt of the second one it adds it to the first and stores the sum. By the same operation, more numbers can be added and the final sum is stored in the accumulator.

Status register

A status register is used to hold a status word which contains information about the conditions of the CPU such as the result of a logical instruction, overflow during an arithmetic operation, and the interrupt state, etc.

Arithmetic and logical unit

The ALU is a hardware unit that consists of some circuitry capable of carrying out arithmetic and logical functions that are specifed in a program. Most ALUs can add and subtract, but there are now some ALUs that can perform multiplication and division, and even other complex mathemetical functions.

The memory, sometimes called main memory, of a computer contains binary storage units which are grouped into bytes of 8 bits length. Then, a number of these bytes are organised into a word whose length depends on the design of the computer. A word is a unit of memory addressable by the CPU and it usually occupies one memory location. The main memory can store information such as program instructions to be executed by the controller and data to be processed by the ALU. This stored information is then made available as required for the use of either the controller, or the ALU, or other units within the computer. Intermediate results can also be stored in memory while the ALU is working on another part of the program. These stored parital answers may later be used to solve other sections of the program.

There are various types of storage medium used for the main memory of a computer. The most common type is the ferromagnetic core often referred to as core storage. A magnetic core memory consists of thousands of elaborately interwoven strands of wire and tiny magnetic doughnut-shaped rings about 1/16th of an inch in diameter. These rings, usually called cores, are organised into groups so that each group represents a unit of data such as a byte or a word. The wires themselves are paths along which signals can be transmitted to energise individual cores so that they adopt a particular polarity. Each core represents a bit and the polarity of the core at any given instant indicates the value of either 1 or 0 for that bit. Core storage is non-volatile, so the data stored in it will not vanish when the power supply is interrupted.

Another common storage medium for computer main memory is the semiconductor storage which consists of electronic devices such as transistors. These transistor circuits are arranged into storage cells, also known as 'flip-flops', each of which represents a bit, as a core does in core stoarage. With the application of Large Scale Integration (LSI), thousands of these transistor memory cells can be etched onto little chips of silicon less than 1/10th of an inch in size. Such a chip can easily contain more than 100 bits of main memory. However, this type of storage medium is volatile which means that the data in each cell will disappear as soon as the power supply is cut off.

2.2.1 Multiple CPUs

Sometimes, a computer system embodies more than one central processing unit. Such a system is often described as multiprocessing. The main purpose of employing multiple CPUs is to improve either the reliability or the performance of the system, or even both. Figure 2.2 illustrates schematically a multiprocessor system with two central processing units that share the main memory and, as a result, can service the same set of programs.

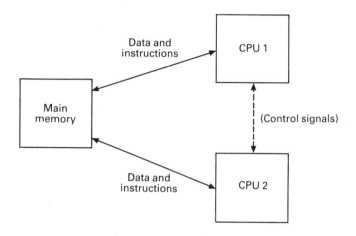

Fig. 2.2 – Organisation of a multiprocessor computer system.

When multiple CPUs are used to increase system performance, the CPUs operate asynchronously and effectively share the processing load. In many computer applications where a rapid response is critical and the smooth running of an organisation may depend very much on a computer, an extra central processing unit is frequently utilised as a backup CPU so that in the event of malfunction in one CPU, the other one can take over the workload of the whole system. The essential thing is that in many operating environments it is preferable to run in a down-graded state than to fail completely.

In this type of multiprocessor system, the central processing units must have access to the same information stored in main memory and mass storage devices. The system must include circuitry for error checking and correction. When an operational error has been detected, the system tries to correct the error condition through some logical procedures. If it succeeds in correcting the error condition, the system will obviously continue as normal. But the error can sometimes occur in a section of main memory or in an input/output or mass storage device, then the system will logically disconnect the faulty component by software and carries on its operation in a less efficient manner. If an error condition cannot really be corrected, a fault signal is transmitted so that one CPU knows the other one has failed and takes necessary remedial actions. For this reason, it is very important that the CPUs in a multiprocessor system are able to exchange control signals about their current status.

There are indeed some multiprocessor systems in which each CPU has its own main memory as shown in Fig. 2.3. These systems are often described as having a multiple-instruction/multiple-data architecture because each CPU is fully programmable and capable of executing its own program. Individual CPU can communicate with any of the other CPUs through the common data bus. The performance of multiple-instruction/multiple-data systems undoubtedly compares favourably with those of single main memory or single CPU, as it can handle simultaneously as many memory references as there are CPUs.

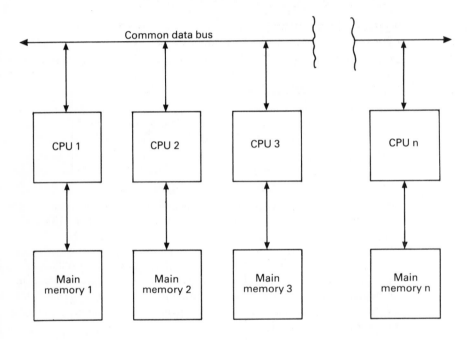

Fig. 2.3 – A multiprocessor computer system with separate main memory for each CPU.

2.2.2 Categories of computer
In the early 1960s, there were only large and fast computers often called mainframes, which were expensive and had a high power consumption. As a result, only large institutions such as government departments, universities, and big commercial and industrial organisations could really afford to use them. During the mid-60s, the market began to see the gradual arrival of the small computers known as minicomputers which satisfied to a certain extent the demand for low-cost computing. The appearance of these minicomputers heralded the ever-growing trend of miniaturisation in computer technology that has currently reached the stage where a CPU, together with a large main memory and sophisticated input/output circuitry, can be squeezed onto a single LSI

circuit chip of several millimetres square. Indeed, the design of many small computers, usually referred to as microcomputers, is based on these chips, together with some input and output devices.

Although the architecture of different types of computers is roughly the same, they can generally be classified as follows, according to their word length which represents the width of the internal data transfer paths of the computer.

(a) The mainframe computer.
(b) The minicomputer.
(c) The microcomputer.

In general, the mainframe computer has a word length of 32 bits or more, and is the most powerful of the three categories in many ways. The minicomputer usually has a word length of 16 bits whereas the microcomputer has a word length of 8 bits and is the smallest of the three categories. However, this classification of computers is gradually becoming more difficult as new mini-computers and microcomputers are developed and introduced on the market. The 16-bit microcomputer, often called the supermicro, has already been with us for some time, and the launch of the 32-bit microcomputer, referred to as the megamicro, has just been announced by some computer manufacturers. With regard to the minicomputer, a 32-bit machine known as the supermini or midi has been in production for several years. Thus, it can be imagined that the division of different categories of computers is not so clear-cut now as new computers become available and their capabilities overlap those of other categories.

As large mainframe computers and large minicomputers are capable of fast processing and have large main memory, they are commonly used to solve complex engineering and scientific problems such as iterative calculations in heat transfer analysis, structural design analysis, or fluid dynamics analysis. In a commercial environment, they are often employed in large-scale data processing operations such as payroll and corporate accounting, production scheduling, stock control, and maintenance of large information databases.

Small minicomputers and microcomputers are obviously less powerful than large mainframe computers and large minicomputers; but, because they are inexpensive, they become very cost-effective in some commercial and industrial applications. Another advantage is that they are small in size, so they are commonly used as a part of some large test equipment, or as the processor of a computer numerical control machine tool, etc. Their applications are numerous, but generally they perform a small specific task within a large overall system for a particular function.

2.3 MASS STORAGE DEVICES

No matter how large the main memory is in the CPU, it is still finite in size and very much limited when compared to the vast quantity of data that the CPU has to process. The purpose of the main memory is essentially to provide a working area for the current program, and it only retains information on a temporary basis until the termination of the program. In other words, the main memory is basically used to contain instructions and data of a program during its execution. As computers may often operate on an enormous amount of data that exceeds what it can hold at one time, backing store is commonly used to supplement the main memory and to save the data on a permanent basis. Information saved in backing store can then be retrieved and quickly transferred to the CPU when it is needed.

There are several types of devices that can provide this kind of backing storage, but the choice will mostly depend on how the information needs to be accessed. Generally, two basic forms of backing store offering two different methods of access are available.

Sequential access storage

In this type of storage, also known as serial access storage, information can only be accessed in the same order that it is stored originally. If a particular piece of information in a file is required, all preceding items will have to be read as well. If additional information is to be written into the file, it is usually only possible to do that at the end of the file, otherwise the rest of the file will have to be rearranged.

Direct access storage

Frequently, the accessing of data needs to be in a more direct manner than sequential storage permits. With direct access storage, individual items of information can be located immediately for read or write without involving any other items in the file. As the information is literally available at random, that is, in any order, direct access storage is also commonly referred to as random access storage.

Clearly, it can be imagined that the random access storage is much more efficient than the sequential access storage, because of the difference in their methods of operation, and, consequently, the random access storage reads or writes data more quickly. However, random access storage is more expensive in terms of cost per bit because the technology involved is more sophisticated. For these reasons, sequential access storage is used mainly to store files as permanent records in an archive where they are very rarely accessed, whereas random access storage is most suitable for storing files that require frequent access.

2.3.1 Magnetic tape

Magnetic tape is a storage medium similar to that used on a tape recorder except that its quality is superior and more durable. It is usually in the form of a continuous strip of plastic material with a magnetic oxide coating on which data may be stored as a series of magnetised and non-magnetised spots. Because of this physical arrangement of data, magnetic tape can only provide sequential or serial access. Computer magnetic tape is commonly 1/2 inch wide and may be up to approximately 2,400 feet long wound on a reel, usually 8 to 12 inches in diameter. There are two main types of magnetic tapes, namely 7-track and 9-track. Most modern systems use 9-track magnetic tape because of its ability to store data at higher densities. Information can be recorded on magnetic tape in different densities typically 200, 556, 800 and 1,600, or even as dense as 6,250, bytes per inch (BPI). Magnetic tape speed ranges from 37.5 inches per second to 200 inches per second, so the data transfer rate of magnetic tape can vary from 7,500 BPI to 1,250,000 BPI.

The major advantages of magnetic tape are that it is relatively cheap when compared with other types of storage medium and that it can easily hold a large amount of data for its size. The capacity of a 450 ft reel may be up to 20 million characters or more. Magnetic tape, unlike punched paper tapes or cards, can be used again by simply overwriting previously stored data.

As magnetic tape is a sequential access storage device, data access may be slow and is not the ideal choice for fast backing storage. However, it is most suitable for some applications such as those that might be required in payroll, personnel management, inventory control and customer invoicing where an immense amount of data is to be processed sequentially.

Magnetic tape may be used not only as a backing storage medium but also as an input/output medium. Information can be read into the computer from the tape for processing and information can be written to tape where it is stored until it is required at a later time − or until it becomes obsolete and redundant, when the tape can be re-used for some other purpose and its contents be over-written. Very often, magnetic tape is used to back up the data in a computer system so that, in the case of a system crash, a copy of the original information is still available and can be reloaded if necessary. Magnetic tape is also widely employed for transferring data between computer installations that are not linked together.

2.3.2 Magnetic disk

Magnetic disk is one of the most commonly used devices for direct access storage. A typical magnetic disk consists of six or more metallic platters that resemble gramophone records, except that the tracks are concentric instead of spiral. Each platter is coated on both surfaces with magnetisable material such as ferrous oxide. The platters are mounted about 1/2 inch apart on a spindle which rotates and spins the platters at speeds of 50 or more revolutions per second. Data are stored as a series of magnetised or non-magnetised spots on tracks on the platter surfaces, and are read or written as the platters rotate. Each track on a platter is further divided into sectors and information is accessed by track and sector address. On each standard-sized platter of 14 inches diameter, there may be as many as 256 tracks and 32 or more sectors per surface, so a platter can easily hold thousands of characters, depending on the recording density which varies from 3,500 to 15,000 characters per track. As a result, a disk is potentially capable of storing as many as 20 to 1,000 Mbytes of data. One Mbyte is equivalent to one million bytes and 1 Kbyte to one thousand bytes. The data transfer rate of disk storage is usually very high and ranges typically from 300,000 to 1.5 million bytes per second.

The stack of platters in a disk is often referred to as a disk volume. Some disk volumes are permanently fixed inside the disk drive unit, while other volumes can be removed and replaced by another easily and quickly. These removable volumes are better known as disk packs. If a disk pack contains only one platter, it is generally known as a cartridge disk. The obvious advantages of using disk packs or cartridge disks are that storage space can be increased without the need to buy another complete device and the total disk storage capacity is not limited by the number of disk units.

Two kinds of read/write head units for magnetic disk devices are in general use. They are the moving-head unit and the fixed-head unit. Removable disk packs are usually associated with only moving-head units which contain a certain number of access arms depending on the number of platters in the pack. For each platter, there is an access arm to control two read/write heads, one for the upper surface and one for the lower surface of the platter. The access arms form a single assembly and move in and out together across the surface of the disk so that it is possible to access each track individually. In the case of fixed-head units, one read/write head is used for each track. As a result, there is no head movement and data is therefore stored or retrieved more rapidly. Because of this difference in the physical arrangement of the read/write head units, a fixed-head device is normally quicker than a moving-head device in accessing information.

Although magnetic disk is a direct, or random, access storage medium, it can be used for sequential access as well. As with magnetic tape, information on magnetic disk can be read again and again, and new data may be stored by simply writing over existing information.

Floppy disk

In the early 1970s, a new disk storage device became commercially available for use as a low cost and faster alternative to storage on magnetic tape. The platter of this disk is coated with a layer of magnetic oxide much as in a conventional disk except that the platter is made of flexible plastic, hence this new disk is widely known as floppy or flexible disk. Floppy disk provides a compact medium for random access backing storage, and it can be used for both input and output operations in a similar way to a hard disk. A floppy disk is normally contained in a plastic or cardboard sleeve, often referred to as a cartridge which protects it from physical damage and dust particles. It is usually removable from the disk drive so the total storage capacity on floppy disk is not restricted by the number of disk units.

Floppy disks often come in two standard sizes: the large is 8 inches in diameter, and the small is $5\frac{1}{4}$ inches and commonly referred to as mini-floppy. The storage capacity of an 8-inch floppy can be anywhere between 250 Kbytes and 1.5 Mbytes, whereas that of a $5\frac{1}{4}$-inch mini-floppy lies in the region between 125 Kbytes and 500 Kbytes, depending on recording density and whether single-sided or double-sided.

In general, floppy disk cannot hold as much data as conventional hard disk and its data transfer rate is less than desirable for use as fast access backing storage. However, because of the fact that it is relatively inexpensive compared to its storage capacity and access speed, it is particularly suitable for use with small and low cost microcomputer systems.

Winchester disk

The Winchester disk is a fairly recent development in disk storage technology. The platter is coated with a magnetic oxide as in the conventional disk and with a special lubricant which reduces the friction between the read/write heads and the platter as it rotates, because the heads actually make contact with the platter when accessing data. A Winchester disk platter is usually contained in a sealed chamber which prevents it from being contaminated with dust and other particles. As greater precision can be achieved in Winchester disk, a greater number of tracks on the platter surface and a higher storage density per track are possible.

There are several standard sizes of Winchester disks. The most commonly used ones are $5\frac{1}{4}$ inches, and 8 inches, but it can be as large as 14 inches. Typical storage capacities are 10, 20 and 40 Mbytes, and can be doubled with dual disk drives. One advantage of Winchester disk is that preventative maintenance required is minimised as the disk units are sealed. Winchester disks provide a fast and reliable storage medium, and yet they are reasonably priced in comparison with conventional hard disk. For this reason, they are often used to support the more powerful microcomputers as well as minicomputers, and are gradually taking over the position formerly held by floppy disks.

2.3.3 Magnetic drum

The magnetic drum is one of the oldest forms of mass storage devices that provide random access. It essentially uses a metal cylinder with a coating of magnetic oxide over its surface as a storage medium. Data access is performed in the same way as for magnetic tapes and magnetic disks except that data are stored in circular tracks on the curved surface of the cylindrical drum. These tracks, often referred to as bands, are generally analogous to tracks on disk. The magnetic drum is usually held vertically in a cabinet and there is a control electronics bay associated with it. Each recording band has a fixed read/write head positioned over it. These read/write heads are staggered round the drum for more efficient utilisation of space and higher rate of data transfer.

Since there is one read/write head for each track of the drum, the time required to store or retrieve one particular item of data on the drum is usually less than that for magnetic disk where read/write heads must normally be moved across the disk surface to reach the appropriate track. As a result of the way in which the heads are arranged, every item of data will pass under a head during one revolution of the drum, so data transfer rates are on average higher than disk storage and range from 300,000 to 1.5 million characters per second. Access time is also faster than disk, whereas storage capacity is comparable and it is possible to have a multiple drum system which can considerably increase the storage capacity. However, a drum storage unit generally costs more than disk storage, and the drum is not removable from the drive unit.

2.3.4 Other storage devices

In previous sections, the mass storage devices widely used in computer systems today have geen discussed. As research and development into storage technology continues, these devices will be further refined to provide more powerful mass storage systems. Indeed, there are already some new storage devices that are based on completely different principles from those of magnetic tape, magnetic disk or magnetic drum. These new mass storage devices are Laser Beam Storage, Videodisk Memory Systems, Bubble Memory, and Electron Beam–Addressable Memory Systems, all of which, although still in their infancy, potentially offer large storage capacity, faster access time and data transfer rate, and may become feasible alternatives to conventional magnetic memory technology in the future.

2.4 INPUT/OUTPUT DEVICES

Within the central processor of any computer, information can be processed at very high speeds, but before the central processor can be set to work, the data and programs must be entered into the computer memory. This is done by means of input devices which provide a vehicle for communications to the computer from the people who are concerned with its operation. Input devices are required to bridge the gap between the language of human beings and the internal code language of the computer because human beings are able to recognise and understand the relationships between numerals, characters and symbols, whereas within the internal store of a computer the various electronic circuits are able to respond only to patterns of electrical impulses. For communications in the opposite direction from the computer to the people, an output device is needed. All the data and programs within a computer are stored as electrical impulses in a coded form according to the machine code system of the particular computer. When data are held in this form, they cannot be readily understood by human beings and therefore output units are employed to translate these data into information that can be used by human beings as and when they require them.

2.4.1 Card reader/punch

The standard punched card measures $7\frac{3}{8}$ by $3\frac{1}{4}$ inches and is 0.007 in thick. The corners of a card may be rounded or square and one corner is normally cut to detect when it is upside down. A card is divided into 80 columns, numbered 1 through 80, with 12 punching positions in each column, numbered 0 through 9 from the third to the twelfth punching position, while the first two at the top of the card are not usually numbered. Information is recorded by punching rectangular holes in the card and one column represents one character.

A card reader is an input device that transfers data from the punched card to the computer system. The reader is designed to read and interpret the codes on a card represented by the punched hole positions, and then generate electrical pulses that correspond to the sequences of bits which the computer can understand. Each card passes between a light source and a set of photoelectric cells.

The presence of a hole causes the light to produce a pulse by triggering one of the cells. Card readers can run at speeds up to 2,000 cards per minute.

A card punch is an output device that transfers data from the computer system onto punched card. The punch outputs the data as a pattern of rectangular holes in a card, and then reads it to verify that the card was punched correctly. Very often, card readers and punches are integrated into one single unit.

Punched cards were once the traditional form of input, storage and even output medium, yet today there are very few applications in which punched cards are still used. The major disadvantage of punched cards as a storage medium is that they are cumbersome to handle and too bulky to keep because they occupy a relatively large space when a considerable volume is involved. Therefore, more direct and convenient methods have evolved and these are now generally preferred by those who used to rely on punched cards.

There is another type of machine, called keypunch, associated with punch cards. Keypunch contains a keyboard like an ordinary typewriter. Blank cards are entered into the keypunch machine, and a corresponding combination of holes are punched when a key is depressed. The cards are then read through a card reader to the computer.

2.4.2 Paper tape reader/punch

Paper tape has been used as a medium for input, storage and output since the early days of computers. The tape is similar in concept to magnetic tape and is basically a continuous strip of paper or material made of thin flexible plastic compounds. The tape is normally 1 inch wide, comes in rolls, and may be used in any length up to several hundred feet. Data is recorded on paper tape by punched holes in rows across the width of the tape. The maximum number of holes in each row indicates the number of channels on the tape. Eight-channel tape is the only one currently in widespread use. Each character is represented on the tape by a combination of holes in a row across the tape. The small holes located between channels 3 and 4, referred to as sprocket holes, run along the length of the tape and are used for feeding the tape through the tape reading and punching device.

A paper tape reader operates in much the same way as a card reader. It translates the data punched in code on paper tape and transmits it into a central processing unit. The tape passes through a reading unit where the presence or absence of holes is detected and converted to electrical pulses. A paper tape punch operates on data in the opposite direction to a paper tape reader, that is, it outputs the information in a central processing unit in the form of punched holes on paper tape. Sometomes, a paper tape reader and a paper tape punch are built into a single unit.

Paper tape, like punched cards, is not so commonly used these days as superior media for input, storage and output have become widely available. The major disadvantages of paper tape are that errors are difficult to correct and that readers and punches are relatively slow, operating at approximately 500 and 300 characters per second respectively.

2.4.3 Terminals

There is a wide variety of terminal devices used in computer systems. One of the earliest types, the teletypewriter, also known as teleprinter, is usually an electromechanical or electronic device that resembles an ordinary typewriter in many ways except that it is connected directly to the computer via data communications lines. The unit works for both input and output, that is, the user can enter data or commands to the computer from the keyboard and may receive from the computer results or responses printed automatically on it. As the user types a character, it is printed at the terminal and transmitted to the computer. When the line is complete, the user presses the carriage-return key to indicate the end of line to the computer. Output from the computer can be under the control of a program. The rate of data communications between a computer and a teletypewriter is often in the range of 110 and 300 baud (bits per second). A teletypewriter prints output one character at a time at rates between 20 to 50 characters per second on continuous rolls of paper of 8 to 20 inches wide or on 'fanfold' paper depending on the application.

As a teletypewriter has mechanical parts, it can be quite noisy in an office environment when it is printing. Output from a teletypewriter is always printed on paper whether or not it is needed as a printed record or 'hard copy', there-fore, it may unnecessarily use up paper and waste material resources. Other major disadvantages of the teletypewriter are its slow data communication rate and printing speed which limit significantly its use as a fast interactive terminal device with a computer. Despite all the above shortcomings, teletypewriters are still used, but only as a console terminal to make enquiries about the status of programs or data being processed by the computer. The information on the print-out paper will form a complete operating log for all the jobs that have been executed.

Owing to the limitations of teletypewriter, a new generation of terminal devices, called visual display units (VDUs), has been developed. A VDU contains a keyboard and a display device which is usually a cathode ray tube (CRT). The display device provides a screen to print information, and essentially performs the same functions as the carriage and paper of the teletypewriter. As each character is typed for input, it is displayed on the screen. At the end of the input line, carriage-return key is pressed and the position indicator, or cursor, is returned to the beginning of the next line for further input. As lines are displayed, preceding lines are successively scrolled up until the top line dis-appears at the top of the screen. For output, lines are displayed in a similar manner. Typically, a VDU is formatted to display 24 lines of 80 characters each.

A VDU generally costs more than a teletypewriter terminal, but it does not have any of the disadvantages described in the previous paragraph. As a VDU is an electronic device with very little in the way of moving parts, its opera-tion is almost silent. There is no hard copy of the output from a VDU, so the problem of wasting print-out paper unnecessarily is eliminated. However, if it is important to have a printed copy of the output, it is possible to attach to a VDU a printing device which can be switched on when needed to produce the hard copy. Other major advantages of the VDU over the teletypewriter are that its data communication rate is much faster, between 110 and 19,200 baud, but 9,600 baud is the most common, and that its printing speed is far superior.

Many VDUs widely used today are intelligent terminals with local memory and a local microprocessor, so they are able to carry out some computing functions. This intelligence is very often used to add or enhance features available on the keyboard such as prompting capabilities, screen formatting and local editing. Indeed, there are now some VDUs built with touch-sensitive screens which enable data to be entered by touching the screen with the fingertip. The screen surface consists of a number of 'touch points' as defined by the program in use. This type of terminal device often works with a menu-driven system. When touched, the terminal sends to the computer the coordinates of the point, and the menu-command corresponding to that position is then executed. Thus, it is not necessary for the user to type in the command from the keyboard, although it is possible if so desired. Instead the command can be more easily issued by pointing at the command in the menu on the screen with a finger.

2.4.4 Printers

A printer is an output device that converts data from a computer and prints it in a readable form on paper. There are a number of different categories of printer used to produce printed output, such as line printers, serial printers and, more recently, laser printers.

A line printer, as the name implies, prints a whole line at a time and operates on the same basic principle as a typewriter. The speed of line printers varies from 60 to 5,000 lines per minute, and there are normally 132 or 136 print positions per line, so it is very suitable for high volume output. To ensure sufficient and uninterrupted supply of paper for these high printing speeds, continuous stationery is used. Continuous stationery is a long piece of paper with sprocket holes on the edges and with perforations to separate the paper into convenient sizes, typically 11 inches long and about 15 inches wide. One disadvantage of line printers is that the quality of printing is not very good, in comparison with that of typewriters.

Serial printers, also known as character printers, are another category of printer which outputs one character at a time, as opposed to one line at a time in the case of the line printer. A serial printing device can normally be operated using continuous stationery or separate sheets, usually A4 size. In general, a serial printer is much slower, but cheaper, than a line printer. The most common type of serial printer is the dot matrix printer. The print head consists of a matrix of tiny needles, typically seven rows of nine needles each, which hammers out characters in the form of patterns of small dots. The shape of each character, that is, the dot pattern, is obtained from information held electronically in the printer. The printing speed of dot matrix printers varies between 45 and 220 characters per second. Another type of serial printer is the daisywheel printer whose name originates from the fact that it uses a daisy-shaped disk made of plastic or metal which holds some 96 characters on its 'petals'. Print heads are interchangeable, so it is possible to use different character fonts. There are normally 132 or 136 print positions per line and printing speeds are typically 25 to 60 characters per second. Daisywheel printers can produce very good quality print and are widely used with word processing systems and other applications where print quality is important. Both the daisywheel and the line printer are based on impact techniques. There is a type of serial printer, called the thermal printer, which employs non-impact techniques. Thermal printers use heat to produce characters in dot matrix form on special sensitised paper. Generally, serial printers are more suitable for low volume output because of their slow printing speed compared with line printers, and are commonly used as output devices for small computer systems.

The laser printer is a relatively new technological innovation in which a combination of electronics, lasers and copier techniques are used. An entire page of different size can be printed at one time by a laser printer. It is capable of producing very high quality print very quickly, and a wide selection of character fonts is also available. Most laser printers operate at speeds between 30 and 250 pages per minute. Despite all these advantages, laser printers are still not in widespread use as very few applications can justify their present high cost.

2.4.5 Other input/output devices

In previous sections, the common input/output devices that are most easily found in general-purpose computer systems have been described. There are many other types of input/output device which are used specifically for some applications. The following paragraphs try to outline and briefly describe some of them.

In data preparation, punched cards were the primary means of inputting original data into the computer, but card reading is relatively slow, so the emphasis today has been switched to magnetic media. An alternative approach to data input is to use a key-to-tape device which enables the recording of information directly on magnetic tape, the data is entered through a keyboard, stored temporarily in some memory and also displayed on a CRT screen for visual inspection of input data for accuracy before being transferred to magnetic tape. Two variations on the key-to-tape data entry method are the key-to-disk and key-to-cassette/cartridge systems. In a key-to-disk system, data input follows much the same procedure as that for a key-to-tape system except that the data are placed directly on magnetic disk. A key-to-disk system usually has a minicomputer, several key stations and one or more disk drives. In a key-to-cassette/cartridge system, data can also be entered via a keyboard direct to small magnetic cassette tapes or cartridges and later copied to standard magnetic tape for processing. Although the cassettes and cartridges are small and not capable of storing a large amount of data, they have their use in some applications because of their compactness and portability, therefore, they are most suitable for the collection of data at source, for example, on sites remote from the computer system.

The process of preparing data for input from a form acceptable to humans to a form acceptable to computers is time-consuming and very often tedious.

The ideal solution would be a method of input that could do away with these data preparation procedures, that is, the use of documents that could serve both as source documents and as an input medium. There are a number of devices that have been developed to achieve this objective. One example is the magnetic magnetic ink character reader (MICR). The recording of data in magnetic ink characters is based on specially designed fonts in which the characters are printed with ink containing ferromagnetic materials. Magnetic ink characters are used principally on bank cheques to indicate the code number of the bank, the customer's account number, and the cheque sequence number. These characters can be read both by people and by a magnetic ink character reader. When a magnetic ink character is being scanned, it induces a current which will be proportional to its area. The patterns of the varying currents can then be compared with, and identified as, bit patterns of the selected character which is translated into the internal machine code of the computer and transferred to the main memory. Two MICR fonts are in common use. One is the E13 B font which is used in the USA and originated there. The other is the CMC 7 font which is used mostly in Europe and originated in France.

Another similar device is the optical character reader (OCR) which works in the same principle as that of MICR. An OCR reads letters, numbers, and special characters from printed, typed, or handwritten documents. The optical character reader scans the document, compares the result with predefined patterns, and inputs corresponding data into the computer. Documents are rejected if invalid characters are detected. New features such as mark reading have been incorporated into OCRs. The optical mark reader can scan a document for marks made by hand or printed by line printers. As the reading head works by detecting the infrared absorption of the mark, the best hand-marking medium is an HB pencil which has a high carbon content and marking errors can be easily erased.

In the retail trade, bar-codes are widely used in supermarkets for labelling goods and shelves, and in stock control. In public libraries, books are labelled with bar-codes so that records of books can be stored in a computer. Bar-code readers use a scanner or light-pen to enter information to a computer through a terminal device. The scanner or light-pen is stroked across the pattern of bars, a sequence of bits corresponding to that pattern is generated and stored in the computer.

In computer graphics and CAD/CAM, graphics input/output devices such as the light-pen, tablet, and graphics display etc., are essential parts of the system hardware. As CAD/CAM is the subject of this book, graphics input/output devices will be described in detail in a separate chapter (Chapter 3) devoted entirely to this topic.

Here, in this section on input/output devices, only the more commonly used ones have been mentioned. Obviously, the list of these devices could go on and on because many different varieties of input/output devices have been developed and modified over the years to suit particular applications. Enough of them have been described here to give the flavour of the topic.

2.5 DATA REPRESENTATION

Information is handled within the computer by electrical components such as transistors, integrated circuits, semiconductors, and wires, all of which can only indicate two states or conditions. The binary number system is thus particularly suitable for mathematically representing the two possible states. The binary number system is based on the number two and involves only two digits, zero (0) and one (1), whereas the familiar decimal number system is based on the number ten and involves the ten digits, zero (0) to nine (9). As both the binary and decimal number systems work on the same principle, it may be useful to explain the binary number system by first describing the decimal number system that we know so well. A decimal number is made up of individual digits, each of which represents a value determined by the value of the digit itself and its position in the number. For example, the decimal number 1,234 can be expressed as:

$$1,234 = 1,000 + 200 + 30 + 4$$

or alternatively,

$$1,234 = 1 \times 10^3 + 2 \times 10^2 + 3 \times 10^1 + 4 \times 10^0.$$

It can be seen from the example that each digit of a decimal number, reading from right to left, indicates a successively higher power of ten, starting from the power 0. Similarly, in the binary system, each digit in a binary number represents a successively higher power of two, again reading from right to left. The digit 1 indicates that a particular power of two is present, and the digit 0 indicates that it is absent. Thus, for example, the binary number 11010011 can be expressed as:

$$11010011 = 2^7 + 2^6 + 2^4 + 2^1 + 2^0$$

therefore, it has a decimal value of:

$$11010011 = 128 + 64 + 16 + 2 + 1$$
$$= 211$$

Apart from the decimal and binary number systems, the octal and hexadecimal number systems are in common use in computers. The octal number system is based on the number eight and involves the eight digits, zero (0) to seven (7), whereas the hexadecimal number system is based on the number sixteen and involves the sixteen digits zero (0) to nine (9) and A to F which represent 10 to 15. Letters A to F are used as digits in the hexadecimal number system because there are no corresponding digits in the decimal number system. All four number systems, that is, binary, octal, decimal and hexadecimal, work on the same principle. The following table shows a list of values and their equivalents in different number systems:

Decimal	Binary	Octal	Hexadecimal
0	0	0	0
1	1	1	1
2	10	2	2
3	11	3	3
4	100	4	4
5	101	5	5
6	110	6	6
7	111	7	7
8	1000	10	8
9	1001	11	9
10	1010	12	A
11	1011	13	B
12	1100	14	C
13	1101	15	D
14	1110	16	E
15	1111	17	F

As eight is equivalent to two to the power three, it takes three binary digits to represent one octal digit, and similarly sixteen is equivalent to two to the power four, four binary digits are needed to represent one hexadecimal digit as shown below using the same previous binary value 11010011 as an example:

$$11010011 = 11\ 010\ 011 \text{ (split into groups of three binary digits)}$$
$$= 3\ 2\ 3_8 \qquad \text{(equivalent octal number)}$$
$$= 3 \times 8^2 + 2 \times 8^1 + 3 \times 8^0$$
$$= 192 + 16 + 3$$
$$= 211$$

also,

$$11010011 = 1101\ 0011 \text{ (split into groups of four binary digits)}$$
$$= D\ 3_{16} \qquad \text{(equivalent hexadecimal number)}$$
$$= D \times 16^1 + 3 \times 16^0$$
$$= 13 \times 16 + 3 \times 1 \ (D_{16} \equiv 13_{10})$$
$$= 208 + 3$$
$$= 211$$

There are generally three common binary coding schemes used in the computer industry. They are ASCII (American Standard Code for Information Interchange), BCD (Binary-Coded Decimal) and EBCDIC (Extended Binary-Coded Decimal Interchange Code). The definitions of these binary coding schemes, if required, can be easily found in most reference texts on computers.

2.6 PROGRAMMING LANGUAGES

In general, there are three basic categories of computer programming language:

(a) Machine languages.
(b) Assembly languages.
(c) High-level languages.

To program a computer to perform a specific function, the program instructions and data have to be expressed in terms that it can understand. As described in the preceding section, data representations in a computer are all in binary numbers, therefore programs and data have to be written in binary code which is often referred to as machine language or machine code. A machine language instruction consists of two parts, namely the operation code and the operand. A short listing of machine language instructions for a certain computer is shown in Fig. 2.4 which illustrates the difficulties involved in writing machine language programs. This process is very often laborious, time-consuming, and, worst of all, error-prone. A machine language is machine-dependent in that it can be recognised and understood only by the computer for which it was designed.

```
000250    004767
000252    000000
000254    012702
000256    000045
000260    012703
000262    000004
000264    017500
000266    000004
000270    010001
000272    042701
000276    110142
000300    072027
000302    177774
000304    077307
```

Fig. 2.4 — An example listing of machine code (compiled from the assembly code in Fig. 2.5).

Because of the problems of writing machine language programs and the difficult task of remembering the numeric machine codes, mnemonic languages, which can help the human brain to memorise the codes and more readily identify them, were developed. They are commonly known as assembly languages. For example, a computer may be designed to interpret the machine code of 110 (binary) as the operation 'add', but it is much easier for a human being to remember it as ADD. A sample listing of an assembly language program is shown in Fig. 2.5. An assembly language is usually designed for one particular computer, so it can only be used on that computer and not any other one, for

```
PLTVEC::  JSR    PC, $SAVAL      ;SAVE ALL REGISTERS
          MOV    #BUFBOT, R2     ;GET PLOTTER BUFFER ADDRESS
          MOV    #4, R3          ;SET NO. OF NIBBLES FOR Y
          MOV    @4(R5), R0      ;GET Y FROM 2ND ARGUMENT
5$:       MOV    R0, R1          ;GET A COPY OF IT INTO R1
          BIC    #177760, R1     ;CLEAR ALL BITS BUT 4 LSBs
          MOVB   R1, –(R2)       ;GET A NIBBLE OF Y INTO PLOTTER
                                 ;DATA BUFFER
          ASH    #–4, R0         ;ARITHMETIC SHIFT TO THE RIGHT
                                 ;FOUR TIMES TO GET NEXT NIBBLE
          SOB    R3, 5$          ;RETURN TO 5$ UNTIL NO MORE NIBBLE
```

Fig. 2.5 – A sample listing of assembly code.

which there may be a different assembly language. Although there exist various versions of assembly language for different computers, they normally include the same kind of instruction sets that perform the four fundamental operations within a computer such as input/output operations, arithmetic operations, movement of data within the CPU, and logical or comparison operations. However, assembly language programs are not in a form directly executable by the computer, therefore, a special program called an assembler is used to carry out the transaction of assembly language programs into machine codes which the computer can understand and execute. Both machine and assembly languages are considered to be low-level languages as they are orientated towards the basic design of computers.

As both machine language and assembly language are machine-orientated, this has led to the development of high-level languages which, by contrast, are problem-orientated and, therefore, are very much independent of the computer on which they are used. Thus, a high-level language program written on one computer can be executed on a different computer without major changes to the program itself. Such problem-orientated languages allow programs to be written using English-like statements and traditional mathematical notations, so it makes the task easier for human beings to think about the problem and to write programs to solve it. The significant advantage of high-level languages is that the need for writing programs in either machine code or assembly language is eliminated to a large extent. Each high-level language statement is equivalent to many instructions in machine code, so a program requiring a large sequence of machine code to perform a specific function can be achieved by a few high-level language instructions. As with assembly languages, high-level languages must also be translated into machine code. A special program, called a compiler, is usually needed to take a high-level language program and perform the necessary translation. A compiler is normally able to detect statement errors in the program, and corresponding error messages are printed in a special program listing to indicate the error conditions. There are many different high-level languages used for various applications. The most common ones found in business and engineering applications are FORTRAN, BASIC, APL, PASCAL, ALGOL, RPG and PL/1 etc. A sample listing of FORTRAN instructions is given in Fig. 2.6.

```
C
C    Read in gear pitch diameter
C
  2      WRITE(5,207)
  207    FORMAT(/, '$ENTER GEAR PITCH DIAMETER AT OUTER CONE DISTANCE
       1 (INS) : ')
         CALL ASKINP(10)
         READ(5,208) D
  208    FORMAT(F10.4)
         IF(D.GE.3.0.OR.D.LE.31.5) GO TO 7
         WRITE(5,103)
  103    FORMAT(' *** – VALUE SPECIFIED OUTSIDE ALLOWABLE RANGE (3.0–31.5 INS)
       1 FOR GENERATING MACHINE')
C
C   Calculate diametrical pitch PD, metric module SIM, and
C   ratio of gear to pionion teeth RATIOM
C
  7      PD = TEETHG/D
         SIM = D/TEETHG
         RATIOM = TEETHG/TEETHP
C
C   Calculate cotangent of approx. value of gear pitch angle
C
         COTTAJ = 1.2/RATIOM
         SINTAJ = 1.0/DSQRT(1.0+COTTAJ**2)
              ,,              ,,              ,,

              ,,              ,,              ,,

         STOP
         END
```

Fig. 2.6 – An example listing of FORTRAN code.

2.7　OPERATING SYSTEM

An operating system may generally be defined as a program which organises a collection of hardware into a consistent and coherent entity so that it can be used for developing and running programs. It can assist users and their programs to make the most efficient use of system resources such as memory capacity and processing power with minimal programming effort by providing software to supplement and enhance the system facilities available. Effectively, an operating system acts as an interface between the user and the computer. There are a number of different processing techniques widely used for implementing this human—computer interface, and each technique has its own characteristics.

Batch processing

In this mode of processing, programs to be run are collected several at a time, and are entered into the computer as a batch of programs, hence the name 'batch processing'. The programs are read in and placed on disk storage to form an execution queue. When one program is completed, the next one in the queue is loaded for processing. In other words, programs are processed sequentially one at a time. Batch processing provides no or very little interaction between the user and the computer. The user usually prepares programs and their associated data on punched cards, submits them to the computer and, some time later, receives a print-out of results which can be examined at leisure. The results may also be output as a graph or drawing produced by the computer's graph plotter.

On-line processing

In an on-line mode of operation, the user communicates directly with the computer through its terminal devices. Information and instructions entered via a terminal are executed virtually immediately, and a response is received as soon as possible, often within seconds. The user can also use all the available devices to their best advantage. On-line processing is normally used where a high degree of user—computer interaction is necessary.

Multiprogramming

There can be only one program at a time in main memory for execution in batch processing. As the CPU is normally the fastest component in the computer system, its expensive memory and the full capabilities of the ALU may be under-utilised because not every program will be large enough to fill main memory and not every program will need to use the ALU all the time. Multiprogramming attempts to balance the CPU's speed with the slower devices by enabling more than one program to reside in main memory at the same time. The objective is to make more efficient use of the CPU by keeping more parts of it busy more of the time. For example, when program A is receiving data or outputting results (I/O operations), program B can then be processed. If both program A and program B are performing I/O functions, a third program, C, can be executed.

Time-sharing

When a computer system operates in time-sharing mode, its resources are shared by several users through terminals which are often connected to the computer by land or telephone lines and are commonly referred to as remote terminals. The basic idea behind time-sharing is to allow all users to have a brief share of the central processing unit in turn. Although the computer actually services each user in sequence, the high speed of the computer makes it appear that the users are all handled simultaneously. Each user is allocated a short interval of time on the computer, often known as a time slice, during which it has sole control of the CPU. Once the interval has expired, the control of the CPU is passed over to another user; thus the CPU cannot be monopolised by one single user beyond a fixed time limit. For example, a computer system can support up to five terminals, and user 1's program is allowed to run. When its time slice is over, or if it requests an input/output function and therefore temporarily does not require the CPU, the CPU is promptly scheduled for user 2's program. User 1's program is then swapped out of memory until its next turn to run has come, and user 3's program is swapped in to be ready for execution. During this time interval, user 2's program is being processed. A program is in main memory only when it is being executed, and it is on backing storage when it is not being executed. As the swapping process, sometimes called 'roll in—roll out', happens many times within a few seconds, the only feasible backing storage devices are disks which transfer data at a much faster rate.

Real-time
The term 'real-time' is used to refer to any system which produces a response almost immediately as a result of inputting data. The processing of these input data must occur virtually at the same time as the event generating them, so that the response is fast enough to affect subsequent actions to be taken promptly within time limits in order for it to be useful in the real world. Real-time systems are often used in process control and airline ticket reservation.

2.8 SYSTEM CONFIGURATION

As is well known by now, apart from the central processing unit, a computer system consists of other devices such as backing storage devices and input/output devices which are collectively called peripheral devices, or simply peripherals. A peripheral unit is defined as any hardware device that is connected to the CPU and performs a specific task such as input/output for the system. There are normally many peripherals in a computer system, and the way in which they are arranged and what they are determine the configuration of the system. Obviously, the number of various possible system configurations is almost infinite, and each configuration produces a totally different performance ,depending on the capabilities of each individual peripheral in it. However, there are in general a number of established types of system configuration as explained below.

Simple configuration
A simple configuration of computer system is illustrated schematically in Fig. 2.7. The system comprises essentially one CPU, various backing storage devices and several input/output devices. As can be seen, such a configuration is rather small and its processing power is undoubtedly limited. Thus, when the number of users increases and their demands for system resources grow, this kind of system configuration cannot really cope and there is very little room to expand by way of introducing additional hardware. Because of the lack of flexibility of this type of configuration, a new concept, often known as network

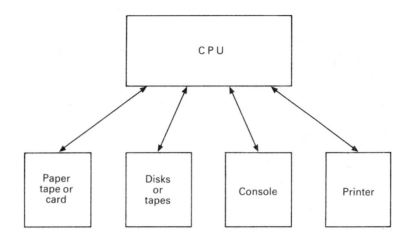

Fig. 2.7 – A simple configuration of a computer system.

configuration, has been developed and has now become widely used in many computer applications.

Network configuration

A computer network system is formed by linking distinct computer installations together through a communications system. The purpose of setting up a network is to share computing power and, perhaps more important, to improve the flow and to allow the exchange of information within organisations and beyond. Taking it further, network systems may be considered as a step towards establishing a means by which people can communicate with each other through a computer. A local area network (LAN) allows organisations operating on one site to link their computers, user terminals and workstations, and often peripheral devices in an efficient and cost-effective way. At the same time, different countries are gradually developing nationwide and worldwide data communication networks to operate as a wide area network (WAN), which links various sites, computer installations and user terminals, and may even permit LANs to communicate with each other.

There are three main classes of configuration for computer network systems:

(a) centralised network
(b) distributed network
(c) ring network.

A centralised network is shown in Fig. 2.8, and it is characterised by a centralised computer complex, a communications system, and a number of users who can access and interact with the computer system via local terminal devices. A centralised network is particularly useful for organisations that require a centralised database or a centralised processing facility. A distributed network is illustrated in Fig. 2.9, and it is characterised by two or more computer

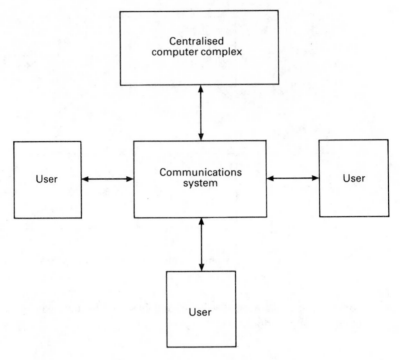

Fig. 2.8 – Centralised computer network.

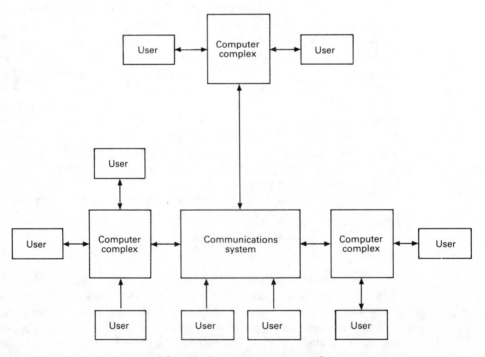

Fig. 2.9 – Distributed computer network.

complexes that are connected via a communications system. Users may access and interact with one of the computer complexes through local communications facilities or may be linked to the communications system directly. The advantage of a distributed network system is to place computing power where the user is. Thus, it eliminates some of the problems associated with a centralised network system. For example, if the computer if a centralised system fails to function, which is not uncommon, all computer operations will have to be suspended. Also, as a result, the system will become overloaded when it is operational again.

Figure 2.10 shows a typical ring network system which is a special case of a distributed network. Every computer system is connected to exactly two other systems in the ring network which is often employed by large organisations to coordinate on an infrequent basis its various branches in different locations.

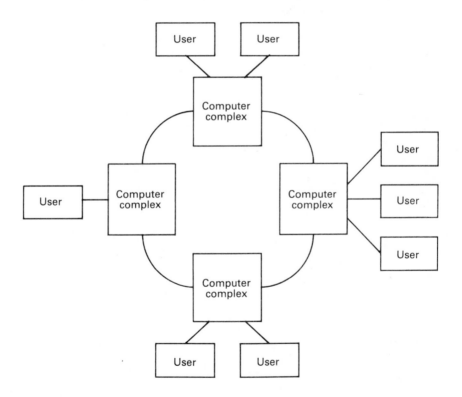

Fig. 2.10 – Ring computer network.

REFERENCES

[1] Besant, C. B., *Computer-aided design and manufacture,* 2nd edn, Ellis Horwood, 1982.

[2] Lui, C. W. K., *The use of computer-aided techniques for hypoid gear design*, Ph.D. thesis, Imperial College of Science and Technology, 1983.

[3] Groover, M. P. and Zimmers, Jr, E. W., *Computer-aided design and computer-aided manufacturing*, Prentice-Hall, 1984.

[4] Eckhouse, Jr, R. H. and Morris, L. R., *Minicomputer systems: organisation, programming and applications*, Prentice-Hall, 1979.

[5] Hunt, R. and Shelley, J., *Computers and commonsense*, 3rd edn, Prentice-Hall, 1983.

[6] ICL, *Introduction to computer systems*, ICL Technical Service, 1972.

[7] Katzan, Jr, H., *Introduction to computers and data processing*, Van Nostrand, 1979.

[8] Chandor, A., Graham, J. and Williamson, R., *The dictionary of computers*, 2nd edn, Penguin Books, 1983.

[9] Computer Technology Review, *The systems integration sourcebook*, West World Productions, Inc., Winter 1983.

3

Computer-aided design system hardware

3.1 INTRODUCTION

In CAD/CAM, some special input/output devices are needed for it to perform the required functions, besides the usual hardware such as the CPU, mass storage devices, printer and alphanumeric terminal devices that can be expected to be found in a computer system. As computer graphics is a function fundamental to CAD/CAM, the special input/output hardware is mainly graphics devices that enable coordinate data to be input and allow graphical information or pictures to be displayed on a screen. Then, if it is required, hard copy of graphical data may be produced. In the following sections of this chapter, the different kinds of input/output device will be individually described in detail, and the various types of computer system configuration will also be considered and discussed with regard to their suitability for CAD/CAM.

3.2 GRAPHICS INPUT DEVICES

Graphics input devices are tools with which the user can interact with the system and specify graphical information such as coordinate data so that graphics software can perform computations on them in order to generate the required results on display devices. In general, there are three main categories of graphics input device commonly used in CAD systems: light-pens, analogue devices, and keyboard devices.

3.2.1 Light-pens

Physically, a light-pen is very similar to a fountain pen in both size and shape. Quite different from what its name implies, the light-pen does not emit light. Instead, it detects light on the CRT screen by using a photodiode, photo-transistor, or some other form of optical sensor. Light-pens can be employed to perform two functions, the principal one of which is to point at graphic objects displayed on a refresh-type CRT. The other function of a light-pen is to locate positions within the displayed area. This latter function is often called tracking in which a cross-hair cursor is displayed on the screen at a position corresponding to that of the light-pen. A light-pen can only be used with a refreshed CRT because the principle on which it works is that a point is recorded at the screen position where the photocell at the pen tip senses the light when the CRT beam reaches that position during one of its cycles to refresh and maintain the image on the display. The resolution of a light-pen is poor when compared with that of a tablet and cursor. This poor resolution is the combined effect of the light-pen's relatively large field of view and the display screen's limited image sharpness.

3.2.2 Analogue devices

Many of the graphics input devices are analogue-to-digital (A/D) converters, which are essentially devices that detect some physical quantity such as speed, acceleration, force, position, direction, rotation, distance, etc., and then trans-late it into a numeric quantity such as a binary value that the computer can accept and understand. Some examples of these analogue graphics input devices are the digitiser, tablet, joystick, tracker ball, and dial.

(a) Digitiser

A common way to take *x, y* coordinates off a drawing is to use a digitiser. The digitiser consists of a table or board similar to a drawing board, and a probe consisting of a pen or cursor which can be moved over the surface. The pen, or cursor, contains a switch that enables the user to register *x, y* coordinates at any desired position. These coordinates can be fed directly to a computer or to some off-line storage device such as industrial compatible magnetic tape.

There are many digitisers on the market, operating on various principles. The Talos digitiser, for example, shown in Fig. 3.1, is based on an accurately set out wire grid mounted on a temperature-stable board. An electric field is electronically switched to portions of the draughting surface in accordance with the movement of the pen or cursor – which, in turn, senses the electric field. Utilising a writing servo principle that can, in fact, resolve much larger areas of active surface, the digitiser surface is driven in increments of only one inch by one inch. When the pen or cursor goes beyond the dynamic range of the one-inch area, the activated area is electronically switched to follow the pen or cursor. The activated area is resolved by circuitry so that interpolation takes in the *x* and *y* directions. This technique does not place any restrictions on the size of the digitiser, since the system resolution is independent of size.

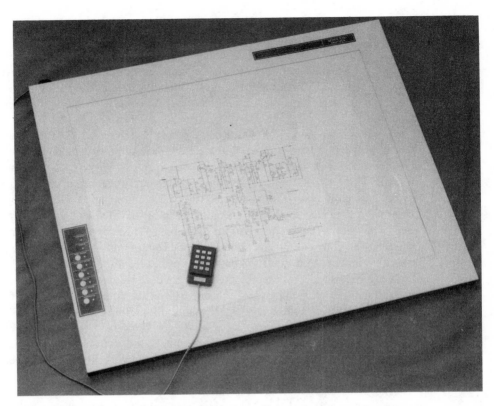

Fig. 3.1 – Complot digitiser with multi-button cursor system.

The system has the advantage that any specific position is defined by a data output that represents the absolute location. There is no need to reference the cursor or pen to a starting position; therefore it can be removed from the surface and replaced, with an immediate output of the absolute position. The Talos digitiser is available in sizes from 11 × 11 inches up to 44 × 60 inches, and an accuracy of ± 0.005 inch is possible.

(b) Tablet
The tablet is a small low-resolution digitiser. It is often used as an alternative to the light-pen or in association with storage tube displays. The tablet serves as a surface corresponding to the CRT but remote from it. The user writes on the surface of the tablet, and the position of the pen is transmitted to the computer. The screen which then displays a cursor is an echo of the position of the pen on the tablet.

The tablet is particularly suitable for interactive design since it allows the designer to work naturally with a pen in his hand while watching the screen to check his actions. The computer can straighten lines and present a clean picture if it is fed with accurate salient points. These points can be input either from an alphanumeric keyboard or from a menu on the tablet. A menu consists of defined squares on the tablet. Any point digitised within a square activates an appropriate instruction within the computer.

The digitiser is often used in draughting applications where information is taken off accurate drawings. It can be useful in working on large layouts. However, tablets are used more and more where increase in computing power, coupled with large and faster display systems, makes use of the digitiser less essential.

A well known tablet, on which many others have been based, is the Rand design. This has a wire grid of 1,024 closely spaced wires in both x and y directions. The wire grid is mounted in a square plate. The wires are pressed in touch with a set of contact pads which are laid out in the margin of the work area and which provide a voltage drive. The pads are arranged in the Gray-code pattern for 10 digits, so that by energising each of the 10 sets of pads in turn, each wire receives a unique pattern of pulses. This contact configuration is applied in each coordinate direction. The voltage pulses are sensed by a capacitive probe which reads the pattern of pulses from the wire nearest to it. The Gray-code signals received by the probe are decoded into binary values, and by subtracting each reading from the one before, the computer can check for errors or pen lift-off, since each signal should differ from the pervious one by 1 bit only.

The wires in the grid must be spaced at intervals of 0.01 inches in a typical 10-inch square tablet. Furthermore, each wire must be accurately positioned, and this can lead to an expensive tablet.

The Talos digitiser in low resolution form (± 0.001 inch) with an 11-inch square working area, makes an ideal tablet, combining low cost with an excellent all-round performance.

There are many other types of tablet such as the sensitive film design developed by Walker at Reading University and the Bell Telephone Laboratories. Another tablet is the Graf/Pen Sonic Digitiser which is based on two linear microphones which sense acoustic signals from a sonic source located in the digitising pen or cursor.

(c) Other digitisers

Apart from the two preceding types of analogue graphics input devices, there are a number of other analogue devices that are sometimes used for graphics input in CAD systems. They are the sonic digitiser, touch-sensitive digitiser, light-detector digitiser and incremental digitiser. A sonic digitiser is based on two linear microphones which sense acoustic signals from a sonic source located in the digitising pen or cursor. A touch-sensitive digitiser uses an ordinary pen or pencil together with a touch-sensitive surface which detects pen movements by sensing the pressure generated when the tip of the pen is pressed onto the surface by the user. A light-detector digitiser, as its name suggests, operates through light detection. A series of tiny light-emitting diodes (LEDs) are arranged along one horizontal and one vertical edge of the digitiser. Opposite the LEDs, a corresponding series of light detectors are installed on the other two edges. Thus, a matrix of infrared light beams is formed by the LEDs. When a finger or a pencil is placed on the active area of the digitiser, the infrared light beams will be interrupted, thus causing the digitiser to generate the coordinates corresponding to the position of the interruption. When compared with other digitisers, light-detector digitisers have an extremely poor resolution, about 0.25 inch, which explains why they are not widely used in applications where positional accuracy is very important. An incremental digitiser utilises a rotary or linear encoder that uses optical, magnetic, mechanical or brush contact means to generate a distance-increment count that corresponds to the total distance travelled by the positional device across the digitiser surface. Incremental digitisers can easily provide a resolution as high as 0.001 inch.

(d) Joystick, tracker ball and dial
Joystick, tracker ball and dial can play the same role as a light-pen in a CAD system, that is, they are used mainly for cursor control. A joystick allows the user to indicate the direction, speed, and duration of the cursor motion. A joystick is a handle approximately 1 to 4 inches high and is mounted vertically in a base. It can be moved with the fingers in any arbitrary direction that comprises forward-backward and right-left components. The joystick gives a faster response in cursor control, but less accurate, than the tracker ball or dial. A tracker ball is a simple rotating sphere operated with the palm of the hand. This device has a ball, typically 3 inches in diameter, mounted on rollers inside a box so that about one-third of the ball is exposed through the top of the box. Cursor movement is controlled by the rotation of the ball. The distance that the cursor travels on the screen is directly proportional to the number of revolutions of the ball, whereas the direction of cursor motion corresponds to the direction of rotation. A dial is a one-axis cursor control device. The dial usually takes the form of an ordinary rotating knob, but in some cases it is a lever that moves back and forth. The direction of cursor movement along an axis depends on whether the knob is turned clockwise or anticlockwise. The distance that the cursor moves varies according to the angle of rotation of the knob. The cursor remains stationary when the knob is released. A familiar example of the dial is the thumbwheel. There are usually two thumbwheels, one each for x-axis and y-axis, mounted at right angles to each other on the side of a graphics terminal keyborad such as the Tektronix 4014. Thumbwheels are most suitable for fast and accurate cursor movement along straight lines, but are not so desirable as a joystick or tracker ball for directing the cursor in a diagonal direction or along a curved path.

3.2.3 Keyboard devices
Keyboards, buttons and switches constitute an important class of input devices that can be used in various configurations to meet different requirements. A keyboard inputs alphabets, numerals, and other characters by sending digital codes to the computer. A button, considered as a special type of momentary switch which rebounds after being depressed, can indicate only one data item, that is, the condition that it has been pressed. Similarly, a switch can send only one piece of information to the computer, that is, whether its current position

is on or off. The interpretation of the meanings of the button and switch input data is performed by the graphics software. A voice data entry system may also be considered as a keyboard device that inputs complete words or phrases from a limited vocabulary by recognising human speech.

(a) Keyboard
Many graphics terminals incorporate some kind of keyboard that has a similar layout to that of a typewriter. Obviously, a keyborad does not provide the most suitable interface for human−computer interaction, but nevertheless it can sometimes be used within its limitations for positioning by typing the actual coordinates of each point and selecting a displayed object by typing its name. As the alphanumeric keyboard is generally the most efficient device for input-ting text, it is frequently used to enter non-graphic data such as part numbers, material specifications, and other textual information associated with graphic objects into the graphics system database. There are now variations of the conventional keyboard for different types of usage. For example, traditionally, the shift key on a keyboard is used to select between upper case and lower case letters, but instead, it is used to choose between upper case letters only and a set of graphics symbol characters which can be treated as building blocks for creating pictures in some limited graphics systems. For applications that involve entering a large volume of purely numeric data, a twelve-key numeric pad that consists of a rectangular arrangement of keys for the digits 0 through 9 along with a minus sign and a decimal point is incorporated into the conventional keyboard, but it can be a completely separate unit. A number of graphics terminal keyboards have a set of five special keys for cursor control. Each of these keys causes the cursor to move up, down, left, right and home, which is a programmed origin position. Diagonal movements of the cursor can also be generated if two keys for perpendicular directions are pressed simultaneously. Some keyboards may have special program function keys which offer a faster and effective way to issue commands to the system with fewer errors because a single key stroke specifies an entire command statement. Program function keys may also be mounted on an independent, self-contained cabinet.

(b) Buttons
Buttons and keys work very similarly, that is, a data item is sent to the system when a button is pressed, and no data item when released. Buttons are often found on the hand-held input device of a digitiser where frequently used commands are assigned so that they can be accessed much more easily and readily.

(c) Switches
A switch is an input device that remains in one of two possible positions until changed by the operator. A data item is transmitted to the computer when there is a change of position by either pressing a key or reversing a switch. The data item indicates that data entry has occurred and it contains the code to identify the specific switch causing that entry. For example, a program function is

invoked by one key strike and is terminated by the next strike of the same key. Active program function switches or keys are often illuminated with a little lamp inside or close to it to help the user select the required function.

(d) Voice data entry system

A voice data entry (VDE) system allows the operator to enter graphics information or system commands using speech, that is, actually saying the input data to the system. For this system to work, a predefined vocabulary of commands must first be established, together with their related words spoken by a particular operator. A separate vocabulary has to be created for and by each operator because no two persons speak in the same way. When an operator issues a spoken command, the VDE system recognises it and searches the operator's own vocabulary for the matching spoken words whose corresponding command entry in the vocabulary are then returned as alphanumeric characters on a visual display for the operator's confirmation. The operator can confirm by saying a code word such as 'go' to mean that the result is correct and can be sent to the computer, or 'cancel' to ignore the result because it is incorrect. VDE systems are still rarely seen in CAD systems as the technology is only in its early stages of development, but it can supplement, or potentially replace, graphics input devices such as an alphanumeric keyboard, a set of program function keys, and a tablet menu. As speech is the most natural human communications method, it is the easiest for the operator to learn. This method of human—computer interaction offers higher accuracy and fewer errors because it eliminates mental encoding by the operator. In addition, the operator has freedom of movement about a work area, resulting in more efficient work flow, and also the hands and eyes of the operator are free so that another task can be performed at the same time.

3.3 GRAPHICS DISPLAY DEVICES

The graphics display screen is the most prominent piece of equipment in a CAD system because the main purpose of a CAD system is to display a computer-generated picture of a design model on a screen so that it can be examined and graphically manipulated according to the operator's commands to make modifications. Nearly all graphics display devices are based on the cathode ray tube, although some are plasma panel displays. Both these two categories of display devices will be described in the following sections.

3.3.1 CRT displays

In CRT displays, a heated cathode emits a continuous, high speed stream of electrons, which are formed into a beam by an aperture in a control grid surrounding the cathode. The electrons are then accelerated and focused to a point on the display surface. The electron beam is rapidly swept across the phosphor-coated face of the tube line-by-line. The beam current is regulated to increase or decrease its intensity in order that brighter or darker points are created along each swept line. The construction and operation of the CRT is shown in Fig. 3.2.

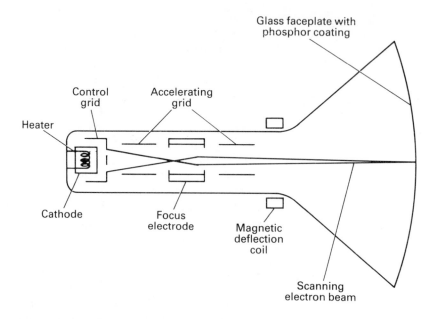

Fig. 3.2 – Cathode ray tube (CRT).

The two major techniques of generating an image are stroke-writing and raster scan. A stroke-writing device draws vectors or lines to create an image, while a raster scan device uses a matrix of closely spaced dots to form a picture. Regardless of whether the stroke-writing or raster scan method was used to create the picture on the screen in the first place, it would have to be maintained by some means as the persistence of the screen's phosphor coating is short and finite. Persistence is defined as the length of time that a phosphor continues to glow after it is excited by the CRT electron beam. There are generally two methods of overcoming this problem of maintaining an image on a screen. A refreshed tube continuously regenerates the image, whereas a storage tube physically retains the picture until it is erased.

Three main categories of CRT graphics displays are commonly used in CAD systems: direct-view storage tube (DVST), directed-beam refresh tube (DBRT), and refreshed raster scan display. Each of these CRT displays employs a combination of the aforementioned techniques to generate and maintain a picture on a screen. The following table summarises the techniques used by these three types of display devices.

Image maintenance method \ Image generation method	Stroke writing	Raster scan
Storage	DVST	
Refreshed	DBRT	Refreshed raster scan

3.3.1.1 Stroke-writing displays

A stroke-writing CRT display uses an electron beam to create a line image on a screen in the same way as when a draughtsman uses a pencil to produce a drawing, that is, only one line at a time. No matter how complex the drawing is, it may be considered simply as a series of straight-line segments. Similarly, an image may be treated as a construction of a number of individual straight lines having any length and orientation. Each line segment is drawn on the screen by specifying the endpoint coordinates which are converted into analogue voltages that control the deflection system in the CRT. These voltages direct the electron beam to trace a path on the phosphor screen that corresponds to the position of the line. Circles and curves may also be produced on the screen if they are approximated by joining a sequence of line segments short enough to give a realistic representation. The method of stroke-writing for generating images is shown in Fig. 3.3.

Direct-view storage tubes and directed-beam refresh tubes come into the category of stroke-writing displays which are sometimes identified by any of a number of synonymous adjectives, such as calligraphic, randon scan, vector writing, line drawing, etc.

(a) Direct-view storage tube (DVST). The most common DVSTs are manufactured by Tektronix who began the production of DVSTs in 1962. Since then the DVST has become so popular that it is a *de facto* standard for graphics displays, against which the performance features of other display devices are judged.

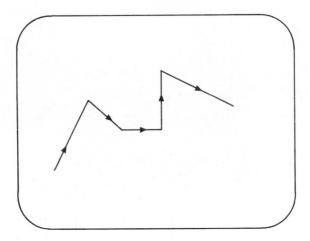

Fig. 3.3 – Stroke-writing method for generating images.

The DVST is similar in construction to the conventional electrostatically controlled CRT, but has an additional grid electrode and a second electron gun called a 'flood' gun. The grid electrode is deposited as an integral part of the phosphor layer on the screen, giving a bistable phosphor.

The operation of the storage tube depends on secondary emission due to electron bombardment. The grid electrode is negatively charged by the application of a suitable voltage before writing. When the main write beam of electrons is moved so as to display a line, the grid becomes positively charged in the area where the tightly focused beam strikes it, so allowing electrons from the flood gun to be accelerated through onto the screen, causing the phosphor to glow. Once the image has been displayed, it is maintained by the energy from the flood gun, and it remains visible without the need for any retracing of the write beam, thus making little demand on the computer for processing. The image can be erased from the screen by recharging the grid negatively. A familiar example of DVST is the Tektronix-4014 shown in Fig. 3.4. The resolution of a storage tube can be up to 4,000 by 3,000 displayable points on a 19-inch diagonal screen, and compares very favourably with that of any other type of display devices. As a result of this high resolution, lines on DVST are fine and sharply focused with crisp edges. Because it is a stroke-writing display, there is none of the 'staircasing' often associated with raster scan displays. However, one limitation is that the lines can only be of one colour. As opposed to a raster image that must be constantly regenerated, the image on a DVST is stored on the surface of the tube, so it is stable and unchanging. Hence, a user does not experience the flicker or character jitter found in some refreshed displays. A trade-off for this advantage of a stable image is that selective erasure is rather difficult with a DVDT. Whenever any part of an image had to be changed or deleted, the stored image would have to be erased completely and the whole image redrawn together with the modifications. Obviously, this is a time-

Fig. 3.4 – Tektronix 4014 Direct View Storage Tube Display. Display sizes
15 in (38 cm) by 11 in (28 cm).

consuming process especially when a complex drawing and a great number of
changes are involved. On DVST, dynamic movements of images are very limited
because it can be achieved only by rapid replacement of the image, and the
drawing speed of a DVST write beam is relatively slow, typically 5,000 inches
per second.

Recent developments in DVST technology have overcome some of the
problems associated with the storage tube by introducing a limited degree of the
capabilities of a refreshed display. An example is the Tektronix 4054 Option 31
desk-top computer with a colour-enhanced refreshed storage tube display
which makes it possible to produce a colour picture and to refresh portions of
a picture – thus selective erasure can be performed.

(b) Directed-beam refresh tube (DBRT). The direct-beam refresh tube is a
stroke-writing and refreshed display device. The DBRT, like DVST, utilises
the stroke-writing approach to generate the image on the CRT screen, except
that the image is not stored and has to be continually refereshed in order to

maintain it in a stable and flicker-free condition. As the phosphor coating on the screen surface has a short persistence, the traced lines are volatile, therefore, the entire image needs to be retraced at a very fast rate so as to avoid flickering of the image. For this reason, there is a maximum limit on the number of lines in the image which can be displayed on a DBRT before the flicker becomes unacceptable. This shortcoming of the need to retrace the image is balanced by several important advantages. Since the image is being regenerated continually, selective erasure and modification of the image is readily achieved. Thus, it is possible to provide dynamic viewing capabilities for an image, and it is highly interactive. Since there is little stored data, memory requirements are low, but expensive and sophisticated circuitry is required to maintain the refresh time. In general, DBRT has a high resolution of up to 4,000 by 4,000 displayable points, and is capable of producing monochrome and colour images. A monochrome monitor can display lines of various brightness levels through control of the electron beam voltage.

DBRT uses the beam penetration method to display colour images. It incorporates a single electron gun and a screen coated with layers of two or more different phosphors separated by a thin layer of dielectric. The colour of each line in the image varies according to the velocity of the beam when that line is stroked. For example, if the screen surface of a DBRT has been coated with a first layer of red phospher and a second layer of green phosphor, the combination of two phosphors provides a possibility of four different colours, that is, red, orange, yellow and green. When the electron beam strikes the screen at the lowest velocity, it can penetrate only the first layer and that phosphor emits a red light. At the next higher velocity, the beam penetrates the first layer as well as a part of the second layer, so some green light combines with the red to give an orange light. The third next higher velocity generates yellow light. The highest velocity forces the beam to penetrate right through the two layers and produce the green light.

3.3.1.2 Raster scan displays

In a raster scan display, the viewing screen is divided into a large number of tiny phosphor picture elements, often referred to as pixels. They are arranged into a matrix, called a raster, that represents the display area of the screen. A raster scan display typically consists of a raster of 256 by 256 pixels, giving a total of over 65,000 pixels, but it can be up to 1,024 by 1,024 pixels, providing a total of over 1 million pixels. Each pixel on the screen can be energised to glow at various brightness levels, and even different colours in the case of colour displays.

Although the construction of CRT units used as raster scan and stroke-writing displays is almost identical, the principle of operation to create graphics is rather different. During operation, the electron beam is magnetically deflected to sweep along a horizontal line on the screen from left to right. The electron beam energises a small phosphor element as it passes a pixel along the line during a sweep. The colour or brightness of a pixel is determined by the momentary intensity of the beam which is continuously changed according to the display data. When the beam sweeps to the end of a line, it is turned off or blanked, quickly moved to the start of the next line, and then turned on to continue sweeping as before. This cycle of scanning a line is repeated at constant rate from the top to the bottom of the screen, where the beam is blanked and swiftly returned to the top again.

Generally, there are two commonly used methods to refresh a raster display: sequential, or non-interlaced, scanning and interlaced scanning.

Figure 3.5(a) illustrates sequential scanning which means sweeping each horizontal line starting from the top to the bottom of the screen sequentially. The problem with this type of scanning is that if the refresh rate is not high enough to match the persistence of the phosphor, the top part of the display appears to fade while the bottom part is still being scanned.

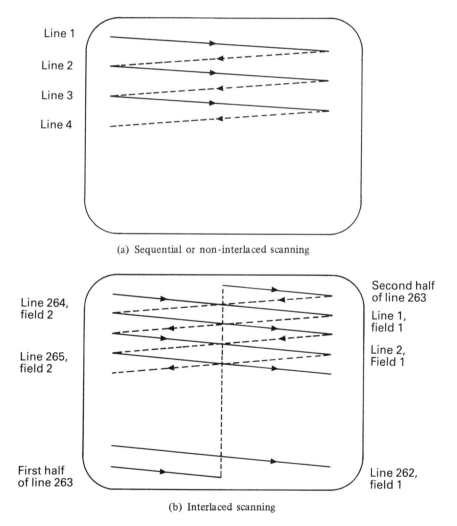

(a) Sequential or non-interlaced scanning

(b) Interlaced scanning

Fig. 3.5 – Raster scanning.

Figure 3.5(b) shows the technique of interlaced scanning which overcomes the problem of flicker in non-interlaced scanning by sweeping all the odd-numbered lines first (field 1) and then sweeping the even-numbered lines (field 2), which means that the screen is scanned twice during each refresh cycle. To prevent flickering, refreshing is usually performed at a fixed rate of between 30 and 60 Hz.

As raster scan display is a refreshed device, the image is regenerated at very frequent intervals, but this necessitates the image to be stored in some memory so that it can be obtained for the purpose of refreshing. The memory used in this way is often known as bit map memory, frame buffer or refresh buffer. For example, if the image on a raster display of resolution 512 by 512 is to be stored in a refresh buffer, and each pixel on the display has two brightness levels, say on or off, the amount of memory required is as much as 512×512 or over 260,000 bits of storage. This calculation is based on the fact that one bit is enough to represent the two beam brightness levels, that is, on or off. As can be imagined, if the resolution is doubled, the total number of pixels on the display will be quadrupled, hence the refresh buffer required to hold the image is increased by four times as well. Now, if it is to enhance the number of brightness levels that can be displayed on each pixel, additional bits will be needed to store the brightness levels on each pixel. To have four different levels, two bits are required. Eight different levels, three bits are needed, and so on. For a colour display, three times as many bits are required to represent the various brightness levels for each of the three primary colours: red, green and blue. Thus, it is not difficult to see that raster scan displays are memory-intensive, and memory costs money.

When compared with DVST and DBRT, the resolution of the raster scan display is relatively low, therefore diagonal lines appear jagged, like staircases. To tackle this problem, a technique, called anti-aliasing, effectively blurs the image so that its edges gradually change from one colour to another.

3.3.2 Plasma panel displays

Although the majority of display devices are based essentially on the CRT, there is one type, called plasma panel display, which employs different display technologies. Plasma panel displays are lighter, less bulky and consume much less power than other displays which are mostly CRT-based, while their performance and price are comparable. No anti-distortion circuitry is required in plasma panel displays because their screens are flat.

Figure 3.6 shows an exploded view of a plasma panel display. The term 'plasma' in the name of the device refers to the neon gas that is sandwiched and sealed between two glass panels. A series of thin conductive strips are attached to the inner surface of each glass sheet, running horizontally on one side and vertically on the other. Each crossing point of two electrodes is a displayable point which emits light when both electrodes are activated. The resolution of a plasma panel display can be up to 512 by 512 pixels (crossing points). A dielectric layer is used to cover the electrodes, and it comprises another sheet of glass coated with magnesium oxide.

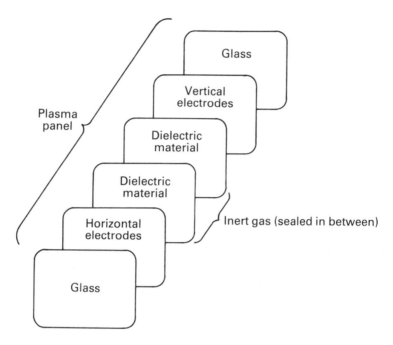

Fig. 3.6 — An exploded view of a plasma panel display screen.

The way in which a plasma panel display creates an image is totally different from that of a CRT. In a CRT, an electron beam is quickly scanned across the phosphor-coated screen of the tube, causing the phosphor traced by the beam to glow and an image to be formed as a result. In a plasma panel display, when a voltage is applied to the electrodes, light is produced at each pixel by electrical excitation of the gas which may glow (active or light-emitting) or change the way that light is reflected (passive). The plasma panel display multiplexes, or time shares, digital circuits to feed signals to one line of pixels at a time. One driving circuit is used for each horizontal row and one for each vertical column, rather than a separate driver for each pixel. A horizontal row of hundreds of pixels can be fed some voltage, causing all the pixels to remain dark. Additional small voltages applied to all the column electrodes light all the pixels along that row. They can be made brighter by using a larger voltage or a longer pulse. The image is created by sequentially painting it line-by-line down the display.

A plasma panel display can be described as a monochrome, single intensity, raster format, stored image device with selective erasure capability. Another unique feature of plasma panel displays is that their screens are transparent, thus additional related information from slides or microfilm can be projected from the rear to superimpose with the displayed image.

3.4 GRAPHICS OUTPUT DEVICES

After graphics information has been entered into a CAD system, the designer will usually examine it on a graphics display and will modify it as and when it is necessary to improve the design. When the design has been finalised, a hard copy is often produced as a record of the design for presentation or archival purposes. To produce such hard copy output for a CAD system a plotter is generally used. Two main classes of plotters are commonly used in CAD systems: pen plotters and electrostatic plotters. This section attempts to provide an introduction to these two types of plotters and other graphics output devices that are sometimes found in CAD systems, and the various modes of operation in which graphics output devices work in relation to the CAD system.

3.4.1 Pen plotters

Pen plotters seem to be the most popular kind of graphics output devices because of their versatility; they may be subdivided into flatbed plotter, drum plotter, beltbed plotter and plot-back digitiser.

Figure 3.7 shows a typical flatbed plotter in which a gantry is supported over a flat surface and can be moved in the x-direction. Mounted on this gantry is a carriage which can hold one or more pens and can traverse in the y-direction along the gantry. Both the gantry and carriage are usually driven by d.c. servo-motors with optical encoders for positional sensing. By a combination of gantry

Fig. 3.7 – A typical A0 flatbed plotter (courtesy of Quest CIL Ltd).

and carriage movements, the pen can be positioned to anywhere within the plotting area. The drawing paper is laid flat on the bed, and is usually held down by vacuum or electrostatic means or simply by adhesive tape. Nowadays, with the advent of microelectronics, a microprocessor is often incorporated into a plotter and acts as a plotter controller. The microprocessor allows certain common geometric shapes, such as circles and ellipses, to be preprogrammed, so that simple instructions to the plotter will quickly produce a complicated shape. The microprocessor is therefore used as a vector generator, and it has the advantage that variations in standard shapes can be programmed into the microprocessor to suit a variety of applications.

A flatbed plotter on average costs more to buy, but it has a few advantages, one of which is that the whole drawing can be seen as it is being drawn. Thus any drawing errors can be spotted immediately as they occur. Another advantage is that it can take any type or size of drawing paper so long as it is properly laid on the plotting surface.

Figure 3.8 shows a typical drum plotter in which a drum or, in fact, a roller is driven by a stepper motor. Paper with perforations along each edge is fed over the drum which has sprockets at both its ends to move the paper when plotting. A pen carriage which can hold one or more pens is mounted on a gantry

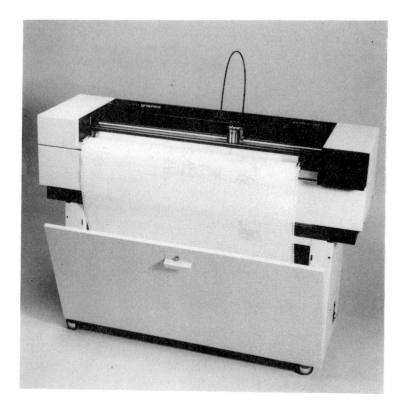

Fig. 3.8 – A typical drum plotter.

above the length of the drum. A second stepper motor is used to drive the pen unit back and forth along the gantry, and the pen is moved up and down by a solenoid. By a combination of drum, hence paper, and carriage movements, a pen can be moved to any position in both the x- and y-directions relative to the surface of the paper.

A drum plotter is inexpensive and occupies less floor space than a flatbed plotter, so it is very suitable for small offices. The only inconvenience of using a drum plotter is that it will accept just one single size of specially made drawing paper. This means that, if the right drawing paper is not available, no drawings can be produced, even when the plotter is in good working order.

Besides flatbed and drum plotters, variations of these to types are also sometimes used. The beltbed plotter is one example and is basically a design that combines the best features of both flatbed and drum plotters. It incorporates a continuous belt that covers both sides of the flat, nearly vertical plotting surface which moves, like the paper on a drum plotter, in coordination with the pen carriage movement to generate a plot. Because a belted plotter has a flat plotting surface, it can accept any kind of drawing paper. Since the plotting surface is almost vertically mounted, floor space is economised.

There is also the plot-back digitiser which is really a graphics input and output device that can function both as a digitiser and as a flatbed plotter. A plot-back digitiser can record drawing data if the operator positions the plotter pen over the desired input point and presses a button. The coordinates of the point are then sent to the computer and stored there. When the input is completed, the whole drawing can be retrieved from the computer and plotted out. This type of device enables the reproduction of some previously plotted drawings whose original plotting data are not available. In addition, a plot-back digitiser can be very useful for comparing and checking drawing revisions if a second plot is produced with a different colour pen.

3.4.2 Electrostatic plotters

Although a pen plotter can produce highly accurate drawings, the time it takes to plot a drawing is usually quite long because it can plot only line by line and only in the order that the data were entered when the drawing was created. Hence, there is a certain amount of unnecessary random pen movement when positioning the pen from one line to the next. As can be imagined, when the drawing is complex and contains a large number of lines, the plotting time can become unacceptably long. In applications where high volume of graphics output is required within a short time, an electrostatic plotter is often used.

An electrostatic plotter, shown in Fig. 3.9, consists of an electronic matrix which can print dots onto charge-sensitive paper. The matrix is held on a bar, and the paper is fed over rollers and under the matrix. The drawing is produced in a raster type format, as the paper passes under the matrix. The dots can be produced with a density sufficiently high, typically 200 dots per inch, for a series of dots to make up a line. It is possible for some electrostatic plotters to produce plots of up to 72 inches in width. Compared to pen plotters, electrostatic plotters can plot much more quickly, sometimes of the order of three to

Fig. 3.9 — A typical electrostatic plotter with 11-inch wide paper, interfaced to a minicomputer (courtesy of Calcomp Ltd).

four times more quickly, but it is over five times more expensive and the quality of the drawing is on average not as good, though acceptable, because it is a raster format output. Special paper and chemicals have to be used for the output, and it is necessary to preprocess the drawing data before plotting, so the high plotting speed is achieved at the cost of extra computational overheads to convert graphics data in the computer database into a raster format. The electrostatic plotter can also be used as a printer to output alphanumeric text.

3.4.3 Other graphics output devices

There is available a whole range of graphics output devices apart from pen plotters and electrostatic plotters. An example of these alternative devices is the dot-matrix impact printer/plotter, also known as dot-matrix impact graphics printer, which is capable of producing both alphanumerics and graphics. It contains an array of needles that hit an inked ribbon against the paper to obtain a pattern of dots. Graphics printers are usually moderate in cost and relatively easy to use. Its major advantage is low cost per copy, but poor image resolution restricts its use in some applications. Another example is the ink-jet plotter which is also a matrix device that produces drawings by forcing droplets of ink directly to the medium, usually paper. Full colour plots can be output through combining dots of red, green and blue pigment. New ink-jet devices, such as the Tektronix 4691 colour graphics copier, offer high quality and reliable production of colour graphics output. Thermal printer/plotters are also used as graphics output devices. An electrical resistor tip is employed to heat temperature-sensitive paper to generate an image. A thermal output device is compact and low in cost. Sometimes, camera/film techniques are applied to generate graphics output. In such a case, superior colour and resolution can be obtained, although camera systems are usually slow, large and expensive.

3.4.4 Modes of operation

A suitable choice of the graphics output device for a CAD system is obviously crucial to the overall performance of the system; but the mode of graphics output operation is just as important to meet the needs of a particular application. Modes of operation, in this case, means the methods used to supply data to the graphics output device, commonly a plotter. In general, there are four modes of graphics output operation: hard copy, on-line, off-line and remote.

(a) Hard copy mode
Figure 3.10(a) illustrates the operation in hard copy mode. If a permanent copy on paper is required of an image on a CRT display screen, it can be produced using the plotter in this mode of operation. The term 'hard copy' distinguishes the permanent physical presence of the image on paper from the 'soft' temporary visual image on the screen. In fact, a hard copy may be simply defined as a plot produced by any output operation mode. Hard copy mode provides a quick way to reproduce on paper the same image on the screen. It is basically a straight copy, or screen dump, of the image. A hard copy is sometimes very useful in

some applications where it is necessary to check rough preliminary or intermediate output before the final version is produced on some other plotter with
the required quality. An electrostatic plotter and a dot-matrix impact plotter
are most suitable for working in this mode because the plot data from the CRT
tube are in a raster format. Pen plotters are not really appropriate for hard copy
mode.

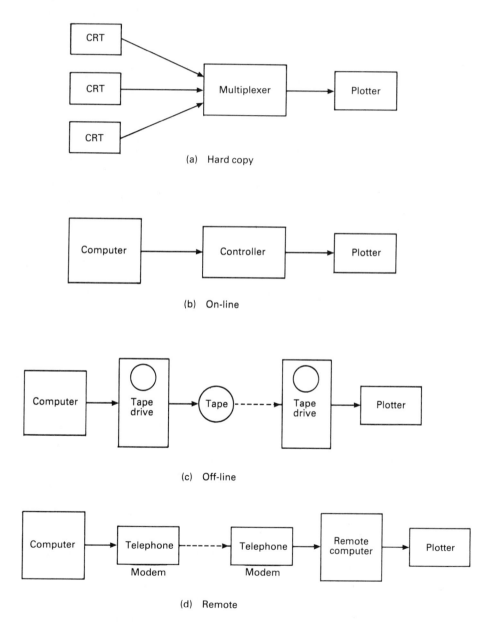

Fig. 3.10 – Modes of operation for graphics output.

(b) On-line mode

Figure 3.10(b) shows the on-line mode of plotter operation. This is the simplest mode in which the plotter is connected directly to the same computer that is running the CAD software. On-line mode has the advantage that intermediate data transfer steps between the plotter and the computer are avoided; therefore, related hardware for this operation is minimised, hence the cost as well. One possible drawback of on-line mode is that additional processing demand is placed on the computer.

(c) Off-line mode

Although on-line mode is simple and involves less hardware investment, it does have its disadvantage. Plotting is usually slow relative to the speed of the computer. In situations where continuous plotting is required, the computer is held up by plotting to the extent that other essential functions of the system cannot really be performed to any acceptable level. Off-line mode is frequently used as a solution to this problem. Figure 3.10(c) illustrates how off-line mode works. In this mode, the graphics data are transferred from the computer to the plotter through some removable storage volume such as a magnetic tape or a disk. The computer outputs the data to a storage volume which is then removed and physically taken by a human operator to the plotter system which reads the storage volume and sends the stored data to the plotter to produce the drawing. With this off-line procedure, it is more efficient and flexible in that there is much freedom to choose whether to produce the plot immediately, keep them for later plotting, or use them repeatedly if multiple copies are required; and the computer is relieved of the task of plotting so that it can perform other functions much more quickly.

(d) Remote mode

The operation of remote mode is shown in Fig. 3.10(d). The transfer of graphics data is achieved through the use of telecommunications. The data transmitted to the remote system may be used for on-line plotting, or they may be stored on magnetic tape or disk for off-line production of plots. Remote plotting is commonly used when the graphics system that generates the plot data is run on a computer that is situated far away in the company's head office, or on a computer that belongs to a service bureau from whom computer time is purchased. The cost of the modems required for sending and receiving the data is relatively low, but the major disadvantage is the high cost of telecommunications time.

3.5 CAD SYSTEM CONFIGURATION

In the early days of computer-aided design, graphics peripherals were connected to the computer system to form a typical configuration of a CAD system as shown in Fig. 3.11. Each graphics device communicates directly with the computer which is very often one of the commercially available general-purpose minicomputers. Traditionally, most CAD minicomputers are 16-bit word machines, but the new minicomputers tend to use 32-bit processors and are also

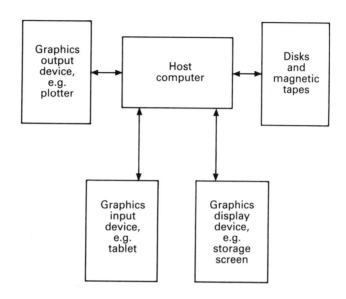

Fig. 3.11 – Conventional CAD system configuration.

known as superminicomputers. The advantages of using a 32-bit minicomputer include higher processing speed, greater accuracy and more main memory. In some powerful CAD systems, a mainframe computer may even be used. For secondary storage, disks and magnetic tapes are commonly employed so that fast access and efficient storage of data are achieved. CAD systems software and its associated database for drawings are generally stored on disks because they provide rapid random access of data. As a result of this feature, the computer can quickly load and swap programs and files between main and secondary memory as an when needed. Since magnetic tape is a sequential access device, it is only used to store programs and files that are not frequently required by the system. In a CAD system, magnetic tape is mainly used for disk backup, permanent archival drawing files, and data transfer to output devices such as plotter or other computer installations.

Clearly, it can be seen that the CAD system illustrated in Fig. 3.11 has a centralised configuration which has the advantage of being simple and straight-forward because each graphics peripheral is connected directly to the computer. However, this configuration does not bring out the best possible performance from the system. As there is only one computer in the system for processing, it cannot support too many peripherals. In addition, graphics functions are usually input/output intensive, so the computer may be unnecessarily held up.

The current trend in CAD systems is to use a workstation in a distributed configuration as shown in Fig. 3.12. A CAD workstation such as the one shown in Fig. 3.13 would consist of a graphics display and graphics input devices which are linked to a local computer with its own secondary storage. The concept of workstation is that a collection of hardware is combined together to form a

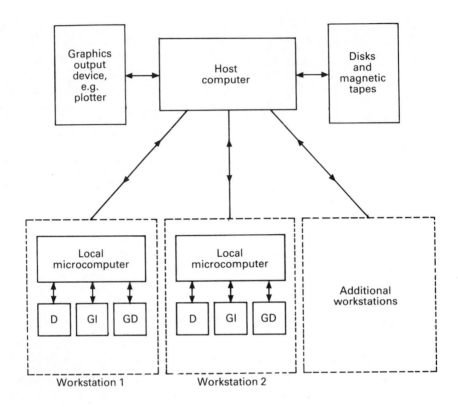

Fig. 3.12 – Current CAD system configuration. D – disk; GI – graphics input
device; GD – graphics display device.

single unit capable of acting as a standalone system to perform computations
and essential functions required for a particular application. As a workstation
has its own computer and secondary storage, it is sometimes referred to as an
intelligent workstation. The local computer in a CAD workstation is usually one
of the microcomputers based on a 16-bit microprocessor, so it is sufficiently
powerful to handle most of the processing demands in CAD. Each CAD workstation is treated basically as a functional entity which is connected to the host
computer through one of the data communication lines. The host computer
takes on the role of central database for storing drawings and other data created
at the different workstations by various users. In addition, the host computer
will run those complex analysis programs with which the local computer does
not have the resources to cope. Sometimes, the host computer acts as an
intermediary for communications between separate workstations.

There are many benefits in arranging devices into workstations and connecting them into a distributed network configuration. For example, if demand for
CAD usage requires that the system be expanded to accommodate more users, it
will simply be a matter of introducing more workstations. Because each user
works on a standalone CAD workstation which is dedicated to provide total

Fig. 3.13 – Interactive CAD workstation (courtesy of Sun Microsystems, Inc.).

support for all interactive graphics activities, the user will not have to compete with other users for system resources and will have a guaranteed consistent response. With the rapid technological improvements in networking capability and the continuing minaturisation and falling costs of computer hardware, the future trend in CAD systems is quite clear. Gradually, the local microcomputer in a workstation will become more and more powerful to the extent that all CAD processing is performed on the microcomputer, and the host computer is relegated to the role of just providing a central database for storage of drawings and other information. It may also be possible for one workstation to communicate directly with the other workstations in the system configuration without the need to go through the cental host computer.

BIBLIOGRAPHY

Scott, J. E., *Introduction to interactive computer graphics*, John Wiley, 1982.

Groover, M. P. and Zimmers, Jr, E. W., *Computer-aided design and computer-aided manufacturing*, Prentice-Hall, 1984.

Newman, W. M. and Sproull, R. F., *Principles of interactive computer graphics*, 2nd edn, McGraw-Hill Kogakusha, 1979.

Besant, C. B., *Computer-aided design and manufacture*, 2nd edn, Ellis Horwood, 1983.

Computer Technology Review, *The systems integration sourcebook*, West World Productions, Inc., Winter 1983.

Mini-Micro Systems, *Special Report: Technology*, Cahners Publication, December 1982.

Smith, W. A., *A guide to CAD/CAM*, The Institution of Production Engineers, 1983.

Systems International, *Distributed CAD/CAM*, June 1983; *Digitiser Survey*, July 1983; Business Press International Ltd.

4

Computer-aided design system software

4.1 INTRODUCTION

CAD hardware, or for that matter, any computer hardware would not be of much use unless there was software to support it. The term 'software' is used to refer to all the programs that run on a computer system to perform various tasks. In this chapter, the software aspects of a CAD system will be discussed. Generally, the software of a CAD system can roughly be divided into three main levels, namely the operating system, the graphics system and the application system. At the lowest level of the CAD system software is the operating system which provides software facilities for the user to develop the graphics system and the application system. The graphics system offers some basic graphics capabilities with which the application system can be built.

4.2 OPERATING SYSTEM

An operating system may be defined as a program which organises a collection of hardware into a consistent and coherent entity so that it can be used for developing and running programs. It can assist users and their programs to make the most efficient use of system resources such as memory capacity and processing power with minimal programming effort for providing software to supplement and enhance the system facilities available. In general, an operating system is usually provided with the computer by the vendor as a complete package, so there is no need for the user to write it. There are numerous different types of operating systems. Most of them are interactive, which means that a command entered by a user is executed immediately and the results are returned as soon as possible. Some sophisticated operating systems provide multi-user and multi-tasking capabilities which allow more than one user to access the system and more than one program to run in the system. Typically, no matter which type of operating system is used, it will invariably have utility programs such as an editor, an assembler and a compiler for some high-level programming language. These are the most basic facilities needed to develop a program. An editor enables a user to create and modify source programs which are just ASCII characters representing instructions in some programming language. An assembler is used to assemble programs written in the system's assembly language into binary code before they can be executed. A compiler serves essentially the same function as an assembler, but for programs written in a high-level language such as FORTRAN or PASCAL. Very often, an operating system provides a system library which contains subroutines to perform basic functions such as input/output and file handling etc.

4.3 GRAPHICS SYSTEM

The function of the graphics software for a CAD system is to provide graphics capabilities so that various applications can make use of them to help solve design problems. As a result of this objective, the graphics software has to be written and organised into a structure that is sufficiently general to meet the requirements of many different and diverse applications of CAD. A graphics system should essentially provide a system for handling user actions, a set of basic graphics functions and utilities, and a system for the operation of applications programs. It is of paramount importance that a graphics system be designed in such a manner as to allow applications systems to be incorporated into the CAD system without the applications programmer having to be concerned with low-level data, detail systems programming, or peripheral handling. So far as is possible, all applications systems should be modular and should conform to a well-defined specification, enabling useful modules in one application to be used in another.

To illustrate the concept and principle of a graphics system, a system, known as TIGER, developed in the CAD section of the Mechanical Engineering Department at Imperial College is described here. Inevitably, the software structure of the TIGER system is somewhat different from those of commercially available systems, owing to the difference in the capabilities of the hardware used; but it is hoped that by using the TIGER system as an example the readers can gain some insight to understand other graphics systems. Maybe, it even provides some ideas as well for them to start developing their own graphics system.

The TIGER system is a general computer-aided design system developed particularly for engineering applications such as draughting, data preparation for finite element analysis and mechanical engineering component designing etc., although it can be applied to any activity that involves three-dimensional graphics. The name 'TIGER' is in fact the acronym for Three-dimensional Interactive Graphics and Engineering Routines. The TIGER system runs within a multi-user environment RSX-11M on a PDP-11/45 minicomputer and is capable of handling more than one graphics workstation operating at the same time.

4.3.1 The overlay system

In general, the software for a large graphics system demands extensive memory for it to run while the computer core size is finite and limited. To overcome this problem of insufficient core memory, an overlay system is used in the TIGER system. The idea is to divide the graphics system into individual modules of code that perform specific functions. Each of these fundamental modules, called an overlay, can be loaded and run independently in memory so that, when a function is required, only the appropriate part of the graphics system need be loaded into memory instead of the entire graphics system which would exceed the memory limit. The following sections will describe how such an overlay system is organised and implemented to control the flow of execution within the graphics system. This technique permits the use of many or even large programs to be executed in a computer with small core storage capacity.

An overlay is basically a program resident on the disk which can be brought into core for use by the CPU and then returned to disk on completion. The principle of an overlay system is that available core memory is divided into two areas called the resident and overlay areas. The overlay area is for the execution of overlays. The resident area is used to handle the calling of overlays and to provide small areas of data storage for the communication of vital information between overlays. The resident area is usually kept as small as possible in order to allow the maximum amount of space for overlays. An example memory utilisation map when the overlay system is running is illustrated in Fig. 4.1.

As all system and user functions are handled as overlays, the overlaying system has to be very fast. Another important requirement of the overlaying system is that overlays should be called just like subroutines, that is, each overlay does not have to specify which overlay is to be called next. This facility can be provided by incorporating an overlay execution stack into the overloading routines. This allows overlay calls to be stacked up ready for execution. A return from one overlay causes the next on the stack to be executed.

By stacking not only the overlay number but also an additional overlay segment variable, overlays can be segmented, giving, in effect, multiple entry point overlays.

The RSX-11M operating system does provide a standard overlay system, but this facility had regrettably been proved to be too slow and restricted for graphics applications, in that it operates on a tree structure where the sequence of overlays to be executed must always be predefined when the graphics system is built and it is not possible to change this sequence during execution. As, in many cases, the order of operation of the graphics system is controlled at random from a menu area, it is important to be able to jump between overlays in any sequence or direction without impairing its execution speed to the extent

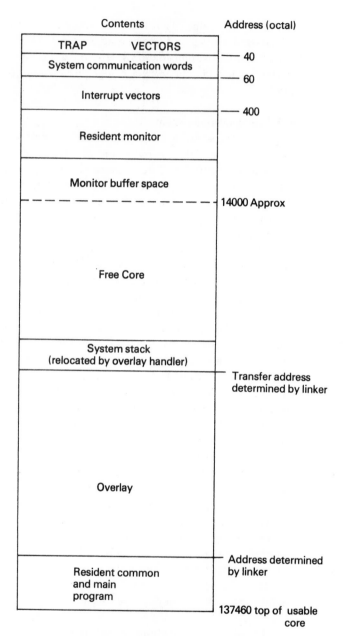

Contents   Address (octal)

TRAP   VECTORS
System communication words   ── 40
                                        ── 60
Interrupt vectors
                                        ── 400
Resident monitor

Monitor buffer space
─ ─ ─ ─ ─ ─ ─ ─ ─ ─ ─ ─ ─ ─   14000 Approx

Free Core

System stack
(relocated by overlay handler)
                                        ── Transfer address
                                           determined by linker

Overlay

                                        ── Address determined
Resident common   by linker
and main
program
                                        137460 top of usable
                                               core

Fig. 4.1 — Memory utilisation map.

of causing irritating delays for the user. Hence, a more efficient and flexible
overlay system, such as the TIGER overlay system, was written to eliminate the
shortcomings inherent in the standard RSX-11M overlay system.

A multi-tasking operating system such as the RSX-11M, allows more than
one overlay running simultaneously in memory. When a new overlay is requested

to be run and there is adequate core space for its execution, the RSX-11M executive will load it from disk. If not, the RSX-11M executive will wait until further memory space is made available when the overlay or overlays currently in core have finished execution and left. Thus, by the time the requested overlay is loaded into memory, the previous overlay or overlays will have almost completed execution, or have finished and gone from the memory. In order to communicate data between overlays, it is therefore necessary to allocate some reserved area of memory which is not affected by this overlaying procedure but can be accessed by any overlay if required. This resident common area in core usually contains the relevant graphics system parameters such as input and output scale factors, grid sizes etc. As the graphics system is designed to cater for more than one user working at the same time, it is necessary to have a resident common area for each user. This is implemented in the following way. When the graphics system is first logged on, a dynamic common region in memory is automatically created for the user and will be initialised with the default graphics system parameters. Each overlay when run will include this common region in its address space by mapping to it so that the graphics system parameters can be accessed and changed if and when necessary throughout the session. These regions are named in the form COMnnn, where nnn is the number of the user station to which it applies. The RSX-11M executive simply identifies this region by a special code which is passed between tasks to enable each to map to the correct region.

The flow of overlay execution is controlled by means of a set of FORTRAN callable subroutines which pass the overlay number as a simple numeric argument. This number can have any value during execution so that any overlay can call any other overlay or even itself if so required. An overlay execution stack is maintained by the graphics system to keep track of the overlay mapping on a 'first in-last out' (FILO) or 'last in-first out' (LIFO) basis on which overlays can be queued for execution. When an overlay returns at the end of its execution, the overlay stack is checked to see if there is any more entry awaiting execution. If there is, the top entry on the stack will be removed and executed. An overlay stack entry consists of two items, that is, the overlay number and the overlay entry point which may be used by the overlay to divide itself into logical segments.

The TIGER overlay system has an overlay library that keeps a list of all the overlays in the system. This library acts as an index file from which a particular overlay, given its overlay number, can be looked up when it is requested to run. The overlay library is maintained by an overlay librarian which updates the library directory. The overlay librarian enables the user to delete, insert and list the overlay entries in the library directory. When an overlay is to be included in the graphics system, the librarian installs the overlay in memory, makes a copy of its 'task control block' into the overlay library created on disk, and then removes it. As far as the RSX-11M system is concerned, an overlay is treated as a task which can be run independently in core. A task control block is a small area of core allocated to an installed task that provides important information about the task such as the task name, execution priority, location on disk etc.

The overlay library basically contains references between user-specified overlay numbers and their corresponding rask control blocks so that overlays can be called by number. Figure 4.2 shows schematically the overlay library format.

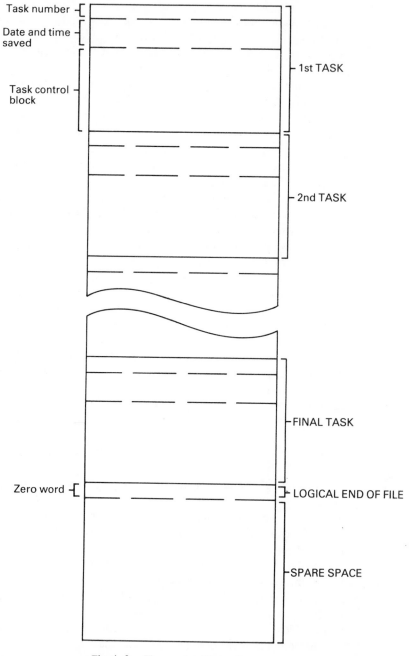

Fig. 4.2 – The overlay library format.

The library can keep a maximum of 32,767 entries which are considered adequate for all foreseeable applications.

The RSX-11M executive can address 32K words of core in the PDP-11/45. With address mapping using the memory management unit, it is possible to address up to 128K words. The RSX-11M executive is generally responsible for the placing of a specific task into an unoccupied location in memory, since several users may have tasks executing simultaneously. This special RSX feature is fully utilised by the overlay system, so eliminating the need to work out the task's start address in memory before loading it. However, a problem exists for the graphics system in a multi-user system. Since it is possible that two or more user stations may simultaneously request the same overlay to be executed, it is very important to identify the overlays with a name unique to the corresponding user station. The graphics system overcomes this by a process known as 'spawning' which is used by the RSX-11M executive to deal with a similar situation when a utility program such as the MACRO assembler is requested at the same time by more than one user. For example, a parent task is installed and known to the executive as ...MAC. When this is requested, an offspring task identical to the parent is generated but with a task name of MACTnn, where the last three characters indicate the number of the terminal that invokes its execution. Thus, MACT11 is a copy of the assembler run from terminal number 11. The graphics system uses the same method except that the name of each graphics overlay task is made up of four characters such as MONT which is spawned to run under the name of MONTnn where nn comes from the user station number.

When a request is made for an overlay to run, its overlay entry in the library is copied and loaded by a special 'loader' task into the RSX-11M system task directory which is simply a list of tasks currently installed in the RSX-11M system. The overlay entry must of course be modified to reflect the user station requesting it. Then the overlay starts execution. When it has finished, the executive can be asked to remove it from memory by setting the appropriate flag in the task control block thus avoiding the need for further action by the graphics software. This special loader task is permanently resident in core and is shared by every user station. When first run at system boot-up, it makes a copy of the overlay library in a condensed format for its own use.

The following sections describe the sequence of operations performed when an overlay is required.

(a) The initiating task ensures the loader is active and sends it the following information:

    (1) The number of the task to be executed.
    (2) The user station which requires it.
    (3) The identification of that user stations resident area, as assigned by the executive.

(b) The loader task then:

    (1) Completes any operation it may already be performing for another user.

(2) Receives the above information.

(3) Locates the appropriate task control block from its library and enters it in the system task directory, modifying the name to suit the requesting user station.

(4) Sets the new task running.

(5) Sends the new task the identification of its resident area.

(c) The new task then:

(1) Receives the resident identification and maps to the correct region in order to achieve access to the common parameters.

(2) Identifies the appropriate terminal numbers corresponding to this user station and assigns them for communicating with the operator.

(3) Continues to perform the required function.

4.3.2 Graphics database structure and handling

The database is a vitally important part of a graphics system. The use of a good and well-designed database determines the general performance and flexibility of that system. It is therefore essential to decide on a suitable database for a graphics system otherwise it will have serious consequences on its ability for future expansion. Obviously, different areas of CAD will benefit from different database structures; for example, the best database structure for a finite element system will be considerably different from that for a gear design system. If a graphics system is intended to be a general system on which other applications can be built, it is of great interest to adopt an efficient and versatile data structure which may be used in many other fields of CAD with no or little modification. Furthermore, it must be simple and easily understood by the applications programmer. Although a sophisticated data structure may seem a better choice as it allows quicker and more efficient manipulation, this advantage is offset by the inherent greater complexity of applications programs which result in longer program development time and higher possibility of errors.

The graphics database in the TIGER system has a list-type structure that permits data records to be stored consecutively within a file. Each data record has a fixed length that comprises four components, i.e. X, Y, Z and I. In general, X, Y and Z define a three-dimensional space coordinate and I indicates the nature of that coordinate. In special symbol mode, a data record may be used to store information such as the symbol number, its size and type. They may also be used to identify complex graphics items such as macros. The principles of symbols and macros will be discussed in detail in later sections of this chapter.

4.3.2.1 Data record

As mentioned above, a normal data record consists essentially of the three-dimensional X, Y, Z coordinates and the I word. X, Y and Z are the coordinates of a point in a drawing and each of them takes up two words to store. The I word is just one word long and is used to indicate the significance of that point.

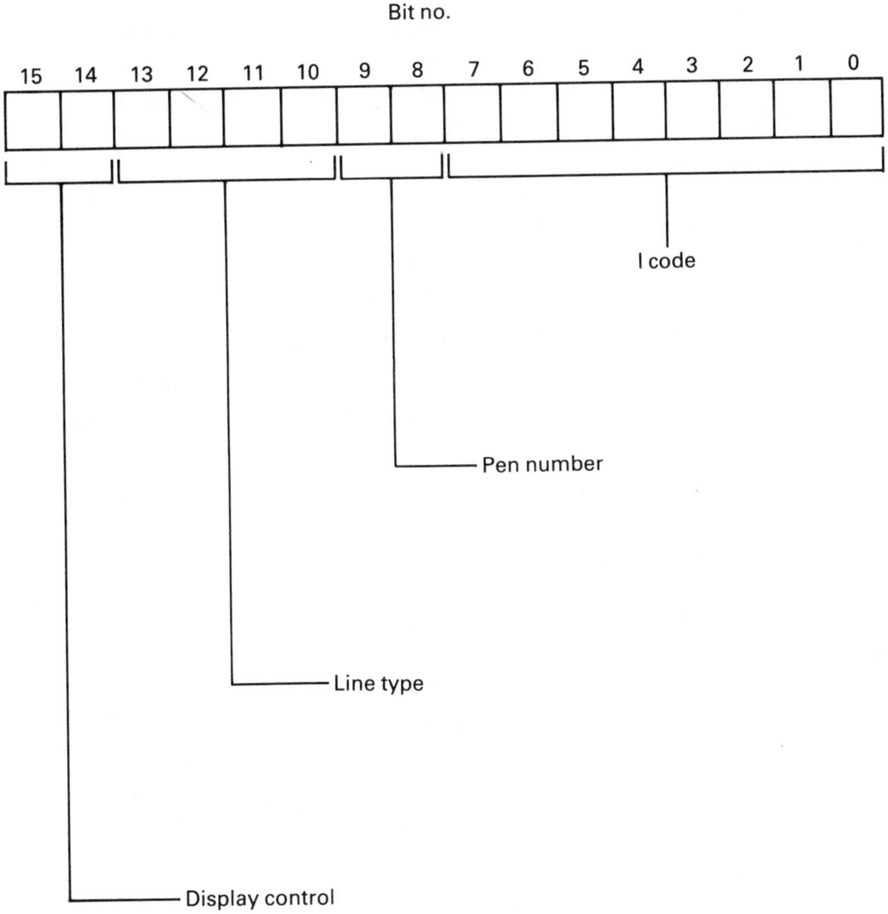

Bit no.

Fig. 4.3 — The I word structure.

Thus each data record has a total length of seven words. Figure 4.3 shows the structure of the I word which is divided into two main parts. The low-order byte of the I word is reserved for I codes which can vary from 0 to 255 which may be considered adequate for most applications. The high-order byte contains three parameters which are encoded into a 'mask' that is put into the I word of each record before it is placed in the database. Typically, the three parameters are the display control, the line type and the pen number.

(a) I code. A full description of the I codes used in the TIGER system is given in the table in Fig. 4.4. Although the I code can have 256 possible values, only 12 of them are assigned to indicate a specific meaning of a record. The remaining values can be used for future expansion of the graphics database.

I CODE	DESCRIPTION OF ASSOCIATED RECORD
1	The start of a line in either two or three dimensions. Plotter/display moves to this point. or A positional reference point if within a symbol block.
2	The end point of a line in either two or three dimensions. Plotter/display moves to this point with pen lowered.
3	The first record and low limits of a symbol block. May be two- or three-dimensional, depending on the symbol.
4	The last record and high limits of a symbol block.
5	An identification record within a symbol block bearing numerical data only.
6	A text record containing 12 encoded ASCII characters.
7	A general record of unspecified format.
13	The first record and low limits of a macro block. May be two- or three-dimensional, depending on the macro.
14	The last record and high limits of a macro block.
15	An identification record within a macro block.
16	A text node within a macro.
255	A null record, to be ignored by all functions.

Fig. 4.4 – The description of I codes.

(b) *Display control*. The function of the display control parameter is to indicate to the graphics system that the data record is true three-dimensional data or two-dimensional information such as annotation. It is important to distinguish between the two different display modes since 3D data will be projected into the x-y, x-z and z-y views and, optionally, an isometric view, whereas 2D data will be displayed directly without any manipulations. This distinction is needed so that the correct geometric transformation is applied on the record. Essentially, all the X, Y and Z coordinates in a three-dimensional record are significant whereas only the X and Y in a two-dimensional record are significant.

(c) *Line type*. There is a choice of up to 16 different line types for use in the TIGER system. The simplest is a solid line and the remaining 15 are constructed from different combinations of marks and spaces. At the start of the graphics system, four line types, namely solid, dotted, dashed and chained dotted, have already been defined for the user. Solid lines are used by default unless otherwise specified by menu command. Each line type is a repeated pattern of two marks and two spaces of varying lengths as shown in Fig. 4.5. The other 12 unspecified

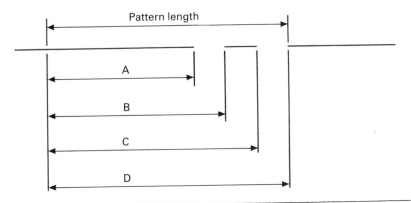

Fig. 4.5 – Line type structure and convention.

Line type number	A%	B%	C%	D%	Overall length (millimetres)	Description
1	100	0	0	0	undefined	continuous
2	25	50	75	100	7.0	dotted
3	25	50	75	100	15.0	dashed
4	70	80	90	100	30.0	chain dotted

line types may be assigned by the user to form a special pattern needed for a particular application.

(d) Pen number. The pen number can have a value of 1 to 4 to refer to the four pen slots available on the flat-bed plotter in the system. Any of these pen slots may be used for pens of different colours or line thicknesses as required by a certain application. For example, different colours are useful for mapping, whereas different line thicknesses are important for engineering drawings. Initially after log on, the graphics system sets the pen number to be 1, but this can be easily changed by use of the appropriate menu command. Obviously, a pen change may also be achieved by physically replacing a pen in a pen slot with the required one.

4.3.2.2 Filing system

As can be imagined, an average drawing contains many vectors. Typically, it has about 40,000 vectors according to one estimate by an enginering company. It is therefore advantageous, if not essential, to have a fast and efficient file handling capability in a graphics system so that long delays for the operator and excessive use of CPU time can be avoided. The TIGER system divides its filing system into three basic levels. There are workfiles for short-term storage, semi-permanent files for temporary storage of incomplete drawings, and permanent files for archival storage of complete drawings. These three levels of filing will be described in detail in the following sections.

(a) Workfiles. Workfiles may be defined as those files with which the user interacts directly during the construction of a drawing. They comprise the work-space which is the area for building up a complete drawing, and any associated 'buffer' files that are commonly used for temporary storage of data before they are processed and then added to the drawing. For graphics applications, it is important to be able to access data records randomly, that is, to have the ability to address records out of sequence, as it is often necessary to modify sections of the drawing without the need to read through the entire database. Unfortunately, the file handling capabilities of the FORTRAN language is limited principally to sequential access of records, that is, data records can only be addressed one by one in a certain order. It is therefore impossible to pick out a particular record without having to read through all preceding records. If a required record is situated somewhere near the end of a large file, it will take a very long time to locate it. Obviously, this is unacceptably slow for an inter-active graphics system which needs a very fast file-handling capability. Thus, a separate random access file handler must be written for the workfiles. True random access operation is only possible with contiguous files, i.e. those in which successive blocks are physically and numerically adjacent on the disk. Thus, the location of any block can be deduced from that of the first or any other block. The filing process for the workfiles can be considered to consist of two stages performed by two independent routines. The first stage packs drawing records into blocks of 256 words long. As each record is 7 words long, there are 36 records in each block with 2 words unused. The second stage is to transfer these blocks to disk storage.

In a multi-user graphics system, a set of workfiles must be allocated to each user currently logged on the system. These workfiles must be given unique names so that the system can distinguish which files belong to which user. It has been decided to divide the workfiles into private and public workfiles. Private workfiles contain information relevant only to a particular user while public ones are for storing read-only information such as character set specifications or menu layout data that are accessible by all users. The system allows each user up to 10 private files and any number of public files. The 10 private files should be enough to meet the present requirements although they can be easily increased with a small software modification if they should prove inadequate. A workfile is named according to the convention described below.

(1) Private file

TIGuuunnn.DAT

where uuu = user station number in decimal
         nnn = file number in decimal (1 to 10 inclusive)

e.g. TIG003002.DAT is private file number 2 for user station 3.

(2) Public file

TIGCOMnnn.DAT

where nnn = the file number in decimal (11 or more)

e.g. TIGCOM014.DAT is public file number 14.

The TIGER system performs a certain amount of checking on these private files. When a user first logs onto the system, it will look up the appropriate directory to make sure that no private files are left from a previous session otherwise they are deleted. Normally, this situation should not happen as private files would be automatically deleted at log-off unless the previous session was prematurely terminated due to system faults. Such a process will ensure that redundant files do not accumulate and waste precious disk space. In general, the number of private files allocated to a user and their lengths are defaulted initially to some suitable values. If this allocation is insufficient, the RSX executive automatically increases them when they become full. The powerful RSX executive provides other special features that make the task of maintaining the workfiles easier. It allows the graphics tasks to access the workfiles quickly by a method called 'opening by file identification' because the executive can open the files directly without the need to search through the directory for its name. This can obviously save a lot of time as many file accesses may take place in an average drawing session. As all I/O requests are double buffered, i.e. two buffers are used in rotation for data handling, it is possible to transfer data asynchronously such that the execution of the calling task and data transfer can happen simultaneously. Furthermore, data may be extracted from one buffer while the other is being read from the workfile and vice versa. However, the execution may be requested at any time to wait for the completion of data transfer before continuing. Since for every open file, RSX will allocate a certain amount of space in memory to store information about it, it is not desirable from space considerations to keep too many files open at all times. The TIGER system imposes a limit of four concurrently open files which are considered adequate for most purposes. When this limit is execeeded and a fifth file is required the least recently used of the four is closed so that the new file can be opened in its place. It is always worth noting that for maximum efficiency, the file opening/ closing overhead must be minimised. A file must be logically opened under RSX before data transfer can take place. When the transfer has finished, the task must close all open files before it exits to avoid leaving them in a 'locked' state. The task achieves this by calling a file closing subroutine as part of its exit procedure.

(b) Semi-permanent files. As the name has already implied, these files are essentially for temporary storage of drawings that are continually undergoing modification. Obviously, they will require high-speed random access to files in this kind of situation and so necessitate the use of disk for their storage. Other bulk storage medium such as magnetic tape cannot possibly be used as it operates on a sequential access basis and becomes too slow when retrieving files stored near the end of the tape.

In the TIGER filing system, each drawing file is given a drawing number which may be any character string of up to 22 characters in length. This is the number that the user specifies to recall the required drawing from storage. The system maintains a drawing index file which keeps general information about the drawing such as when it was saved and by whom, how many times it has been accessed etc. This index file enables drawings to be quickly located and basically contains a correlation between the drawing number and an internal sequence number used to identify the drawing file whose name in the disk directory is derived from this internal sequence number. When a drawing file is recalled, the system searches through the index for this file and reads it directly without referring to any other drawing files.

The contents of a drawing file can be divided into two parts. The first part starts from the first block of the file and stores all the relevant system information about the drawing such as scale factors etc. which are decoded on its recall. The second part contains the actual 3D data of the drawing. There is no limit to the total size of the file which obviously varies according to the complexity of the drawing.

The system provides other facilities to the user apart from the basic storage and retrieval of drawings. For example, all the files currently in the directory can be listed out and entries of drawings may be selectively deleted. When a drawing is requested for deletion, the system automatically checks whether the current user's name matches that recorded when the drawing was saved. If the check fails, the file will not be deleted. This check, however, may be overridden by users with the appropriate privilege status. An index file records only those drawings that are available on the same disk as itself. This, in effect, gives some sort of protection against unauthorised access as a user can access only those drawings on the particular disk allocated to the project in which he is involved.

(c) Permanent files. Permanent files are those drawing files that have been completed and are not likely to be accessed in the foreseeable future. One way to deal with them is to leave them on disk just as semi-permanent files so that they can be quickly accessed if necessary. However, this will fill up valuable disk space with infrequently used files. A disk may be able to store a large amount of data, but there is still a finite limit to its storage capacity. Furthermore, disks are expensive and it is simply uneconomical to use up many disks in this way. An alternative way is to use magnetic tapes as they are much cheaper and smaller than the average disk. However, they are restricted by the limitations usually associated with a sequential device. They take a long time to read files stored some way into the tape. A new file can only be added to the end of the information already on the tape and selective file deletion is very difficult if not impossible. Despite all these disadvantages, magnetic tapes can still be used since permanent files are kept for reference only and are not updated frequently.

The permanent file handling routine works in much the same way as that for semi-permanent ones except that the index file is not stored on the tape as it is subject to frequent changes which can be time-consuming if done on tape. The index files for all tapes are maintained on the system disk and additionally record the label of the tape on which the drawings are stored. Thus, a given permanant drawing can be quickly retrieved in this way as the system indicates which tape is to be loaded.

4.3.3 Operating features

The following sections describe the actions needed to log onto the TIGER graphics system, use it and then finally log off when finished. It explains the use of the various means of interaction between the user and the computer to create a drawing. At appropriate points, the principle and the design of a particular function will be outlined to give an insight into the way in which it works.

4.3.3.1 Log-on

In order to access the TIGER graphics system, the user must first log onto the RSX-11M operating system by the usual method from the alphanumeric terminal at the user's workstation. A typical command to do this is 'HEL n,n/ PASSWORD', where HEL is the RSX command HELLO for logging on, 'n,n' and PASSWORD are respectively the user's identification code and password in RSX. The graphics system is started by entering the command 'LOG' on the alphanumeric terminal. Then the TIGER system title page is displayed on the alphanumeric terminal to indicate to the user that it is running. Before the user can really start, the system requires the following information for accounting purposes:

(1) User's name — up to 14 characters
(2) Contract number — the job to which the session is to be charged
(3) The drawing number — up to 22 characters
(4) User's time sheet number
(5) Cost code section number
(6) Cost code activity number

The user can provide this information on the alphanumeric terminal at each log-on, but this can be tedious as there are six entries each time. Also, most of this information is the same for a particular user, therefore, the system allows the user to specify it in a file called GRAPHICS.LOG. When the LOG command is issued to invoke the graphics system, the user's directory is searched for this file. If it exists and is found, its contents will be used as the default information. GRAPHICS.LOG contains six pieces of information, one on each line, in the same order as above. If no default is available for a particular piece of information, as indicated by a single '?' character on the corresponding line in this default file, the user will be prompted to supply that piece of information from the alphanumeric terminal. Comments may be included in the file either on a line of their own or after a piece of information provided it is preceded by an exclamation mark, The following is an example of a GRAPHICS.LOG file:

```
! An example log-on file showing
! how defaults are established.
!
LUI ! user name
? ! prompt for contract number
? ! prompt for drawing number
45 ! timesheet number
7 ! section number
10 ! and activity number
```

The system has another feature which permits the user to specify in a file the default data that are not always true, but true only for a particular job. If, for example, the user logs onto the graphics system with the command LOG GEAR, it will look through the user's directory for the file GEAR.LOG first. The contents of GEAR.LOG will then be used as the default information. If any required data are not defined there, the default file GRAPHICS.LOG is tried. If even this fails to produce the information, the user is prompted to specify it manually from the alphanumeric terminal.

Once this log-on formality is complete, the system checks whether or not the specified drawing number matches one that has already been saved. If so, the drawing is automatically recalled, otherwise a new drawing is assumed and the default system parameters are set up. When this operation is finished, the log-on procedure is over and control is passed to the 'Background'.

4.3.3.2 Background mode

Background mode may be defined as the state in which the graphics system is idle and awaiting a command from the user. When background mode is first entered after log-on, the graphics screen displays the default view assignment indicated by dashed lines as shown in Fig. 4.6. Each of these views is completely independent of the others, and may be disabled and changed to a view of different projections, origins, axes and scale factors. The user may even optionally select the 2D mode in which the entire screen represents the 2D *X-Y* view.

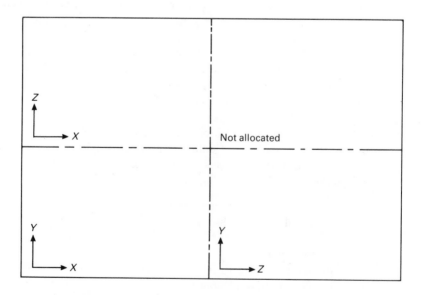

Fig. 4.6 – Default view allocations.

The display screen corresponds to the drawing area in the digitiser which can be treated as a drawing board. In background mode, a non-store cursor is displayed on the graphics screen to indicate the current position of the digitiser stylus within the drawing area. There are two different modes of cursor display available. The first, which is the default at log-on, is the standard mode where the positional relationship between the cursor and the drawing are maintained regardless of the size of the display window, i.e. the part of the drawing that is currently being displayed. The second is the non-scale mode in which the cursor always traverses the width of the screen irrespective of the size of the display window. In standard mode, accurate cursor positioning is difficult as small movements of the stylus result in large movements of the cursor. This problem does not occur in non-scale mode where the positioning accuracy is much better. This mode is particularly useful for digitising small items in a drawing.

The digitiser surface is divided into three areas, namely the drawing area, the command menu area and the macro menu area as shown in Fig. 4.7. The drawing area is that area of the digitiser over which drawings or rough sketches may be placed for digitising. The command menu area provides the user with a means of issuing commands from the digitiser. The macro menu area serves the same purpose as command menu area except that it is used to recall macros instead of tasks. The theory and operation of the macro facility will be dealt with in a later section.

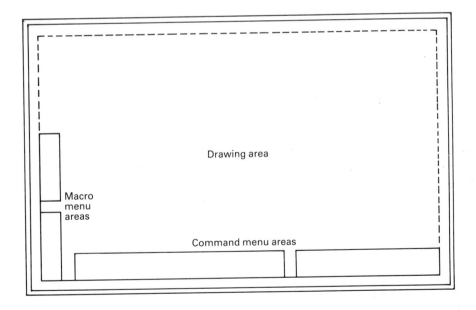

Fig. 4.7 – Layout of the digitiser.

In background mode, the system constantly monitors the digitiser to see if a button has been pressed. When a button is detected to have been pressed, the system records that button number and the current position of the stylus on the digitiser. The system then decides what to do with this information according to whether the returned position of the stylus lies within the menu area or the drawing area. The following two sections describe the actions that the system takes for each area.

(a) Menu area. The menu is placed horizontally along the bottom of the digitiser. It is divided into 250 squares of 1.5×1.5 cm each. They are arranged into 50 columns of 5 squares each. A function within the system is assigned to each square and is activated simply by digitising over that square. Each square contains a short description to indicate to the user of the function assigned to it. For convenience to the user, the squares are grouped according to functions, for example, symbol generators, system parameters, or editors. This grouping is illustrated by the section of a typical menu in Fig. 4.8. As it is envisaged that a large graphics system such as TIGER will need many more than just 250 commands, a system of menu pages has been developed. Each menu page is completely independent of the others and has the same size and specifications as mentioned above. The default at log-on is page 1 and it will remain page 1 until the system is informed of a new page selection by means of a menu command. After a menu change has been made known to the system, the user simply replaces the menu page on the digitiser with the appropriate one and the new page will then come into effect. This system of menu pages has the additional benefit of allowing each user to have his own menu page. Four menu pages have been defined in the TIGER system although only two can be used at any one time owing to the limitation of space on the digitiser.

If the returned position of the stylus at the press of a button lies within the menu area, the system computes the menu sqaure on which it lies. The button pressed in this case has no significance except buttons A and B whose functions will be explained in section 4.3.3.3. The system then jumps to execute the task corresponding to the menu square digitised. At the beginning of the task execution, a message identifying that task is displayed on the alphanumeric terminal so that the user can be certain that the right task is running. Depending on the operation of the task, the user may be prompted for various types of information that are needed to run the task. The prompt messages and questions are very self-explanatory and the user merely has to follow the actions suggested in them. Generally speaking, the requested input data may be classified into two main types. There are those input from the alphanumeric terminal and those from the digitiser. For example:

(1) Input from the alphanumeric terminal

    *Enter the required scale factor
    Range 1.0 to 100.0, default = 2.0
    ⟨NUM⟩:

	CIRC	SEMI CIRC	MIN ARC	MAJ ARC	ELLI PSE	POLY GON	RECT -GLE	SPH ERE	CYLI -NDR	CUR VE	HAT CH	FILL ET	DIMS	ARR OW	TEXT	
	DEFAULT RADIUS	DEFAULT RADIUS	DEFAULT RADIUS	DEFAULT RADIUS	BY CENTRE	ON SIDE	DIA-GONAL	DEFAULT RADIUS	DEFAULT RADIUS	BETWEEN POINTS	IN BOUNDARY		HORI-ZONTAL	PLACE	LEFT JUSTIFY	RIGHT JUSTIFY
	TYPED RADIUS	ON DIAMETER	TYPED RADIUS		MAJOR AXIS	CENTRE AND VERTEX	CENTRE AND CORNER	TYPED RADIUS	TYPED RADIUS		BLOCK FILL		VERTICAL		CENTRE JUSTIFY	FROM FILE
	DIGITISED RADIUS	DIGITISED RADIUS	DIGITISED RADIUS	DIGITISED RADIUS			CENTRE AND SIZE	DIGITISED RADIUS	DIGITISED RADIUS				PARALLEL			
	THREE POINT	THREE POINT	THREE POINT	THREE POINT			CORNER AND SIZE				SET SPACING		ISOMETRIC		SET HEIGHT	SET ANGLE
	SET DEFAULT		CHORD AND HEIGHT	SET DEFAULT				SET DEFAULT	SET DEFAULT	SET ACCURACY	SET ANGLE	SET DEFAULT	SET ISO PLANE	DEFINE	SET ASPECT	SET FONT

MENU NUMBER 2

Fig. 4.8 — Section of a typical command menu.

The first line informs the user of the variable whose input is requested. ⟨NUM⟩ indicates that the input is to be a numeric value ranging from 1.0 to 100.0. The default of 2.0 is taken if a carriage return is entered in reply.

> *Please enter the drawing title
> ⟨TEXT⟩:

The first line prompts the user for the variable to be entered. ⟨TEXT⟩ indicates that the input is to be a text string.

> *Do yout want to contiune? [Y/N]:

This question requests the user's confirmation by typing Y for 'YES' and N for 'NO'. If a carriage return is entered, 'NO' is assumed to be the answer.

> *What type of valve do you require?
> 1–Gate
> 2–Ball
> 3:–Butterfly
> ?

This question requires the user to choose an option from a list of options. The user can specify his choice by entering the number next to it.

(2) Input from the digitiser

Prompt messages for input from the digitiser usually have the same format as those examples above. The only difference is that the required information is generally the coordinates of a point on the drawing or a button number to indicate a certain action to be taken by the system.

Once all the required information has been provided, the task starts to perform its function. Depending on the nature of the output, the task sends them to either the graphics screen or the alphanumeric terminal. When the task has finished its execution, control is returned to background mode which is ready once again to accept another command.

(b) Drawing area. If the returned position of the stylus at the press of a button lies within the drawing area, the way in which this point is processed depends on the button pressed. The buttons on the digitising stylus can be treated as an extension of the command menu. Each button can be assigned a function as an ordinary menu square. The advantage of using a button in this way is that frequently used functions may be conveniently made available 'at the user's fingertips'. The button assignments are described in some detail in section 4.3.3.3. When the point has been processed according to the button pressed, control is then passed back to background mode which awaits another command from the user.

When a menu task is running, background mode does not actually exit. It merely passes into a semi-dormant state in which it checks every few seconds that the graphics system is functioning correctly. If an error of any kind has occurred and caused the task to exit prematurely or the graphics system to crash, background mode informs the user of such a failure and activates an error-handling routine to take remedial action. This error processing and recovery facility will be discussed in section 4.3.3.5.

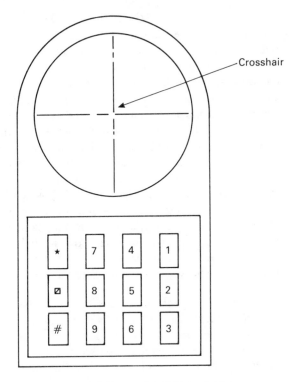

Fig. 4.9 – Layout of buttons on digitiser stylus.

4.3.3.3 The assignment of digitiser buttons

As shown in Fig. 4.9, the digitiser stylus has only 12 buttons, namely 0 to 9, * and #, but there are altogether 16 button assignments. To overcome this shortage of buttons on the digitiser stylus, it is arranged that the assignments of buttons 0 to 9 correspond exactly to the buttons on the stylus whereas the assignments of buttons A to F correspond to those pseudo-buttons each of which is formed by combining button # and one of buttons 1 to 6. Each button assignment is explained below.

(a) Button 0 – Verify current positon. This button is used to indicate to the user on the graphics screen the position of the point that is actually recorded in the workspace if a button is pressed at that position of the cursor. Normally, if the graphics system is in ordinary digitising mode, the position of the point as a result of pressing button 0 will be the same as that of the cursor. However, when control mode or grid round-off are active, the point displayed shows the position after correction and not the one that reflects the cursor on the screen.

Both grid round-off and control mode can be activated by buttons 5 and 6 respectively, and are explained under those headings. Button 0 provides the user with the facility to verify that the position of the point is right before it is entered.

(b) Button 1 – Input 3D coordinates. This button is used to instruct the system to read the current digitiser position and convert it to true 3D co-ordinates. As shown in Fig. 4.6, there are four views for use on the graphics screen. The conversion from a digitiser position to true 3D coordinates depends on which view the digitiser position lies within. When digitising over any one view, it is possible to define only two of the three space coordinates. The third coordinate is usually defaulted to the last value given for that coordinate. For example, if the digitiser sends a position in the $X-Y$ view, the x and y space coordinates are explicitly defined but the z coordinate is defaulted to the z coordinate of the last point. Similar conversions occur when the digitiser position lies in the $X-Z$ and $Z-Y$ views. This third or 'trailing' coordinate may be changed to some other value using button 3 which will be described in section (d). This point in 3D coordinates is usually added to the workspace as the next sequential entry, and the line joining it to the previous point entered is displayed on the graphics screen. If a symbol generation routine is running, the point entered will be passed to that routine for further processing. Button 1 is the most basic way of entering graphics data to the TIGER system.

(c) Button 2 – Break line. This button works in exactly the same way as button 1 except that the point is stored in the workspace with the 'pen up' attribute rather than 'pen down'. This means that the line joining this point and the previous one will be drawn with the vector write-beam switched off, that is, the vector is not visible on the screen. This indicates that the point is the start of a new line rather than a continuation of the last line.

(d) Button 3 – Define trailing point. As mentioned in section (b), button 3 is used to set the trailing coordinates. When this button is pressed, the system displays the message 'Set Trailing Point' and returns to background mode. The user then enters a point from the digitiser in the usual way to change the latest trailing point. If the specified point lies in the $X-Z$ view, the trailing X and Z values are defined for subsequent 3D input. If the succeeding input points are over the $X-Y$ view, X and Y are explicitly specified whereas Z is obtained from the new trailing point.

(e) Button 4 – Window selection. A window is essentially an imaginary rectangle through which all or part of the drawing can be examined. Any area of the drawing lying within the window will be displayed on the screen. The window used in the TIGER system has a fixed horizontal to vertical ratio of $4:3$ which corresponds to the shape of the display screen. A window may be specified by digitising the left and right bottom corners. In fact, only the left bottom corner and the x value of the right bottom corner are necessary to uniquely define a window as it must lie horizontally, i.e. parallel to the x-axis.

The user is allowed to define up to 15 different windows so that certain areas of interest on the drawing can be zoomed in and out for detail inspection. Each window can be stored under any button on the digitiser stylus except button C which has been assigned as the 'Cancel' function. To recall a window, the user only needs to press button 4 to enter 'windowing' mode and then press the corresponding button assigned to the required window. Button 0 is always assigned to the original window, that is, the window first entered after the system is started.

(f) Button 5 – Enable or disable grids. This button enables the user to define a grid of nodes in 3D space to which the coordinates of any point entered are rounded. The grid facility can handle uniform and non-uniform nodes. The user can specify different grid spacings, types and limits for each of the three major axes, and round-off on each axis may be enabled to disabled separately. This facility is very useful for drawing items which are spaced out at equal and fixed intervals. Button 5 acts as a flip-flop type switch that, when pressed, enables grid mode when it is disabled and vice versa. When any grids are enabled, each node in the grid is marked with a small '+' symbol to indicate its position. Grids can work in either standard display or boundary display mode. When standard display mode is active, all grid nodes in the pattern are displayed. However, a problem will arise when the grid spacings become too small, in which case grid nodes will obscure the drawing and thus render further digitising difficult. When grid node density in standard display mode exceeds a certain predefined limit, the system automatically switches to operate in boundary display mode which shows only the outer nodes on each for the grid edges. This mode solves the problem of grid node overcrowding and yet any non-displayed nodes can easily be located by lining up the arms of the cross-hair cursor on the screen with the boundary nodes.

(g) Button 6 – Enable or disable control mode. This button provides the user with the means of constraining a point to lie in a specified direction from the previous point entered. This feature is extremely useful for generating items that consists of lines at a fixed and known angle to each other. Button 6 is a flip-flop type switch just like button 5 except that it is used to enable or disable control mode depending on its current state. Each time button 6 is pressed, the user is informed on the alphanumeric terminal of the new state and the control angle currently in force. When a point is entered while control mode is enabled, the point is rotated so that it lies in a direction that is nearest to a multiple of a previously specified angle from the last point entered. This control angle is defaulted to 15 degrees initially and may be changed to some other value by menu command. As the point displayed on the graphics screen in control mode does not coincide with the position of the cursor, the user is strongly advised to use button 0 to verify the corrected position before entering the point.

(h) Button 7 – Find point mode. This button provides the user with the means of entering a point that coincides exactly with a point already in the drawing.

This find feature is very useful for closing up polygons or other geometric figures where the last point must coincide with the first one entered. It is very hard to do this manually as the resolution of the graphics screen is poor. The entered point and the desired point may appear very close on the screen, but in fact they are not coincident or near enough for the required accuracy. Button 7 functions in much the same way as button 1 except that the point entered is passed to the find routine which searches for the point that lies closest to the digitised point. The point thus found will be put into the workspace as if it was entered with button 1 instead of the original digitised point.

(i) Button 8 – Find line mode. This button works in much the same way as button 7 except that the system searches for the line nearest to the point digitised with this button. Then the new point entered in the workspace is the point that lies on the line found and is closest to the user-specified search point. It is necessary to compute this new point from the end coordinates of the line found rather than just assume the coordinates of a point already existed as in the case of button 7.

(j) Button 9 – Insert fillet mode. Button 9 enables the user to insert a fillet between two straight lines. There are two types of operation in fillet mode. One type assumes a default fillet radius which may be changed by menu command. In this case the user enters button 9 once which is preceded and followed by either button 1 or 2. The point digitised with button 9 is taken as the intersection point of the two lines where a fillet of default radius is constructed. The other type of fillet mode computes the required fillet radius. The sequence of action is essentially the same as the previous case except that button 9 has to be entered twice successively. The lines ending and starting at the two button 9 points respectively are calculated and the smaller radius fillet is inserted such that it is tangential to both the lines calculated. It is important to note that the function of button 9 does not start until all the required points have been received. For the first type, the three required input points are digitised with buttons (1 or 2), 9, and (1 or 2), whereas for the second type the four input points are entered with buttons (1 or 2), 9, 9, and (1 or 2).

(k) Button A (buttons # and 1) – Floating assignment. Occasionally there are certain functions that are required very often when doing a particular type of drawing. It is obviously convenient to have them available at the user's fingertips. Button A is reserved exactly for this purpose. This button allows the user to assign any of the menu or macro commands to itself. The assignment can be made by placing the stylus over the required command square and pressing button A. The command remains assigned to button A until the user changes this assignment by entering a different command square with button A. The user can easily invoke an assigned command by pressing button A over the drawing area. This button has a further advantage apart from convenience, that is, it enables the user to run a command that is not on the current menu or macro page.

(l) Button B (buttons # and 2) – Floating assignment. Button B functions in a similar way to button A, that is, the button can be assigned to any menu or macro command, except that it is initially defaulted to the assignment of 'Break line'. This button is different from button 2 in that it indicates that the line is to broken at the next point without actually entering that point. This facility is useful for breaking a line before using Find mode to locate the next point.

(m) Button C (buttons # and 3) – Cancel operation. This button provides the user with the means of aborting a function prematurely. It is extremely useful for leaving an active function after a wrong command or some bad information has been entered. When button C is pressed, a general tidy-up of the system is performed and control is passed back to background mode. Button C may be used with most functions, but it must be emphasised that some functions are irreversible and do not respond to this button. In this case some other means of correction has to be used after the function has been completed.

(n) Button D (buttons # and 4) – Unassigned. This button has yet to be assigned a function.

(o) Button E (buttons # and 5) – Numeric entry. This button allows the user to enter the coordinates of the point numerically. Such a facility is extremely useful for specifying accurately known coordinates that cannot be achieved with any one of the correction methods such as grids or control mode. When button E is pressed, the user is asked on the alphanumeric terminal to enter the X, Y and Z coordinates of the point. The coordinates may be specified as absolute or relative values. The latter case is obviously convenient for entering lines of known length. The system then displays the position of the point as if it has been entered from the digitiser. With this method of data entry, a point lying outside the current drawing area may also be entered.

(p) Button F (buttons # and 6) – Finish input. There are certain functions such as duplication at random that accept an undefined and variable number of points for their execution. To mark the end of input for these functions, the user need only press button F. This button is not normally recognised by any function except those that use it.

4.3.3.4 Help mode

In the TIGER system, there is a help mode available on-line to outline to the user the operation of menu tasks. This facility is very useful in that the user can obtain information about a particular task without the need to refer to the manual. The user can enter help mode by menu command just as most other graphics tasks. Once in help mode, information about a given menu task may be obtained by digitising over the menu square to which that task has been assigned. This help information is usually output to the alphanumeric terminal and can be expected to include all the major points about the operating procedure of the task. Help mode continues to take control of the system until button C is pressed whereupon it exits and background mode becomes active again.

The information output from help mode has to be entered manually into a text file named according to the format 'Tnnnnn.TXT', where nnnnn is the task number without leading zeros. The contents of a text file may be divided into sections by lines starting with an *, where each section corresponds to an entry point of the task. When a help request is made, the appropriate section is selected from the file and sent to the alphanumeric terminal. Any text before the first * is assumed to be the task's title information and is output for every entry point. The following illustrates an example of such a text file:

> This is a title which is always output
> * 1
> This is the information about entry point 1
> * 2
> This is the information about entry point 2
> * 3
> This is about entry point 3
> etc.

4.3.3.5   Error processing and recovery

In any systen, it is highly desirable to include some sort of error detection and correction facilities so that errors of any kind can be detected as soon as they have occurred and remedial action may be taken immediately. This function ensures that the user is informed of such a failure and vital data of any type are protected. If at all possible, and attempt is made to recover the system. There are basically two categories of errors, namely operator errors and system errors. The following sections explain how both categories are dealt with by the system.

(a) Operator errors. Operator errors often occur when the user does the wrong thing at the wrong time. For example, a certain graphics task requires from the user a numeric input to be within a given range of values, but the actual value entered is outside this range, or the input contains some illegal characters which make it completely meaningless to the graphics task. Such error conditions are normally detected and handled locally by the graphics task itself since they seldom have a detrimental effect on the system. As can be imagined, there must be numerous such situations as the above example in a sophisticated graphics system. The TIGER system incorporates a special error message issuing task that enables common error messages to be stored in a single file and accessable by different tasks, therefore, the problem of having to include error messages within each task is eliminated. To ensure a fast response on an error condition, the error task is normally run in parallel with the task in which the error occurs. The error message file must be created by the programmer and each entry in the file consists of the message number and the corresponding error message. Comment lines are allowed in the file provided they are preceded by a ';' character. It is also possible to substitute values and chracter strings into an error

message. This capability is useful in that a single error message can be easily modified to indicate different error conditions, thus the size of the error message file may be substantially reduced. Finally, it is worth noting that the error task does not attempt to make an error recovery which is considered to be the responsibility of the task that originally detected the error.

(b) System failures. A program error can sometimes cause a serious system failure such as a system crash. Such an event happens quite frequently when new functions are being developed. It is important that the effect of such a failure is minimised by preventing the loss of any valuable work. As mentioned in section 3.3.3.2, background mode does not exit after control is handed over to another task. It just enters into a state in which every 5 seconds it monitors the activities of the task to see whether it is still running correctly. Background mode simply checks that some task, apart from itself, is active for the user station. If no other tasks are found to be running, it is assumed that the system has crashed and background mode initiates the system crash recovery task. This task basically cleans up the system and returns it to the state before the crash. The operations performed are shown below.

(a) Unlock any workfiles that are locked because a task opens it for access but fails to close it properly because of the unexpected crash.
(b) Reset system parameter flags, such as the symbol mode flag, to their correct state.
(c) Clear the system execution stack.
(d) Check the workspace to ensure that indices and pointers are set with the correct values.
(e) Generate a crash dump into a special file which may be output later for checking. A crash dump contains a full octal dump of the resident common area and a number of pieces of information about various other aspects of the state of the system at the time of the crash.

In general, a successful recovery will usually result in most cases and normal background mode takes over control once again. If the recovery attempt is unsuccessful for any reason, the system is shut down and the user is asked to log on again.

4.3.3.6 Log-off

When a graphics session has come to an end and the user wants to log off the system, he can do so by initiating a log-off routine from the menu. This log-off routine deletes all temporary files that have been allocated to the user during the graphics session. This effectively means that the drawing in the workspace is destroyed as well. To make sure that the user does not accidentally lose any vital information during log-off, a question is asked to remind the user of this consequence and to confirm the decision to log-off. If the user answers yes to this question, the system is shut down, otherwise control is returned to background mode and the graphics session continues.

4.3.4 Symbols

A typical engineering drawing contains not only straight lines but also many common geometric figures such as circles and polygons. Since these geometric symbols follow a certain known mathematical definition, the TIGER system provides a 'symbol' mode which enables the user to specify through the digitiser or the alphanumeric terminal a number of pieces of key information about the symbol which is then constructed and displayed on the screen based on this information. To illustrate the principle of symbol mode, a very simple example is a circle in 2 D space. Such a symbol is fully defined when its centre and radius are given. The actual points that make up the circle can easily be calculated using the well-known circle equation. The required circle is then constructed by joining all these points.

A symbol task is normally invoked by menu command. When it is first started, a message identifying its function is output on the alphanumeric terminal. Then the task places itself in the task execution queue with a different entry point from its menu entry, sets a flag within the resident common area, and exits to background mode so that the user can enter the points and other information needed to construct the symbol. As the points are entered in background mode, all its facilities such as grid round-off and control mode are available to correct the point before it is placed in a data block within the resident common area. Background mode then checks the state of the above flag and, if set, takes the top entry in the task queue for execution. The task at the top of the execution queue will be the original symbol task which fetches the point entered and the button number from the resident common area where it was placed by background mode. This input process may be performed for as many points as are required for a particular symbol. When all necessary data have been received, the symbol task constructs the requested symbol and then displays it on the graphics screen prior to returning to background mode. The data entered are arranged into a 'symbol' block in the workspace which contains only the minimum amount of information necessary to re-construct the symbol for displaying or plotting at some later time. The general format of a typical symbol block is given below.

```
1,3,XMIN,YMIN,ZMIN
1,5,STSKNO,BKLENG,-
1,5,-,-,-
1,1,X1,Y1,Z1
1,1,X2,Y2,Z2
1,1,X3,Y3,Z3


  . . .

1,1,XN,YN,ZN
1,4,XMAX,YMAX,ZMAX
```

A symbol block is basically a string of records of fixed format. The first value in each record is a 1 to indicate random access file 1 which is the workspace. The second value in each record is the I code which indicates the significance

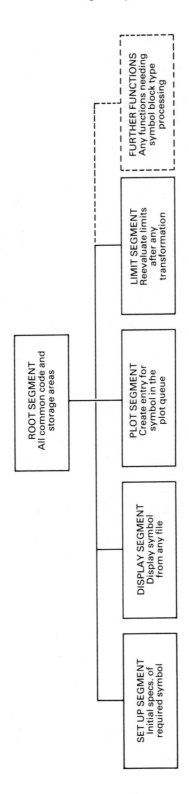

Fig. 4.10 — Symbol task overlay structure.

of the values in the rest of the record. I codes 3 and 4 define respectively the start and end of the symbol block. The start record contains the lower limit (XMIN,YMIN,ZMIN) in space of the symbol whereas the end record contains its upper limit (XMAX,YMAX,ZMAX). The second and third records are the identification records as indicated by I code 5. They usually contain the number of the symbol task (STSKNO) responsible for creating it, and the total length (BKLENG) of the symbol block, as these pieces of information are essential for the system to correctly identify and process symbols. Other additional information can be put into the vacant spaces (−) of these identification records, as may be required by some symbols.

Each symbol task is arranged into some kind of overlay structure as shown in Fig. 4.10. An overlay in a symbol task corresponds to a function performed on its own symbol block. Each overlay is essentiall a subroutine that is called by the root segment to perform the required function. The most common functions handled by a symbol task are to display the symbol on the graphics screen or to generate it on the plotter. Each function is specified by a unique entry point to the symbol task. This arrangement has the advantage of efficient core usage and allows additional symbols to be incorporated into the system without the need to modify the format of the symbol block or other symbol tasks.

4.3.5 Macros
A macro may be considered as a standard component or item that appears repeatedly on one or a number of different drawings. It may contain only straight lines or only geometric symbols or a combination of both. The main benefit of this macro facility is to enable the user to create a frequently used component only once, store it in a library of macros, and subsequently recall it as many times as needed onto any part of the current drawing. Thus the macro facility saves the user from having to generate a complicated but common item every time it is required, and this is the real power of a graphics system. In fact, when a comprehensive library of macros has eventually been built up, it is actually possible to produce a drawing by using macros only. The use of macros may be divided into three separate stages which are described in the following sections.

4.3.5.1 Creating macros
A user can start creating a macro by digitising over a special menu square which puts the system into a macro-creation mode. Once in this mode, the system checks to see if any data are present in the workspace. If there are, the user is warned of this situation and is given the choice of either continuing or aborting the process because all the data in the workspace will be overwritten by the macro. If there are not, the system stores a number of macro parameters and pointers, and then background mode is entered in order that the new macro can be digitised. The user can employ all the graphics facilities available in the system to create the macro which may include any combination of simple lines and symbols, and even other existing macros.

4.3.5.2 Saving macros

When the macro has been digitised, the user issues another menu command which invokes the macro-saving task. This task requires three types of information about the macro before it is saved in the macro library. First, three positional reference points have to be digitised to specify the orientation and size of the macro so that when it is recalled onto a drawing these three points can be used to rotate and scale the macro according to the three positional reference points given during the recall process which will be described in the next section. These three points should be suitably chosen so that they have some significance in relation to the macro, such as points at which connections are made to other items. Second, a code name for the macro has to be specified. This name may consist of up to six alphanumeric characters that can be treated as the identification code for the macro. The code name is used to form the filename of the disk file that actually contains all the data of the macro; so this code name is very important when recalling a macro. Third, the user must enter two numbers to indicate on which page and in which square the macro is to be assigned. This is necessary because in a large graphics system a paging system of macros is commonly used to cope with the many macros that may be needed. In the TIGER system, macros are grouped into pages of 50 each. Each macro page can then be mapped onto the macro area of the digitiser hence allowing the user access to every macro.

When the above three inputs have been entered, the system first of all checks to ensure that the specified square is not currently occupied by another macro. If it is, the present macro must be deleted before a new macro assignment can be made to that square. This serves as a safety measure against the occurrence of an existing macro being accidentally overwritten. If all is well, the system copies the workspace and the three positional reference points directly to a macro file on disk whose name is derived from the six-character code name. There is also an index file which keeps a list of all the macros currently stored in the macro library and other information about them such as when they were created, last accessed, etc.

4.3.5.3 Recalling macros

There are generally two methods of recalling a macro. The first and commonest method is simply to digitise over the appropriate square on the macro menu on the digitiser. The system then works out the square and page numbers of the macro required, and uses them as pointers in the macro index file to obtain the macro code name which enables the disk file containing the macro to be located. Such recalling process is simple and easy for the user to perform. All the user has to do is to choose the correct macro from the menu and digitise over it without the need to know its code name. An additional advantage of this method is that if the macro appears frequently on a drawing, button **A** may be assigned to the square of that macro and this enables further recalls to be made without having to refer to the menu. The second method is to recall the macro by its code name. This recalling method is invoked by a standard menu command which prompts the user to enter the macro code name on the alpha-numeric terminal. The system checks the specified code name against the macro index file for its validity before access can be made directly to the corresponding macro file.

Once the macro-recalling process has been started by either of the two methods mentioned above and the correct macro has been identified, the system requires the user to enter some points from the digitiser to specify the positioning of the macro. Three points are usually needed that correspond to the position of the reference points stored within the macro file. The macro is rotated and scaled so that the first two reference points are aligned with the two points entered by the user, and the third point is used to calculate the space orientation. When these inputs are completed, the system is ready to copy the macro file into the workspace. It first of all modifies the associated parameters such as display control, line type and pen number to some values corresponding to the current drawing. Then it arranges the data into a macro block very similar to the symbol block described in section 4.3.4. A typical macro block is given below as an example:

    1,13,XLIM1,YLIM1,ZLIM1
    1,15,AMACNO,BKLENG,−
    1,1,X1,Y1,Z1

    . . .

    1,1,XN,YN,ZN
    1,14,XLIM2,YLIM2,ZLIM2

where the first and last records indicate the limits of the macro, and the second record contains identification data.

Apart from the above three functions, the macro facility also provides some maintenance operations initiated by standard menu commands. They are included to enable the user to delete old macros that are no longer needed, and to list a directory of all the macros in the library on the alphanumeric terminal or line-printer.

4.3.6 Editing facility
In any respectable graphics system, an editing facility is vital. It enables the user easily and efficiently to modify drawings because of changes in design specifications or perhaps just errors that occur when the drawings are being digitised. The following sections outline the various editors available in the TIGER system.

4.3.6.1 Line, symbol and macro editors
The TIGER system divides its editing facility into separate editors according to the different data types that it handles. There are altogether three distinct categories of data, namely simple lines, symbols and macros as described in previous sections. The editing facility is arranged in such a way that there is an independent editor for each data type. This arrangement is considered more efficient because each editor can perform its function much more quickly as it

operates on only the particular data type that it recognises. The editor for each data type can operate in either last entry mode or random search mode.

Very often, during a drawing session, a situation may arise in which the user has realised that a point has just been digitised at the wrong place, or an inappropriate symbol or macro has been selected. The best way to correct this error is to use the last entry mode of the relevant editor. In this mode, the editor starts searching for the error backwards from the end of the drawing in the workspace. When the specified data type is first encountered, it is assumed to be the error and automatically deleted from the workspace. This method of removing the error is fast and efficient in that the user does not have to identify specifically the error from the digitiser and only a small part of the drawing need be searched to locate the error. Despite this advantage, the use of this mode of operation is rather limited as it is valid for only one, and must be the last, entry of the data type in the drawing.

Another mode of editing operation is the random search mode. As the name implies, this mode enables the user-specified data type to be located from anywhere within the drawing and to be removed from the workspace. When this mode is in operation, the user is asked to indicate the error in the drawing by digitising a point close to it. The editor then uses the point to search from the start of the drawing in the workspace for the specified data type that lies closest to this point. Once the data of the correct data type are located, they are displayed on the graphics screen for verification. At this stage, the user is given a choice of three buttons to indicate whether to delete the located data if they are correct, or to repeat the search procedure if they are incorrect, or just to verify again to be sure about them. When the error has eventually been found and deleted, the random search mode loops internally, to give the user an option to delete further items of the same data type until exit is requested by pressing the appropriate button, Random search mode is better than last entry mode in that the item to be deleted can be anywhere in the drawing and more than one item may be deleted from the drawing, albeit only one item at a time. However, there is one shortcoming inherent in random search mode, that is, it takes much longer to locate the error as the whole drawing has to be searched.

4.3.6.2 *Editor for selected data*

Sometimes it is necessary to perform editing on data that are neither simple lines, nor pure symbols nor macros. For example, when a design change has made a substantial part of a drawing obsolete, it is certainly useful to be able to delete the entire section all at once regardless of its data type and composition. There is in the TIGER system an editor which allows the user to do just that, provided the data concerned have been previously selected by one of the few selection methods described in section 3.3.7. This powerful editor simply deletes from the drawing any data that have been marked as selected. This process is very fast because of the way in which data are removed from the drawing when they are deleted.

This paragraph explains how deleted data are removed from the workspace. The method by which actual deletion of data is achieved applies to all the editors previously described. When the user indicates that certain specified data are to be deleted, the editor simply marks them as deleted but they are not yet physically removed from the workspace. Their removal is left to be done automatically at a later time by any other function that needs to rewrite the entire drawing in the workspace. This arrangement can eliminate the need to rewrite the whole drawing in the workspace every time a deletion is made, which is a time-consuming process if a large drawing is being produced. In addition, this method has the advantage that if the user for any reason has deleted the wrong data and realised the mistake early enough, it is still possible to recover them before they are removed completely from the workspace.

4.3.6.3 Data patcher
The TIGER system includes one more editor which permits the user to modify any piece of data within the database via the alphanumeric terminal in an inter-active question and answer session. This facility is useful for examining problem areas or patching errors that have arisen. However, an in-depth understanding of the data structures used is needed in order to use the facility effectively and efficiently, and the operation is generally rather slow as one command can only change one parameter of a record. It must be stressed that the use of this data patcher is intended for, and limited to, programmers only; and they have to use it with caution since it is possible to corrupt the entire database beyond recovery.

4.3.7 Data selection
From experience it has been found convenient and useful to have in a graphics system the facility to select particular parts of a drawing that may be made available for editing and graphical manipulation such as rotation and transla-tion without affecting the rest of the drawing. The TIGER system has some data selection facility that allows the user to do just that. This data selection facility works in much the same way as an editor except that in this case the data are marked as selected instead of deleted. This method enables the user to have total control over which parts of a drawing are effected by any of the graphical functions that operate on selected data. As a general precaution, these graphical functions will normally check if any data have been selected for the function. If not, an error message is issued to inform the user of this fact. The following sections will describe the various ways of selecting data in the TIGER system.

4.3.7.1 Select lines, symbols or macros
The arrangment for data selection facility is very similar to that for editing. There is a data selector for each data type and each selector can operate in both last entry mode and random search mode.

Last entry mode provides the user with a quick and easy method of select-ing an item in a drawing. Very often during a drawing session the last entry of a particular data type is needed for further processing. For example, if the same geometric symbol is required at many places on a drawing, it is obviously much quicker to create the symbol once, select it as the last entry, and then duplicate it at the required locations, because this method needs less input from the user and less processing by the system than generating the symbol separately each time.

The operation of the random search mode of the data selector follows a similar procedure to that of the random search mode of the editor, except that the data selector in this mode allows the user to specify up to 50 different search points before processing commences. Input of search points is normally terminated by pressing button F. The data selector then searches for the item that is closest to each of the specified search points and marks the item as selected. The item searched may be a line, symbol or macro depending on the data type active at the time. This mode of operation is very fast, as the work-space needs to be searched only once; and when the data selector is used together with the editor, it can provide an effective way of deleting at once a large number of items from the drawing. This is certainly more efficient than using the editor repeatedly for deleting each item.

4.3.7.2 Select within a digitised boundary

There is one further data selector that offers the most powerful and flexible means of selecting data of interest in a drawing. A maximum of 30 lines may be digitised to indicate the boundary that surrounds the data to be selected. The data selector simply checks which item lies within the area enclosed by the boundary and marks it as selected irrespective of its data type. The selected area may consist of any number and combination of different data types. It is very important to exercise care when defining a boundary that intersects data items. When the boundary intersects simple lines, the selector simply ignores them. When the boundary intersects symbols and macros, the situation becomes more complicated. The data selector in the TIGER system handles items of these two data types as a complete entity and so must be either selected in their entirety or not selected at all. If the symbol or macro falls principally within the boundary, the entire item will be selected. If, on the other hand, it does not, the item concerned will not be selected at all. If the user is not sure, the 'display selected data' menu command may be used to verify the data selection.

When data have been selected by any of the methods previously described, they remain so until specifically de-selected by the user or released by default upon completeion of one of the graphical functions. The selector can operate in either select mode or de-select mode to offer the user maximum flexibility in selecting awkwardly shaped or overlapping sections. The operation of the data selector is cumulative in that a number of selections and de-selections may be performed to build up the required selected data if this cannot be achieved with one operation alone. For example, the user can initially select a large area of interest in a drawing, then de-select within it smaller parts that are not needed. Thus the user has total freedom as to which data item is to be selected.

By default, any selected data will be automatically released at the end of the graphical function that operates on them. This is necessary because of the cumulative operation of the selector, otherwise subsequent graphical functions will affect data that are no longer of interest. However, this default may be over-ridden if the user specifies from the menu that the selected data are to be retained as selected after the graphical function. This feature can be very useful in situations where it is necessary to apply more than one function on the same selected data. For example, when a part of a drawing is to be duplicated at a different angle at another place in the drawing, this action may best be per-formed by two separate graphical functions. The way to achieve this is to select the data in the normal way, specify retention of selected data, duplicate the selected part at the same angle as the original, then rotate them to the required angle. Finally, the selected data must be manually released by the user using the appropriate menu command. This last operation is very important and must be performed at the end if the problem of cumulating selected data is to be avoided.

4.3.8 Graphic transformation

The heart of any graphics system is its capability to perform graphic transforma-tions on existing data. The TIGER system incorporates a large number of transformations, a typical example of which is the rotation of an item in a drawing through a certain angle about a specified axis. Each transformation can be applied to selected data only, thus the user has the power to specify which items are to be affected by a transformation and which are to be ignored. As a necessary precaution, every transformation routine automatically checks to make sure that data have been previously selected before starting its operation. If data have not yet been selected, the user is issued an error message to inform of this mistake. In general, any graphic transformation is made up of three basic operations, namely rotation, scaling and translation which are explained individually in the following sections. It is absolutely essential to have a thorough understanding of these basic operations in order to appreciate more complex transformations.

4.3.8.1 Rotation

The simplest case is the rotation on a two-dimensional plane about an axis that is perpendicular to that plane. By geometry, it can easily be deduced that 2D rotation may be represented mathematically by the following matrix equation:

$$\begin{bmatrix} \cos A & \sin A \\ -\sin A & \cos A \end{bmatrix} \begin{bmatrix} X \\ Y \end{bmatrix} = \begin{bmatrix} X' \\ Y' \end{bmatrix}$$

or, alternatively, $TP = P'$

where A = angle of rotation, T = the transformation matrix, P or (X, Y) = the original point, P' or (X', Y') = the new point after transformation.

In the case of three-dimensional rotation, a similar mathematical represen-tation can be deduced, and it can also be shown that multiple transformations may be compressed into one single transformation by multiplying together the

transformation matrices. As the order and direction of applying these transformations affect the final result, the TIGER system adopts the following
convention. Generally, multiple transformations are always applied about the
X, Y and Z axes in that order. Figure 4.11 shows the system of orthogonal
axes used for transformation and the direction of positive rotation. If the $X-Y$
plane lies flat in front of the observer with the Y-axis vertical, the Z-axis is
always positive in the direction towards the observer. The direction of positive
rotation about any axis is defined as that of anticlockwise rotation when the
observer looks along that axis in a negative direction towards the origin.

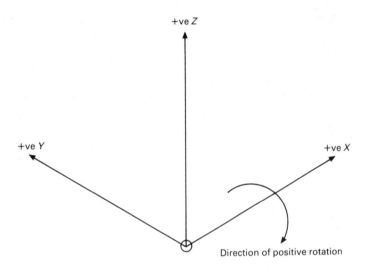

Fig. 4.11 – The system of orthogonal axes used for transformation.

The TIGER system incorporates several FORTRAN-callable subroutines
that perform three-dimensional rotations. Each rotation routine must be able
to apply the transformation on the three different data types. In the case of
simple straight lines, the transformation is relatively easy. However, transform-
ing symbols and macros is not entirely trivial. As described in section 4.3.4,
the first and last record of both symbols and macros are their low and high
limits in space, i.e. they define the bounding cuboid. When the symbol or macro
is rotated, it is obvious that these limits have to be completely re-evaluated.
Each symbol block contains positional data as well as numeric data. For example,
a symbol block defining a polygon contains coordinates specifying its position
and a numeric value defining the number of sides. It is essential that the routine
is capable of distinguishing the two types of information and applies the transformation to the positional data only, so that other records in the block are not
corrupted.

4.3.8.2 *Translation*

Translation is the simplest of all transformations and easiest to handle mathematically as it does not have some of the problems associated with rotation. The orientation of data remains unchanged before and after translation, therefore, the low and high limits on symbols and macros do not have to be re-evaluated. They may be transformed in exactly the same way as positional data by adding the correct displacements on each axis.

4.3.8.3 *Scaling*

Scaling is basically the changing of the size of an object. It is slightly more complicated than translation, but the principle involved is still relatively simple. Scaling can be defined by a scale factor and a reference point about which the object is to be scaled. In the case of magnification, the scale factor is greater than 1 whereas for diminution the scale factor is a fraction of less than 1. As the orientation of data is not changed before and after scaling, the low and high limits in symbols and macros do not have to be completely re-evaluated, provided they are transformed similarly to the positional data by multiplying the appropriate scale factor on each coordinate. The reference point is vitally important in that it allows the user to specify the final location of the object after the transformation has been applied. If, for example, an object has been magnified by a scale factor of 2 to twice its original size, it still does not provide the information to determine the location of this magnified object. The system gives the user full control over its location by specifying a reference point, sometimes called relative origin, which must lie at the same position both before and after the transformation.

4.3.9 Plotting

In the TIGER system, the plotting process of a drawing is divided into two separate stages, each of which will be explained in details in the following sections. Very briefly, the first stage involves converting the drawing data in the workspace into plotter commands in a plot file. Then the second stage simply opens the plot file to read the plotter commands and sends them to the plotter to produce the required drawing. As plotting a drawing is time-consuming and, to a certain extent, expensive in terms of computer and human resources, users are advised to output a drawing only when absolutely necessary.

4.3.9.1 *Requesting a plot*

The poltting process is invoked by menu command which requests the current drawing in the workspace to be plotted. If it is necessary to plot a drawing currently stored on a disk file, that drawing must first be recalled to the workspace. The system then enters into a short question and answer session with the user on the alphanumeric terminal. The user is essentially asked to specify the pens, paper type and size etc. which are needed to set up the plotter and load the correct paper at the actual plotting stage. The user is also requested to digitise a point on the drawing to indicate the bottom left-hand corner of the paper on the final plot, and then two more points to specify the bottom left- and top right-hand corners of the plot window which works in much the same way as the display window on the graphics screen. When all the required information has been received, the system asks one more question to confirm that the user is satisfied with all the specifications. If the confirmation is positive, the system proceeds to convert the drawing into a plot command file. In addition, an entry corresponding to this plot file is placed in a master plot queue index file which is used to locate the plot file when despooling. Each entry in this index file comprises a reference to the associated plot file, paper type and size, and all other information specified by the user when the plot was requested. When all these have finished, the system returns to background mode and the user may carry on with the current drawing.

4.3.9.2 The plot file despooler

As described above, every time the user requests a plot of a drawing, the plot request is spooled in a queue contained in the master plot queue index file. Each plot file is despooled by an independent task called the despooler which may be used to obtain a listing of the plot queue and plot any spooled drawings when the plotter next becomes free. The despooler task normally removes plot files from the queue on a 'first-in—first-out' basis. However, if an important drawing in the middle of the queue is needed urgently, this rule may be easily overridden. Similarly, if a drawing is for any reason no longer required before it is despooled, its entry may be deleted from the queue. When the plotter is ready for operation, the despooler simply reads the commands from the plot file and sends them to the plotter to produce the drawing. At the end of each plot a question is asked to confirm whether the drawing is satisfactory before the entry is deleted from the queue. This confirmation is needed in case one of the pens has run out of ink unexpectedly, or some other problems which make the plot unacceptable, and therefore it has to be repeated. If the confirmation is positive, the despooler will automatically continue to indicate the pens and paper for the next plot, and then wait until it is informed that the plotter is ready again for plotting. This process is repeated until the plot queue becomes empty, or it is prematurely terminated by user request. One added advantage of storing the plot queue on disk is that the drawings may be despooled at any time as the user desires and it is possible for drawings to be waiting overnight when the system is powered down. More importantly, the queue does not disappear in the event of any sudden system failure.

4.4 GRAPHICS STANDARDS

Graphics software needs standards; without them users find that, having purchased new computer graphics hardware, the further cost of modifying their existing software to run on the new equipment is prohibitive and the procedure is time-consuming. There are three major benefits of introducing standards for basic computer graphics. The first and obvious benefit is 'program portability' which permits application programs involving graphics to be easily transferred to and executed on almost any graphics installation. Graphics standards can help applications programmers in the understanding and use of graphics techniques. This gives rise to the second benefit of 'programmer portability' which means a programmer can leave one graphics application project for another without the need to learn a new set of graphics commands again. The third benefit is that graphics standards serve as a guideline for manufacturers of graphics equipment in providing useful combinations of graphics capabilities in a device, thus save programmers from having to create their own set of graphics commands and to write the interpreter for them. This advantage greatly helps to speed up the development of projects.

4.4.1 GKS and CORE

Through the 1960s and into the 1970s, computer graphics became a very versatile tool for many applications that employ graphics. However, most of these graphics applications ran on machines of different sizes and types, ranging from small personal computers to the top level CAD/CAM systems. It eventually seemed clear in 1974 that the computer graphics field could benefit from standardisation. When the need for graphics standards was first officially recognised, the Graphics Standards Planning Committee (GSPC) was formed by SIGGRAPH, a special-interest group of the US Association for Computing Machinery (ACM). Its objective was to produce a draft proposal that could be used as a basis for a computer graphics application interface standard. In 1977, the GSPC published the so-called CORE specification which many assumed was going to become the groundwork for this international graphics standard. The major problem with this CORE specification was that it was defined too early, when vector graphics were still the dominant display device, so it did not support raster graphics in any way. In 1979, after over two years of implementation and review, the GSPC released another CORE specification which included raster graphics as well as the machine/device interface and the need for metafiles to store images before displaying them.

At about this time, the West German standard-making body, Deutsches Institut für Normung (DIN), came up with the Graphical Kernal System (GKS). Since this adoption of GKS by DIN, the International Standards Organisation (ISO) and American National Standards Institute (ANSI) have cooperated to refine GKS in an attempt to make it a truly acceptable graphics standard. Indeed, ISO has already accepted GKS as a draft international graphics standard in 1984 against the opposition of CORE, and ANSI are likely to follow suit.

GKS was designed in accordance with the following three requirements. First, GKS has to provide all the capabilities that are important for the whole range of graphics, from simple passive output to highly interactive applications. Second, different types of graphics devices, such as vector and raster devices, microfilm recorders, storage tube displays, refresh displays and colour displays must be controllable by GKS in a consistent way. Finally, GKS must include all the capabilities required by a majority of applications without becoming excessively large. Details of the specifications of GKS can be found in references [11] and [12].

4.4.2 GKS-3D and PHIGS

One of the differences between CORE and GKS is that CORE allows coordinate data to be specified in either two-dimensional or three-dimensional form, whereas GKS accepts only two-dimensional data. Work is being carried out to extend GKS into a full 3D standard. ISO is now receiving two proposals, GKS-3D and PHIGS (Programmer's Hierarchical Interactive Graphics Standard) which are considered as complementary to, rather than competing with, each other.

The objective of GKS-3D is to enhance GKS to 3D by introducing such capabilities as:

(a) the definition and the display of 3D graphical primitives,
(b) mechanisms to control viewing transformations and associated parameters,
(c) mechanisms to control the appearance of primitives including optional support for hidden line and/or hidden surface elimination but excluding light source, shading and shadow computation,
(d) mechanisms to obtain 3D input.

The condition is that GKS-3D retains the general style of capabilities provided in GKS (2D) so that existing GKS application programs could execute without any modifications.

The objective of PHIGS is to provide capabilities that support such functions as:

(a) definition, display and modification of either 2D or 3D graphical data,
(b) definition, display and manipulation of geometrically related objects,
(c) rapid dynamic articulation of graphical entities.

The functionality of PHIGS uses essentially the same terminology and form as those in GKS, but in some areas PHIGS aims to achieve far beyond the scope of GKS or GKS-3D. As there is an important need for the ability to display an object more than once with different attributes, PHIGS supports a multi-level, hierarchical data structure which can be modified and allows the values of attributes to be inherited from the object's context, whereas GKS binds primitives and attributes together in a much more permanent manner. Both GKS-3D and PHIGS are still in the early stages of development, so they are not likely to be adopted very soon, although ANSI has been reviewing the PHIGS document for some time and ISO has shown interest in it as well.

4.4.3 IGES

The Initial Graphics Exchange Specification (IGES) is a standard data format for product design and manufacturing information created and stored in a CAD/CAM system in digital form, and is designed in such a way that it is independent of all CAD/CAM systems so that geometric and manufacturing data can be readily transferred between different systems. IGES was first proposed in 1979 by the US National Bureau of Standards (NBS) as a result of recognition of the need for exchanging data between CAD/CAM systems. NBS published its first report on IGES in January 1980, and the Version 2.0

specification [14] [15] was released in July 1982. This latest version is the product of drastically improving the initial specification from the first report.

IGES format can be used to describe draughting and geometric entities such as:

(a) geometric — point, line, arc, spline, etc.
(b) annotation — dimensions, drawing notes, etc.
(c) structure — properties, associations, groups, etc.

There are new entities introduced in Version 2.0 that include finite element, node, ruled surface (parameterisation), tabulated cylinder (more general form), surface of revolution (related to spheres and cones), rational B-spline surface, rational B-spline curve, etc.

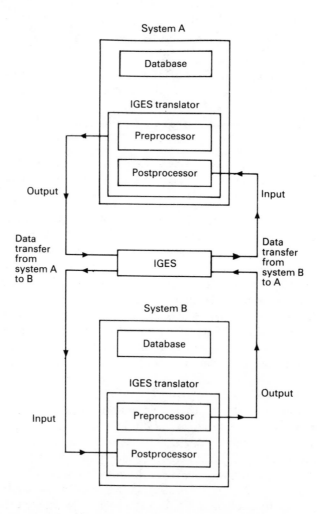

Fig. 4.12 — Data transfer between CAD/CAM systems.

The transfer of data between CAD/CAM systems using IGES format requires two separate stages as illustrated in Fig. 4.12. For example, if data are to be sent from system A to system B, they must first be translated into IGES format by using the IGES preprocessor in system A. Then, the resulting IGES entities are translated into the internal format of system B by its own IGES postprocessor. If data are to be transferred in the opposite direction, that is, from system B to system A, the process involved is essentially similar except that the IGES preprocessor of system B is used to convert its data into IGES format which is subsequently converted into system A data by its IGES postprocessor.

4.4.4 Other graphics standards

GKS, CORE, and IGES are just some of the graphics-related standards. There are other areas in which standardisations are being proposed. ANSI are now considering also Virtual Device Interface (VDI) and Virtual Device Metafile (VDM), which have recently been renamed respectively Computer Graphics Interface (CGI) and Computer Graphics Metafile (CGM). In addition, there is North American Presentation Level Protocol Syntax (NAPLPS) which has already been accepted by ANSI.

Figure 4.13 shows schematically the role played by the different standards in conceptual graphics system. At the highest level, the already approved IGES is a standard database format for the exchange of data between different CAD/CAM systems.

GKS and CORE type standards may be considered as functional standards for software and firmware implementations of computer graphics utilities. They basically act as a software interface between the application program and the graphics functions. Typically, they will provide a library of subroutines or procedures such as DRAW, PLOT, WINDOW, etc. which can be called from an application program.

At the next level, there are VDI and VDM standards which describe the format for transferring graphics data between the application programs and different graphics devices. These standards aim to overcome the problem of graphics device dependence. Many devices exist for graphics output and input which may seem similar, but in fact use different protocols for the production of graphics. VDI is a high-level bidirectional interface, or set of commands, for creating picture primitives such as lines or circles and for reading from graphics devices that an application program requests. The key advantage of VDI is that it is not necessary to modify the application programs as new graphics devices are added to a system. All that is required is to write a device driver for the new unit. VDM is a definition for storing graphics images in computer memory. The purpose of VDM is to provide a means for transferring images from one system to another. Under VDM standard, an image can be stored in one of three ways: clear-text encoding, character encoding or binary encoding.

Clear-text encoding is made up of commands that resemble VDI commands, so that a user familiar with VDI can understand a VDM file and can make changes to it if necessary. Character encoding is a more compact method of storing images. Finally, binary encoding uses a compression scheme to take

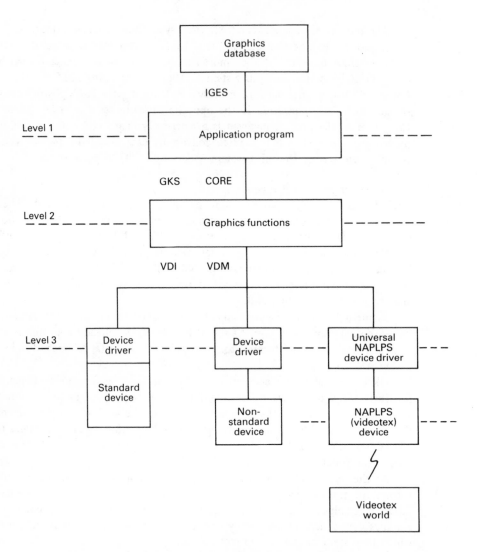

Fig. 4.13 – The role of different standards within a conceptual graphics system (reproduced from *Electronic Design*, 12 July 1984).

up the least possible space in memory or on magnetic disks. An additional advantage of compressing data in this way is that it will cost less to transmit them via a telephone line. The three different levels of encoding in VDM standard are shown in Fig. 4.14.

NAPLPS is a videotex standard which covers the presentation of graphics information via Prestel-type communications. NAPLPS provides a very compact coding scheme for transmitting graphics images over a communications link. As a result of its compactness, the coding scheme permits low-cost but relatively rapid transfer of images.

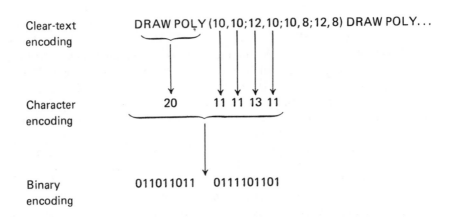

Fig. 4.14 – Three different levels of encoding in VDM standard (reproduced from *Electronic Design*, 12 July 1984).

Having discussed the various approved graphics-related standards and those under consideration by ISO and ANSI, it is not at all difficult to imagine that the interaction between different areas in the world of graphics standards is extremely complex. It is absolutely vital that there should be standards to cover all aspects of graphics interfacing such as for software, device, data and even character coding.

REFERENCES

[1] Newman, W. M. and Sproull, R. F., *Principles of interactive computer graphics*, 2nd edn, McGraw-Hill, Kogabusha, 1979.

[2] Foley, R. D. and Van Dam, A., *Fundamentals of interactive computer graphics*, Addison Wesley, 1982.

[3] Besant, C. B., *Computer-aided design and manufacture*, 2nd edn, Ellis Horwood, 1983.

[4] Groover, M. P. and Zimmers, Jr, E. W., *Computer-aided design and computer-aided manufacturing*, Prentice-Hall, 1984.

[5] Lui, C. W. K., *The use of computer-aided techniques for hypoid gear design*, Ph.D. thesis, Imperial College of Science and Technology, 1983.

[6] Panasuk, C., Software Standards will Usher in the Age of Graphics, *Electronic Design*, 12 July 1984.

[7] Warman, E., Graphics – A European Standard, *Systems International*, April 1983.

[8] Geller, R. D., Two Graphics Standards, CORE and GKS, vie for U.S. Market Acceptance, *Computer Technology Review*, Winter 1983; *The Systems Integration Sourcebook*.

[9] Sturridge, H., Standard Solution to a Portability Poser, *Computing*, 15 November 1984.

[10] Osland, C. and Hopgood, B., Standard Progress, *Systems International*, January 1985.
[11] BS 6390:1983, A Set of Functions for Computer Graphics Programming, the Graphical Kernal System, British Standards Institution, 1983.
[12] Hopgood, F. R. A., Duce, D. A. and Gallop, J. R., *Introduction to the GKS*, Academic Press, New York, 1983.
[13] Fong, H. H., Interactive Graphics and Commercial Finite Element Codes, *Mechanical Engineering*, June 1984.
[14] IGES, Version 2.0 (NTIS. Order No. PB 83-137448), National Technical Information Service, Springfield, VA, July 1982.
[15] Smith, B. M., IGES: A Key to CAD/CAM Systems Integration, *IEEE Computer Graphics and Applications*, November 1983.
[16] Smith, W. A., *A Guide to CAD/CAM*, The Institution of Production Engineers, 1983.
[17] Boniwell, S., How ANSI Sees Graphics, *Systems International*, May 1983.

5

Transformation systems

In the previous chapter we encountered various two- and three-dimensional transformations. They include scaling, rotation, translation and general homogeneous matrix transformations. The purpose of this chapter is to discuss how these transformations are made for those who might wish to write certain graphics software.

We can use hardware as well as software to perform transformations. There are a number of displays that use hardware to perform all the most common transforms but these displays are expensive. Most displays require a level of software to perform transformations and sometimes this software is located in a microcomputer display processor system.

The most important aspect of software-driven picture transformations is the speed of the transformation. In practice the calculation of transformations can be time consuming and prove to be a bottleneck in the performance of a display system. In order to optimise the time in calculating transformations it is often necessary to concatenate two or more transformations into a single transformation. The concatenating of transforms will be discussed in this chapter.

5.1 DISPLAY

CAD involves displaying a large number of pictures and much time can be consumed in converting structured data into display signals. The display file, as previously discussed, may be regarded as a table of instructions to be executed by the display processor. The display file, with hardware consisting of a digitiser and storage tube, may be the same as the workspace for a two-dimensional graphics system, since there is no need for a conventional display memory. In the case of a three-dimensional system, intermediate files may be used to advantage and this will be discussed later.

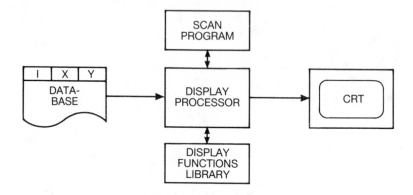

Fig. 5.1 — One-pass display processor.

The display processor consists of a program and a number of subroutines built into a library which are loaded to execute the instructions contained in the display file. This is shown in flow chart form in Fig. 5.1. In the simplest data structure, the instructions are stored in $[I, X, Y]$ records with I representing the command code with respect to the display procedure and X, Y the data. Compiling of the graphics language is simple, for instructions are usually generated in a sequential manner. The instruction code I may have a wide range of meanings and the following examples are now given:

I = 1   drive beam from last position to position $[X, Y]$ with beam switched off.

I = 2   drive beam from last position to position $[X, Y]$ with beam switched on.

I = 3   stop display.

Efficient algorithms for converting display instructions to signals driving the CRT are particularly important for fast display of data.

5.2 WINDOWING AND CLIPPING

When it is necessary to examine in detail a part of a picture being displayed, a window may be placed around the desired part and the windowed area magnified to fill the whole screen. This involves scaling the data which lies within the window so that the window fills the entire screen. Data that lies out-side the window must be eliminated so that only the data required for display is processed. This process is known as clipping. Some hardware devices have automatic scissoring in which the window and the display vectors may be larger than the display raster. For some systems, as with the present DSV system, hardware clipping is not available and so software clipping must be employed.

It is necessary to define two viewing areas: a viewport and a window. It is usual to make the viewport equal in size to the screen to take advantage of maximum screen area.

The limits of the window are determined by the coordinates on the bottom left-hand corner taken as [0,0] and the dimensions of the required frame. The window is set up by digitising the [0,0] coordinates and moving the cursor on the digitiser to the right until the window encloses the desired area. Once the window is defined the data outside the window is clipped before scaling to the screen coordinates, This considerably reduces the amount of data before display signals are generated.

The Tektronix 4014 storage tube used in the Imperial College system contained 4,096 points horizintally by 3,072 points vertically. The windowing transformation is given by:

$$X_s = 4,096 \frac{X_p - X_w}{D_x}$$

$$Y_s = 3,072 \frac{Y_p - Y_w}{D_y}$$

where (X_w, Y_w) are the coordinates of the bottom left-hand corner of the window and D_x and D_y are the frame length and height of the viewport defined on the table (see Fig. 5.2).

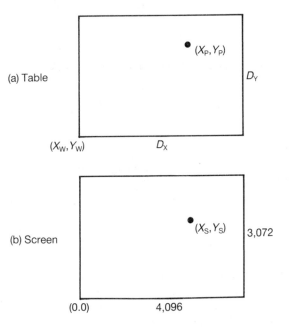

Fig. 5.2 – Windowing transformation.

The clipping algorithm tests for the position of lines in relation to the window. First it tests for the trivial cases where lines lie entirely within or without the window as shown in Fig. 5.3. If these tests are not satisfied then a line is assummed to intersect the window. The intersecting lines are trimmed so that the external points lie on the edge of the window.

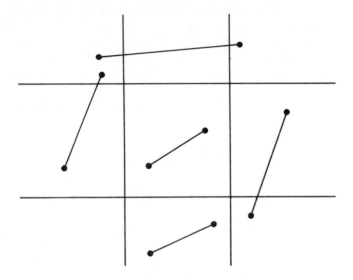

Fig. 5.3 – Trivial cases of clipping.

5.3 TWO-DIMENSIONAL TRANSFORMATIONS

Every picture manipulation performed in the system can be represented by a matrix transformation:

$$\mathbf{X_T} = \mathbf{XT}$$

where $\mathbf{X}$ is a vector in the data file and $\mathbf{X_T}$ the transformed vector contained in a display file. $\mathbf{X_T}$ can be written back in the data file to represent a transformed new data source.

For two-dimensional data $\mathbf{X}$ and $\mathbf{X_T}$ are represented by:

$$\mathbf{X} = [x, y]$$

$$\mathbf{X_T} = [x'\, y']$$

Thus T is a 2×2 matrix:

$$T = \begin{bmatrix} a_{11} & a_{12} \\ a_{21} & a_{22} \end{bmatrix}$$

where the terms a_{ij} can be defined for each transformation

$$[x \; y] = [x' \; y'] \begin{bmatrix} a_{11} & a_{12} \\ a_{21} & a_{22} \end{bmatrix}$$

T can be a single matrix or concatenated by multiplying two or more matrices. The coordinate values in X are normally relative to a reference point X_r. Therefore, in general:

$$X_T = TX + X_r$$

where $\qquad T = T_1 * T_2 * T_3 * \ldots$

The order of the individual matrices T_i determine the resultant transformation. However, using 2×2 matrices it is not always possible to concatenate the matrices. For example, to rotate a point $[x \; y]$ through an angle θ about an arbitrary point at a distance $[R_x \; R_y]$ from the origin, conventional operation in 2×2 matrices would give:

translate $\quad [x' \; y'] = [x \; y] - [R_x \; R_y]$

so that $[R_x \; R_y]$ becomes the origin.

Rotate $\quad [x'' \; y''] = [x' \; y'] \begin{bmatrix} \cos\theta & -\sin\theta \\ \sin\theta & \cos\theta \end{bmatrix}$

translate again to old origin

$$[x''' \; y'''] = [x'' \; y''] + [R_x \; R_y]$$

The transformation becomes:

$$X''' = (X - T)\,R + T$$

The matrices cannot be combined into one single transformation. This is made possible by using homogeneous matrix representation [1], where each transformation in two dimensions is represented by a 3×3 matrix.

Although there is no practical advantage in storing data as homogeneous coordinates it is useful to perform transformations with 3×3 matrices. Adding a third element we obtain:

$$[x' \; y' \; 1] = [x \; y \; 1] \begin{bmatrix} a_{11} & a_{12} & 0 \\ a_{21} & a_{22} & 0 \\ a_{31} & a_{32} & 1 \end{bmatrix}$$

The extra dimensional element represents a scaling factor and is always kept as unity because a single common scale is used which does not require to be constantly redefined and stored as data. This also reduces the data storage requirement by 25 per cent.

The multiplication of the last column can be eliminated, reducing the computation of the matrix from 9 multiplications and 6 additions to 6 multiplications and 4 additions:

$$[x' \ y'] = [x \ y \ 1] \begin{bmatrix} a_{11} & a_{12} \\ a_{21} & a_{22} \\ a_{31} & a_{32} \end{bmatrix}$$

The above translation and rotation becomes:

$$\text{translation} \quad T = \begin{bmatrix} 1 & 0 & 0 \\ 0 & 1 & 0 \\ R_x & R_y & 1 \end{bmatrix}$$

$$\text{and rotation} \quad R = \begin{bmatrix} \cos\theta & -\sin\theta & 0 \\ \sin\theta & \cos\theta & 0 \\ 0 & 0 & 1 \end{bmatrix}$$

$$\text{and} \quad \mathbf{X} = [x \ y \ 1]$$

The matrices can be combined as

$$\mathbf{X}''' = X(-T) \, RT$$

where

$$(-T) \, RT = \begin{bmatrix} \cos\theta & -\sin\theta & 0 \\ \sin\theta & \cos\theta & 0 \\ -R_x \cos\theta - R_y \sin\theta + R_x & R_x \sin\theta - R_y \cos\theta + R_y & 1 \end{bmatrix}$$

Again the last column can be eliminated from the final matrix computation. It is, however, necessary for the three matrix operations.

5.4 THREE-DIMENSIONAL TRANSFORMATIONS

Three-dimensional transformations are performed on data by matrix operations similar to those described in the previous section.

It is necessary to define a system of reference axes and adopt a convention for the direction of rotation before one can consider any transformations in three dimensions. The conventional right-handed reference set of orthogonal axes is used here and is shown in Fig. 5.4. The $x-y$ plane is chosen to correspond to any flat working surface such as the digitiser or the viewing screen. The z-direction is always forward, towards the observer.

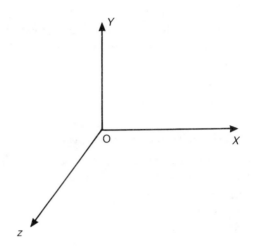

Fig. 5.4 — Right-hand reference set of axes.

The angle of rotation θ about any axis is taken to be positive where the rotation is anti-clockwise and negative when clockwise. Thus the rotation is said to be positive when in the sense of a right-handed corkscrew as shown in Fig. 5.5.

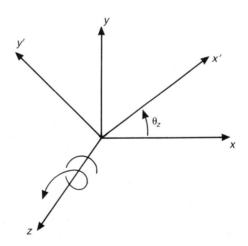

Fig. 5.5 — Rotation about z-axis.

The rotation is given by the matrix

$$\begin{bmatrix} \cos\theta & -\sin\theta \\ \sin\theta & \cos\theta \end{bmatrix}$$

with θ measured in the anti-clockwise direction. The rotation refers to the set of axes on the plane of rotation. For rotation of objects in a space with a fixed set of reference axes the opposite applies and the matrix becomes

$$\begin{bmatrix} \cos\theta & \sin\theta \\ -\sin\theta & \cos\theta \end{bmatrix}$$

5.5 LINEAR TRANSFORMATIONS

Three-dimensional transformations are similar to those in two dimensions. We will concentrate on the following:

Scaling
Translation
Rotation (about one or more axes)

For three-dimensional data $[x, y, z]$ the transformation can be represented by

$$\mathbf{X'} = \mathbf{X}T$$

where $\mathbf{X} = [x, y, z, 1]$ the original coordinates

$\mathbf{X'} = [x', y', z', 1]$ the transformed coordinates

and T can be represented by a single or concatenated 4×4 matrix:

$$T = \begin{bmatrix} r_{11} & r_{12} & r_{13} & 0 \\ r_{21} & r_{22} & r_{23} & 0 \\ r_{31} & r_{32} & r_{33} & 0 \\ t_1 & t_2 & t_3 & 1 \end{bmatrix}$$

where r_{ij} are the terms of the rotation matrices and t_j the translation offset.

The r_{ij} terms normally consist of a single or a combination of rotation matrices.

(a) Rotation about z-axis (Fig. 5.5).

$$R_z = \begin{bmatrix} \cos\theta_z & -\sin\theta_z & 0 \\ \sin\theta_z & \cos\theta_z & 0 \\ 0 & 0 & 1 \end{bmatrix}$$

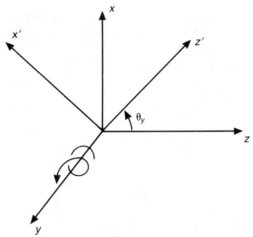

Fig. 5.6 – Rotation about y-axis.

(b) Rotation about y-axis (Fig. 5.6)

$$R_y = \begin{bmatrix} \cos\theta_y & & \sin\theta_y \\ 0 & 1 & 0 \\ -\sin\theta_y & & \cos\theta_y \end{bmatrix}$$

(c) Rotation about x-axis (Fig. 5.7)

$$R_x = \begin{bmatrix} 1 & 0 & 0 \\ 0 & \cos\theta_x & -\sin\theta_x \\ 0 & \sin\theta_x & \cos\theta_x \end{bmatrix}$$

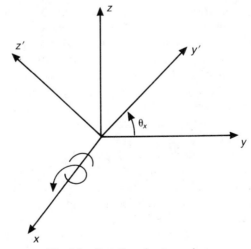

Fig. 5.7 – Rotation about x-axis.

R_x, R_y, R_z can be concatenated to give a general rotation matrix R, however, the order in which the individual matrices are combined will affect the results thereby achieved.

The scaling transformation is given by:

$$S = \begin{bmatrix} S_x & 0 & 0 \\ 0 & S_y & 0 \\ 0 & 0 & S_z \end{bmatrix}$$

For equal values of S_i the scaling is linear, otherwise distortions will be introduced into the system.

These transformations allow data to be repositioned anywhere and in any orientation in space. The translation, rotation and scaling effects can all be combined into a single transformation matrix by concatenating the individual matrices. However, the same effects can also be achieved by varying the viewing parameters used to define the object—observer geometry in a perspective transformation.

The equivalent effects: pan, rotation and scaling or zooming can be obtained by varying the observer's position and orientation relative to the object. This eye-movement technique represents a more economical way, in terms of computing time, of obtaining movements from a static object. The database remains unaltered and only the viewing transformation, which is always required irrespective of the parameter values, is performed. The joystick function described in section 5.9 uses this technique to provide a very quick means of manipulating the display.

5.6 DISPLAY FILES FOR THREE-DIMENSIONAL DATA

The introduction of a third dimension results in an increase in the number of processes involved in converting data from a basic structure to drive a display terminal. Two major stages are involved: compilation and processing (Fig. 5.8).

In the compilation stage the data file is first scanned by a viewing algorithm, an output routine which generates function calls to the appropriate subroutines (scaling, perspective, windowing, clipping). Then, in response to these function calls the display compiler interprets the stored data and creates a display file. The cumulative effect of these two processes is to generate a second data structure from the first.

In the second stage, the display file, which contains simple commands for the display, is scanned by the display processor, which then generates the drive signals for the CRT. The instructions are simple x–y drive signals, but the processor may incorporate some hardware character or symbol generation facilities.

These two stages are very similar, both involving a scanner and an interpreter, and can be combined into one, so that a set of data serves both purposes.

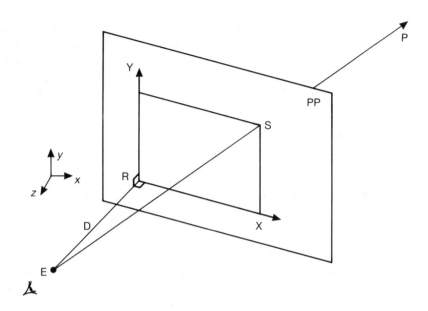

Fig. 5.8 — Projection on plane PP.

With a storage tube a conventional memory for storing the display file is not essential, since the image is only displayed once. Because of the lack of selective erasure facilities the whole screen must be redisplayed to incorporate any changes. However, the user of a separate display file does have the advantage of enabling the use of the concept of segmented display files, which reduces the amount of compilation involved each time the picture is selectively modified. For example, in macro manipulation only the affected macro or modified segment is recompiled.

It is usual to work with two files, so that the data and display files are stored for example, in FILE 1 and FILE 2 respectively. FILE 2 does not fall within the exact definition of a display file because it contains more than simple instructions for the display processor and it also contains data that lies off the screen area. The main purpose of this file is to permit faster display at different window sizes without the need to repeat viewing transformations. It also permits interactive editing of the database by a parallel search technique and it useful in visualisation tests for three-dimensional viewing.

5.7 VISUALISATION OF THREE-DIMENSIONAL DATA

The presentation of three-dimensional graphic data on a plane is important since many people have great difficulty in understanding engineering drawings. When an object is rotated, for example, if the new view is not clearly presented then the viewer could be confused. Such confusion might arise from a 'wire cage' drawing where it might be difficult to distinguish between lines representing the front and the rear.

In order to help the CAD system user to interpret static two-dimensional projections of three-dimensional objects an range of visual aids may be adopted.

There are several ways in which the visualisation of objects can be improved:

(a) Perspective transformation.
(b) Brightness modulation.
(c) Hidden-line removal.
(d) Shading.
(e) Movement.

The first two techniques mainly apply to wire-frame models while (c) and (d) are associated with solid objects. These techniques are now discussed since the use of the compter in this area is particularly significant.

(a) Perspective transformation

Perspective is one of the most valuable techniques for visualising objects. The transformation is often used when displaying three-dimensional data but also it can be used with two-dimensional data to represent a plane lying in three-dimensional space. Perspective is easily represented in mathematical form since it is based on elementary optics.

Consider Fig. 5.8 where x, y, z represent the axes corresponding to the spatial cartesian coordinates. The system of reference axes used here is in the conventional right-handed reference directions. If the $x-y$ plane is made to coincide with the screen the positive z direction is out of the screen towards the observer.

For the purpose of the projection, a plane PP is defined with coordinates (X, Y). The projection maps the point $\overline{P}$ to a point $\overline{S}$ on the projection plane PP. The projection plane in this case corresponds to the viewing screen, with X to the right and Y upwards.

The observer's eye is situated at the point $\overline{E}[E_x, E_y, E_z]$ and the line of sight makes angles α, β and γ with the spatial x-, y- and z-axes respectively. The angles α, β, γ can be specified by their direction cosines or by a point defined as the centre of attention, from which the direction cosines can be calculated. Normally the centre of attention is made to coincide with the origin or the volumetric centre of the object being observed.

Consider the plane of projection PP at a distance D from E, normal to the line of sight. For an observer with the eye position at $\overline{E}[E_x, E_y, E_z]$, direction of line of sight at angles $[\alpha, \beta, \gamma]$ to the major axes, the point $\overline{P}[P_x, P_y, P_z]$ in space is mapped onto the projection plane PP, by the following relationships:

$$X = \frac{(S_x - R_x) \cos \gamma - (S_z - R_z) \cos \alpha}{\sin \beta}$$

$$Y = \frac{(S_y - R_y)}{\sin \beta}$$

where $[X, Y]$ are the coordinates of the projected position of the point $\bar{P}$. $\bar{S}[S_x, S_y, S_z]$ is the point of intersection of the line of sight on the projection plane and $\bar{R}[R_x, R_y, R_z]$ the foot of the normal from the observer eye to the plane.

Isometric or orthogonal projections are often preferred to perspective drawings in engineering because measurements can be related more easily to the drawing. However, from aesthetics the perspective view looks right and, since it is easy to produce from a CAD system, its use may well increase.

(b) Brightness modulation

With this technique parts of the picture near to the observer are bright while those far away are dim.

When this is required on a view an extra routine in the program is entered just before the vector generation and this selects the required brightness levels for the display file as it is being constructed. When a picture has been constructed, the maximum and minimum z coordinates are noted. The z range is then divided into n regions where n are the visible brightness levels available in the display system. The picture is then displayed with the appropriate brightness level corresponding to its z region.

This technique is easy to implement and is very effective when displayed on the screen. It is difficult to obtain a hard copy version unless an electrostatic plotter is used.

(c) Hidden-line removal

If a complex three-dimensional drawing is fully displayed then the large number of lines usually render the picture impossible to perceive. The main problem lies in that the lines which are normally hidden by the object, are all displayed and this can lead to confusion. The hidden-lines can be removed by the computer but large amounts of computing time are usually required. Computation approximately increases as the square of the number of edges. Therefore for moderately complex situations computation can become prohibitive on a small computer.

It is not easy to establish reliable algorithms to identify the lines to be removed. In general, the geometric calculations are straightforward if objects are convex polyhedra. But if the three-dimensional bodies are not rectilinear the problem can be very difficult. A number of successful algorithms have been devised [2, 3] to perform hidden-line removal.

(d) Shading

Shading techniques have been developed extensively in work related to hidden-line removal, particularly at the University of Utah [4, 5] and at the CAD centre at Cambridge [6]. The technique is based on the recognition of distance and shape as a function of illumination.

The technique is similar to finite elements. The surface of a solid is divided into patches and in regions of large curvature, the patches are decreased in size. Each patch or element is then tested for visibility and the degree of shading required. It should be remembered that hidden-line removal is a prerequisite for any shading algorithm.

The amount of shading required is determined by calculating the angle between the normal to the plane of the element and the vector direction of the propagation of the light. The normal vector can be calculated from the cross-product of two vectors on the plane or from the equation of the plane. The angle of incidence is given by the dot-product of the normal and the line of incidence. Brightness and visibility increase as the angle of incidence increases from 0 to 90°. For the visibility test only the sign of the dot-product is required to determine whether the plane is fairing the light. This is a preliminary test to eliminate all planes facing in the wrong direction. Gourand used a method where the intensity at the point where the elements meet is calculated. The intensity is then interpolated to provide smooth shading of the surface. For illumination purposes a point or parallel beam source of light may be used. The surface can also be given a reflective index to make it shiny, or dull or even transparent.

The output is obtained on devices that can scan at different intensity levels, either by drawing a series of parallel lines or by overwriting. Results obtained by Parker, Gourand and Newell are impressive but they do rely on powerful, special-purpose hardware. The work at Imperial College by Yi on shading was based on the use of a Calcamp microfilm plotter.

(e) Movement
Movement leads to improved recognition of displayed objects. As an object is rotated or translated, ambiguities that arise due to the superposition of points are eliminated and the geometrical properties of the object are revealed by the interaction of the points defining the object.

The storage tube is not really suitable for displaying movement since the whole picture must be erased before a new one is displayed. This can take time so that real time movement is not possible. However, it is possible to photograph a series of pictures from the screen of a storage tube so that they can be replayed to give real-time movement. This is particularly useful for a simulation exercise since it is an inexpensive method for obtaining a simple animated line test.

The use of movement in CAD will become increasingly important as new cheap refreshed displays are developed. The use of moving blueprints for the simulation of engineering dynamics is an exciting prospect for engineers. Movement, coupled with colour, will be an extremely powerful tool in CAD.

In summarizing the discussion on displaying three-dimensional objects, the main drawback of most of these techniques is the large amount of computing power required together with the requirement for specialised hardware in some cases. Many of these techniques place considerable strains on mini-computer-based systems and skilled programming is required in order to obtain a viable system. Probably the most suitable technique for the minicomputer and storage tube combination is that of perspective.

The advent of microprocessors and new types of refreshed displays is bringing near the time when three-dimensional graphics in colour with real-time movement will be a practical reality in low cost CAD systems.

5.8 EYE COORDINATES SYSTEM

In the last section a method whereby a perspective view of a three-dimensional object could be generated was described. The final drawing was made up of $[X, Y]$ points on the projection plane. However, it is not always desirable to represent an object as a flat drawing. Sometimes it is essential to preserve the depth information in order to determine the spatial properties of the solid, which would have been lost with a plane representation.

What is needed is a transformation which converts a three-dimensional object as viewed in perspective into another object which will give the same view when projected orthogonally (Fig. 5.9). The transformation in effect moves a local observer to infinity and distorts the object appropriately so that it still looks the same on the viewing screen. The transformation preserves all the spatial qualities of the object so it is always possible to apply the three-dimensional perspective transformation before doing any visualisation computation like brightness modulation and hidden-line removal. It is much easier to perform hidden-line removal computation from an orthogonal projection than a perspective projection.

Before the distortion transformation is applied it is necessary to define the object in terms of a local system of reference axes with a different orientation.

The new set of axes x', y', z' is taken with the origin at the eye position and with positive z' along the line of vision in the opposite direction (Fig. 5.10). x' is in the plane parallel to the $x-z$ plane and y' is upwards. The angles the line of sight makes with the x-, y-, z-axes are given by their cosines $\cos \alpha$, $\cos \beta$, $\cos \gamma$ respectively.

The transformation is a combination of a translation and a rotation.

$$\mathbf{X'} = R(X - T)$$

$\mathbf{X} = [x, y, z]$ point defined in the old coordinate system.

$\mathbf{X'} = [x', y', z']$ point defined in the eye coordinate system.

$T = $ translation vector given by $[E_x, E_y, E_z]$.

Matrix R is given by

$$\begin{bmatrix} \lambda_{x'x} & \lambda_{x'y} & \lambda_{x'z} \\ \lambda_{y'x} & \lambda_{y'y} & \lambda_{y'z} \\ \lambda_{z'x} & \lambda_{z'y} & \lambda_{z'z} \end{bmatrix}$$

where $\lambda_{x'x}$ is the direction cosine between x' and x and so on.
From above

$$\lambda_{z'x} = \cos \alpha$$

$$\lambda_{z'y} = \cos \beta$$

$$\lambda_{z'z} = \cos \gamma$$

and x' is parallel to X, y' is parallel to Y.

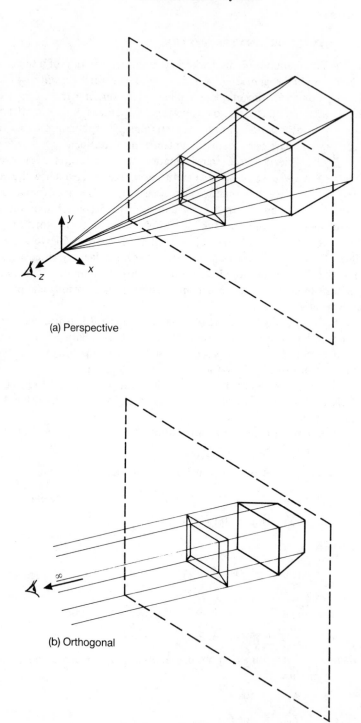

(a) Perspective

(b) Orthogonal

Fig. 5.9 – Perspective – orthogonal projections.

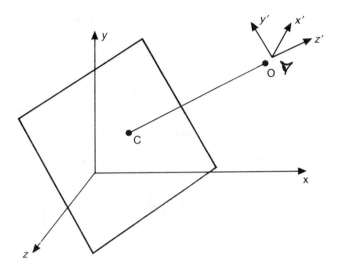

Fig. 5.10 – Eye coordinate system.

$[X, Y]$ are the axes on the projection plane.
Now

$$\lambda_{x'x} = \cos \gamma / \sin \beta$$

$$\lambda_{x'y} = 0$$

$$\lambda_{x'z} = -\cos \alpha / \sin \beta$$

and

$$\lambda_{y'x} = -\cos \alpha \cos \beta / \sin \beta$$

$$\lambda_{y'y} = \sin \beta$$

$$\lambda_{y'z} = -\cos \beta \cos \gamma / \sin \beta$$

thus

$$R = \begin{bmatrix} \dfrac{\cos \gamma}{\sin \beta} & 0 & \dfrac{-\cos \alpha}{\sin \beta} \\[2mm] \dfrac{-\cos \alpha \cos \beta}{\sin \beta} & \sin \beta & \dfrac{-\cos \beta \cos \gamma}{\sin \beta} \\[2mm] \cos \alpha & \cos \beta & \cos \gamma \end{bmatrix}$$

Alternatively, R can be derived by regarding the transformation as a combination of two rotations (Fig. 5.11). The first R_θ about the y-axis in the $x-z$ plane through angle θ anticlockwise (object rotation).

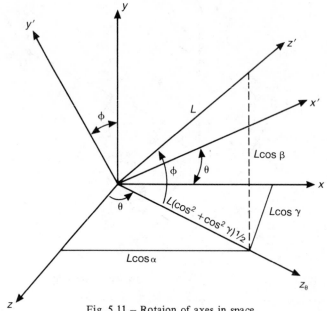

Fig. 5.11 – Rotaion of axes in space.

Thus

$$R_\theta = \begin{bmatrix} \cos\theta & 0 & -\sin\theta \\ 0 & 1 & 0 \\ \sin\theta & 0 & \cos\theta \end{bmatrix}$$

from Fig. 5.11

$$R_\theta = \begin{bmatrix} \dfrac{\cos\gamma}{\sin\beta} & 0 & \dfrac{-\cos\alpha}{\sin\beta} \\ 0 & 1 & 0 \\ \dfrac{\cos\alpha}{\sin\beta} & 0 & \dfrac{\cos\gamma}{\sin\beta} \end{bmatrix}$$

where $\sin\beta = (\cos^2\alpha + \cos^2\gamma)^{\frac{1}{2}}$;
and

$$R_\phi = \begin{bmatrix} 1 & 0 & 0 \\ 0 & \cos\phi & -\sin\phi \\ 0 & \sin\phi & \cos\phi \end{bmatrix}$$

$$\begin{bmatrix} 1 & 0 & 0 \\ 0 & \sin\beta & -\cos\beta \\ 0 & \cos\beta & \sin\beta \end{bmatrix}$$

therefore $R = R_\phi R_\theta$; as before

$$R = \begin{bmatrix} \dfrac{\cos\gamma}{\sin\beta} & 0 & \dfrac{-\cos\alpha}{\sin\beta} \\ \dfrac{-\cos\alpha \, \cos\beta}{\sin\beta} & \sin\beta & \dfrac{-\cos\beta \, \cos\gamma}{\sin\beta} \\ \cos\alpha & \cos\beta & \cos\gamma \end{bmatrix}$$

The above rotation matrix is valid for all directions of z' in Fig. 5.11 except when it is vertical. In either of the cases shown in Fig. 5.12 the direction cosines of the axis z' with respect to the reference axes can be determined by inspection and the rotation matrix becomes

$$R = \begin{bmatrix} 1 & 0 & 0 \\ 0 & \cos\beta & 0 \\ 0 & 0 & -\cos\beta \end{bmatrix}$$

The matrix is valid for both cases in Fig. 5.12. The direction cosine $\cos\beta$ is $+1$ for the first case and -1 for the second.

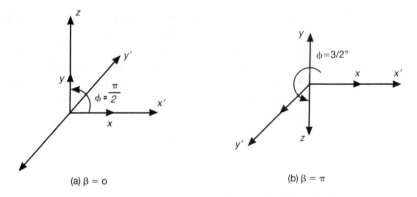

(a) $\beta = 0$ (b) $\beta = \pi$

Fig. 5.12 – Orientation of axes for a vertical line of sight.

In the new coordinate system the perspective projection becomes a simple linear relationship because the $x'-y'$ plane is parallel to the projection plane. From Fig. 5.13

$$[X \ Y \ Z] = [x' \ y' \ z'] \begin{bmatrix} D/z' & 0 & 0 \\ 0 & D/z' & 0 \\ 0 & 0 & 1 \end{bmatrix}$$

where D is the distance from the eye position to the projection plane. The z term is the depth coordinate and is stored in the z location of the display file.

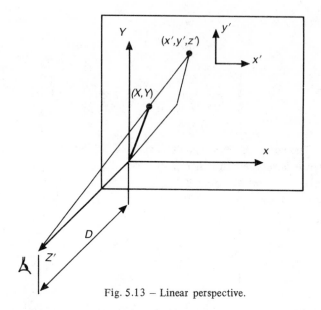

Fig. 5.13 – Linear perspective.

5.9 JOYSTICK FUNCTION

The variations in perspective parameters, as described earlier, allow a very convenient solution to a joystick controlled display.

The joystick function allows the operation of a joystick to be emulated on a patch menu. By pressing the pen button on the menu the user can control two functions: rotation and scale change or zoom. Rotations are possible about two axes: y-axis and an axis on the $x-z$ plane determined by the pan and tilt angle respectively. (θ and ϕ in Fig. 5.14.)

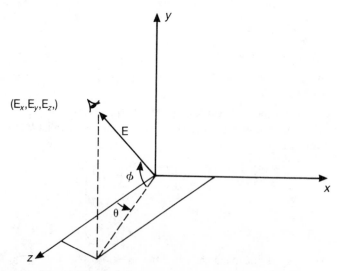

Fig. 5.14 – Eye position in terms of pan and tilt angle.

The eye position is defined by $[E, \theta, \phi]$ where E is the distance from centre of attention, θ is the pan angle and ϕ the tilt angle.

The coordinates of the eye $[E_x, E_y, E_z]$ with respect to the centre of attention are given by

$$E_x = E \cos\phi \sin\theta$$

$$E_y = E \sin\phi$$

$$E_z = E \cos\phi \cos\theta$$

and the direction cosines

$$E = (E_x^2 + E_y^2)^{\frac{1}{2}} \qquad \cos\alpha = E_x/E$$

$$\theta = \tan^{-1}(E_x/E_z) \qquad \cos\beta = E_y/E$$

$$\phi = \sin^{-1}(E_y/E). \qquad \cos\gamma = E_z/E.$$

Conversely

The observer is assumed to be always facing the origin of the coordinates system. For cases where the object is positioned so that it falls out of range, it can be repositioned by using the centering function, which automatically places the object with the centre of volume at the origin.

Movement of the object may be created by varying the observer viewpoint rather than the actual object position. With the centre of attention inside the object, the object is always visible on the screen. The coordinates of the object in the data file remain unchanged; only the perspective parameters are updated. Since the perspective transformation is always requested before display and the time factor involved in the updating of parameters is relatively insignificant, the resultant increase in display time is minimal.

Rotation is obtained by varying the two angles θ and ϕ. The relationships between θ and ϕ and the direction cosines $\cos\alpha$, $\cos\beta$ and $\cos\gamma$ are given by (see Fig. 5.11).

$$\theta = \tan^{-1}\left(\frac{\cos\alpha}{\cos\gamma}\right)$$

$$\phi = \tan^{-1}\left(\frac{\cos\beta}{\sqrt{\cos^2\alpha + \cos^2\gamma}}\right)$$

$$\cos\alpha = \cos\phi \cos\theta$$

$$\cos\beta = \cos\phi$$

$$\cos\gamma = \cos\phi \cos\theta$$

The zoom factor in the display is determined from the ratio between the distance of the plane of projection and the observer distance (D/E).

It can be shown that magnification is proportional to the above ratio, i.e.

$$\frac{H_2'}{H_1'} = \frac{R_2}{R_1}$$

where H_i' is the image size for ratio $R_i = D_i/E_i$ and $D_i = E_i - P_i$, P_i being the distance of the projection plane from the object. Any of the distances E, P and D may be varied independently, but the effects achieved are different.

5.9.1 Distortion

Another ratio needs to be defined, the distortion, given by H'/H'' which is the ratio of the image sizes for two objects of the same size at a fixed distance apart.

(a) Lens zoom (constant P)

By fixing the plane position and varying the eye position we obtain a 'lens zoom' effect. The distortion varies with the magnification in the same way as the image obtained through a zoom lens, when changing the focal length of the lens, with the camera in a fixed position relative to the object.

Take Fig. 5.15(a). From similar triangles

$$\frac{H}{E_1} = \frac{H_1'}{D_1}$$

$$\frac{H}{E_2} = \frac{H_2'}{D_2}$$

Magnification, M, is given by

$$\frac{H_2'}{H_1'} = \frac{D_2}{D_1} \cdot \frac{E_1}{E_2}$$

therefore

$$M = \frac{R_2}{R_1} = \frac{D_2}{E_2} \cdot \frac{E_1}{D_1}$$

From Fig. 5.15(b).

$$\frac{H}{E_1} = \frac{H_1'}{D}$$

$$\frac{H}{E_1 + \ell} = \frac{H_1''}{D_1}$$

where H' and H'' are the image sizes of two segments of same length H, distance ℓ apart. E is the distance from the first object to the eye positions. The same relations apply for E_2, therefore distortion is given by:

$$\frac{H'}{H''} = \frac{E + \ell}{E}$$

For a fixed P_1 given E_1 and D_1

$$P = E_1 - D_1$$

$$R_1 = D_1/E_1$$

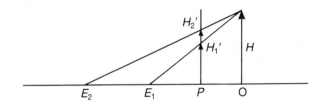

(a) Constant *P*

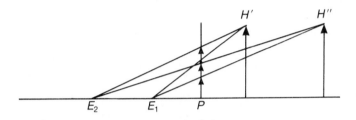

(b) Distortion

 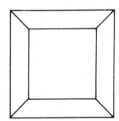

(c) Less distortion

Fig. 5.15 – Lens zoom.

to obtain a magnification

$$M = \frac{R_2}{R_1} = \frac{D_2}{D_1} \cdot \frac{E_1}{E_2}$$

The new eye position is given by

$$E_2 = \frac{P}{1 - R_1 \cdot M}$$

and $D_2 = E_2 - P.$

(b) *Object zoom* (constant *D*)

The second alternative is to move the eye and the plane together by keeping D constant. The size change is given by Fig. 5.16(a).

$$\frac{H_2'}{H_1'} = \frac{E_1}{E_2}$$

$$\text{size } \alpha \ \frac{1}{E}$$

again

$$M = \frac{R_2}{R_1} = \frac{D}{E_2} \cdot \frac{E_1}{D} = \frac{E_1}{E_2}$$

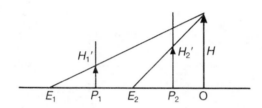

(a) Constant *D*

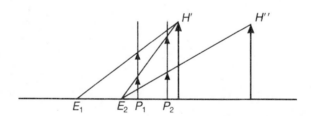

(b) Distortion

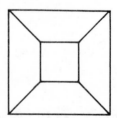

(c) More distortion

Fig. 5.16 – Object zoom.

Therefore, for a certain magnification M,

$$E_2 = E_1/M$$

$$D_2 = D_1$$

The distortion given by

$$\frac{H_1'}{H_1''} = \frac{E_1 + \ell}{E_1}$$

actually increases when the magnification increases. The zoom has the same effect as closing in with a lens of fixed focal length on a fixed object, or conversely moving the object towards the observer. This is called an 'object zoom'.

(c) Combined zoom (constant E)

If the eye position is now fixed and the projection plane is moved we have a 'combined zoom'. This is equivalent to looking through a zoom lens and closing in as it is zoomed out. The distortion remains the same as the size increases.

$$\frac{H'}{H''} = \frac{E + \ell}{E}$$

is a constant now, because the eye position is fixed.

From Fig. 5.17(a).

$$M = \frac{H_2'}{H_1'} = \frac{D_2}{D_1} \quad \text{since } E \text{ is constant.}$$

size α D

again magnification $M = \dfrac{R_2}{R_1} = \dfrac{D_2}{D_1} \cdot \dfrac{E_1}{E_2} = \dfrac{D_2}{D_1}.$

For a combined zoom:

$$D_2 = D_1 M$$

$$E_2 = E_1$$

The effects of the three types of zoom on a cube are depicted in Figs. 5.15(c), 5.16(c), 5.17(c). The three can be combined to give almost any type of zoom effect required.

For any eye-plane combination

magnification α R α $\dfrac{E - P}{E}$ α $\dfrac{D}{E}$

When performing a zoom effect it is essential to ensure that the projection plane P lies between the eye position and the object centre, i.e. D must always be positive or unpredictable results, such as inversion, partial inversion or very large-scale magnification, will occur.

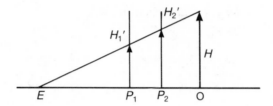

(a) Constant E

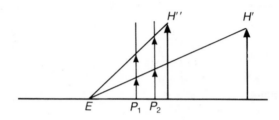

(b) Distortion

 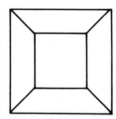

(c) Same distortion

Fig. 5.17 – Combined zoom.

The control of the parameters θ, ϕ and R is via a patch on the menu (Fig 5.18). It is the same as the patch used for drive mode consisting of a square 9×9 cm in size divided into 81 smaller squares, and a rectangle 6×9 cm divided into 9 rows. The square controls the tilt and pan angles and the rectangle the zoom magnification. The angles are in increments of 1, 5, 15 and 45 degrees, or combinations of these. Rotations can be accumulated before display which is activated by the 'enter' switch. The zoom zone is divided into three parts, each controlling a different type of zoom. The upper rows are for zoom-ins and the bottom ones for zoom-outs.

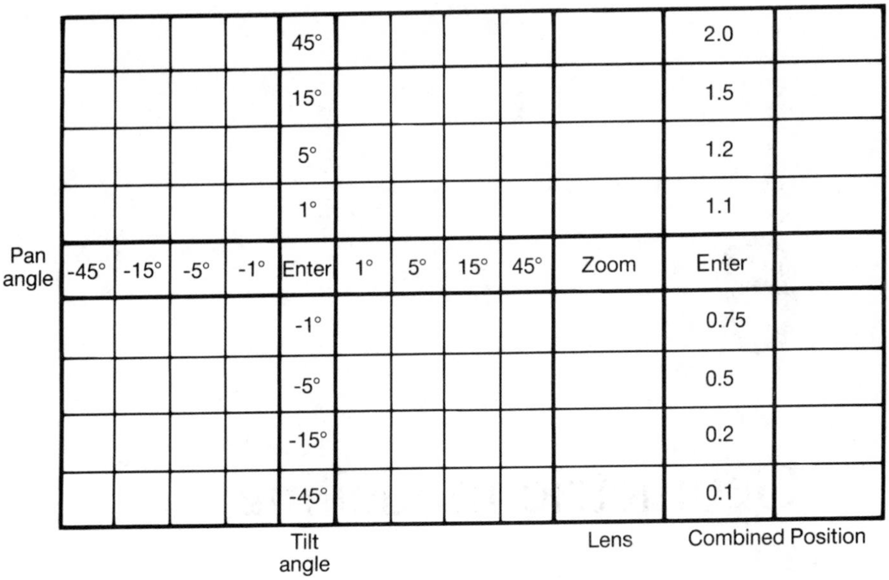

Pan angle											
				45°						2.0	
				15°						1.5	
				5°						1.2	
				1°						1.1	
-45°	-15°	-5°	-1°	Enter	1°	5°	15°	45°	Zoom	Enter	
				-1°						0.75	
				-5°						0.5	
				-15°						0.2	
				-45°						0.1	
				Tilt angle					Lens	Combined Position	

Fig. 5.18 – Joystick patch.

A point within the small squares must be digitised to obtain the desired orientation and magnification. The values of these are displayed on the screen.

The joystick control facility enables the user to select quickly the best orientation of an object for inspection. Because the display is linked point by point to the object data, it provides a convenient means of isolating data for construction and editing purposes.

REFERENCES

[1] Roberts, L. C., Homogeneous matrix representation and manipulation of N-dimensional constructs, *Computer Display Review,* 1969.

[2] Sutherland, I. E., Spinill, R. F. and Schumacker, R. A., A characterisation of ten hidden-surface algorithms. *Computer Surveys* 6, 1974.

[3] Watkins, G. S., A real time visible surface algorithm. Ph.D. thesis, Dept. of Electrical Engineering, University of Utah, 1970.

[4] Parke, F. I., Measuring three dimensional surfaces with a two dimensional data tablet. *Computers and Graphics* 1, 1, 1975.

[5] Gourand, H., Computer display of curved surfaces. Ph.D. thesis, Dept. of Electrical Engineering, University of Utah, 1971.

[6] Newell, R. G., The visualisation of three dimensional shapes. Proceedings of the Conference of Curved Surfaces in Engineering, Cambridge, March 1972.

6

Geometric modelling

6.1 INTRODUCTION

The first step in using a CAD/CAM system to design and manufacture a component is geometric modelling. It is the process in which a geometric model is created to represent the size and shape of a component in computer memory. Obviously, geometric modelling is a very important part of CAD/CAM because it is the starting point of the product design and manufacture process. Another crucial reason for its importance is that the accuracy of the model and the way in which the model is structured and stored in computer memory will have far-reaching effects on other CAD/CAM functions such as finite element analysis, draughting and NC part programming. For example, the geometric model may be used to define a finite-element model of the component for stress analysis. The model may also be used as a basis for automated draughting to produce engineering drawings. In addition, when CAD and CAM are interfaced, the geometric model may serve as an input for generating NC tapes to make parts on automated machine tools, or for producing process plans to specify the sequence of steps required to manufacture the component.

Geometric modelling is often performed with the assistance of graphics systems because interactive graphics enables users easily to enter, manipulate and modify data for the construction of geometric models, although it is not absolutely necessary.

6.2 DIMENSIONS OF MODELS

A geometric model representing a component in computer memory may be a 2-dimensional, 2½-dimensional. or 3-dimensional type depending on the capabilities of the CAD/CAM system and the requirements of the users. It is not difficult to understand that a 2D model represents a flat part and a 3D model provides representation of a generalised part shape. For a 2½D model, it may be slightly tricky to visualise. A 2½D model can be used to represent a part of constant section with no side-wall details. To give a simple example of a 2½D model, consider a cuboid which may be defined by specifying its uniform cross-section A and length ℓ as shown in Fig. 6.1. The major advantage of a 2½D model is that it gives a certain amount of 3D information about a part without the need to create a database of a full 3D model.

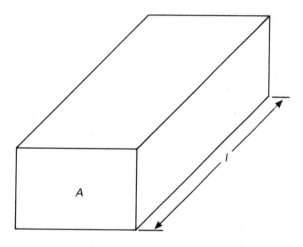

Fig. 6.1 – An example of a 2½D model.

A solid object may be represented as a 2½D model. For example, in the case of a sphere, it can be thought of as consisting of a series of disks with constant circular section as illustrated in Fig. 6.2. The thickness of each disk is exaggerated here, but if it is much reduced, this 2½D model can provide a very close approximation to the original sphere.

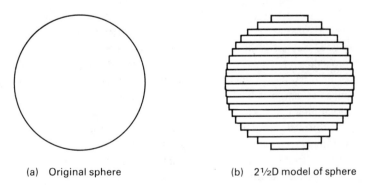

(a)   Original sphere   (b)   2½D model of sphere

Fig. 6.2 – Sphere represented as a 2½D model.

6.3 TYPES OF MODELS

In general, there are three types of model in common use to represent a physical object in CAD/CAM systems. They are wireframe, surface and solid models as illustrated in Fig. 6.3.

Wireframe models, also called edge-vertex or stick-figure models, are the simplest method of modelling and are most commonly used to define computer models of parts especially in computer-assisted draughting systems. The reasons are that wireframe models are simple and easy to create, and that they require relatively little computer time and memory so even inexpensive computers can cope with the processing needed.

Although wireframe models provide accurate information about the location of surface discontinuities on the part, they do not give a complete description of the part. Very often, the model will subsequently be used as a basis for automatically generating cutter paths to drive an NC machine tool to manufacture the part. As wireframe models contain little if any information about the surfaces of the part, nor do they distinguish the inside from the outside of part surfaces, they are inadequate for this purpose. From Fig. 6.3, it can be seen that wireframe models are ambiguous especially in representing complex 3-dimensional parts, and sometimes can be interpreted in many different ways.

An example of the procedures for creating a wireframe model of a rectangular block is depicted in Fig. 6.4. Procedures for creating models may vary from one CAD/CAM system to another depending on its capabilities and the individual technique of the user. In general, most CAD/CAM systems use a split-screen approach, as shown in Fig. 6.4, in which multiple views, usually three orthogonal and one isometric view, are displayed and manipulated simultaneously. In Fig. 6.4(a), four points are indicated on the top view to represent the vertices of one of the faces of the rectangular block to be modelled. The points are then joined up as straight lines to represent the edges of the top face of the rectangular block as shown in Fig. 6.4(b). Then, in Fig. 6.4(c), the image is projected into the other three views. Finally, the face is projected into the third dimension to give the model a thickness producing a rectangular block as a result in Fig. 6.4(d).

A higher level of sophistication in geometric modelling is the surface model which can overcome many of the ambiguities of wireframe models. A surface model can be built by defining the surfaces on the wireframe model. The procedure of constructing a surface model is analogous to stretching a thin piece of material over a framework. As surface models precisely define part geometry such as surfaces and structure boundaries, they can help to produce NC machining instructions automatically.

Surface models may be constructed using a large variety of surface features often provided by many CAD/CAM systems. The plane is the most basic feature to represent a surface element. More complex shapes can be defined by tabulated cylinders, ruled surfaces of revolution, sweep surfaces and fillet surfaces.

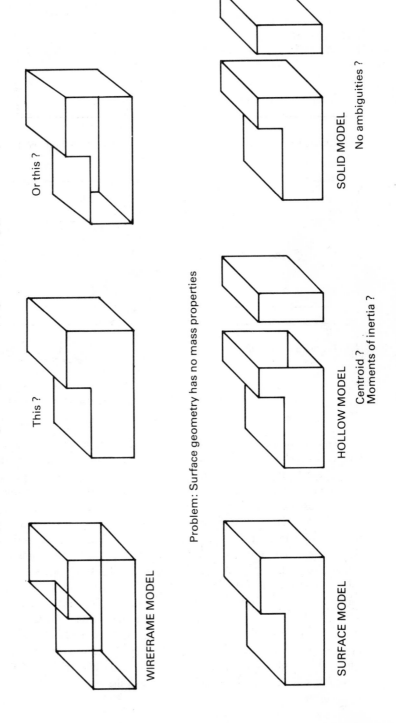

Fig. 6.3 – Types of geometric models.

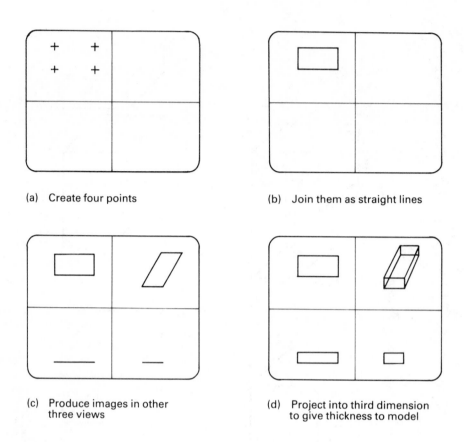

(a) Create four points

(b) Join them as straight lines

(c) Produce images in other
     three views

(d) Project into third dimension
     to give thickness to model

Fig. 6.4 – A typical procedure for creating a wireframe model.

The most general and complex surface representation are sculptured surfaces which are also known as curve-mesh surfaces, B-surfaces and cubic-patch surfaces. A sculptured surface may be considered as a surface produced from combining two families of curves that intersect one another in a criss-cross manner, creating a network of interconnected patches.

A model composed of surface elements can easily be made to appear as if it is solid through the technique of hidden-line removal, although surface models represent only an envelope of part geometry in computer memory. As surface models do not actually represent the solid nature of parts because they contain no information describing what lies within the part interior, they cannot be used as a basis for engineering analysis programs such as finite element and modal analysis for stress and strain predictions. These programs often require such properties as weight, volume and moments of inertia which cannot be derived from a surface model alone. Interpretation of a surface model can still be ambiguous, as shown in Fig. 6.3, because the model may represent a totally solid object or is merely a thin-walled structure made, for example, of sheet metal.

The highest level of sophistication in geometric modelling is 3D modelling. Solid models often resemble wireframe or surface models with hidden lines removed. However, the difference is that wireframes have severe limitations when used as mathematical models of the parts they represent, because for modelling purposes parts need to be represented in the computer in a way that enables mass properties such as centre of gravity or weight to be associated with the part. The mass properties of parts models are important for the prediction of such properties as weight, stability, moments of inertia and volume, etc., of finished products containing the parts. Solid models are better in the sense that they allow the solid nature of an object to be defined in the computer and thus help to calculate mass properties. Also, sections can readily be cut through solid models to reveal internal details.

Solid models are recorded in the computer mathematically as volumes bounded by surfaces rather than as stick-figure structures. As a result, it is possible to calculate mass properties of the parts which are often required for engineering analysis such as finite element methods and kinematic studies for interference checking.

To give an example such as a cylinder, a wireframe cylinder is defined in a computer as two circles connected by two line segments, whereas the solid model of a cylinder is represented as a 3D object that contains a volume. If it is to determine the volume of the wireframe cylinder, the formula for cylinder volume can be used. However, this formula is only valid for cylinders. For other types of volumes, different formulae would have to be programmed into the computer. Obviously, it would be difficult to calculate volumes in this manner for complex shapes because each volume would have to be determined using a different formula. Another drawback is that the computer programmer would have to know the shape of the part in advance before the appropriate formula could be used to program the computer. On the other hand, to calculate the volume of a solid model, it is much easier because the computer can employ a general numerical integration that can be applied to solids of any shape. For example, the volume of a solid may be determined by dividing one face of the solid into a rectangular grid, then tracing the rectangular shapes back through the solid until it reaches the back edges of the model. The volume of the part is then obtained by adding the volumes of all the parallelepipeds. This technique is commonly used in solid modelling and is often referred to as the approximating sum integration method.

Despite the benefits of using solid models, they have their disadvantages. For one thing, they use much more memory to represent a model than either wireframes or surface models; for another, they require extensive processing for their manipulations owing to the more complicated data structure and associated mathematics.

6.4 CONSTRUCTION OF SOLID MODELS

Most commercially available solid modelling systems use one of two common approaches to construct solid models. One approach is to use simple geometric shapes such as cubes, spheres and cylinders, etc. These elementary geometric shapes are often called primitives. The idea is to combine a number of these primitives to create complex solid models. This constructive solid geometry (CSG) approach is known as primitive or building-block modelling, if one imagines that the primitives are just building-blocks which can be put together to build up a complicated model. Another approach is called boundary or perimeter modelling in which 'elastic' lines are stretched to form the outlines to define the boundary of the part to be modelled.

Primitive modelling is essentially based on the principle that any part, no matter how complex, can be designed by adding or subtracting elementary shapes such as cubes or cylinders, and putting these primitives in the appropriate position. Each elementary shape may be considered as a solid model with associated mass properties. Hence, a finished model of a part consisting of a series of three primitives may similarly be treated as a solid model.

Primitives can be combined to construct a solid model by Boolean or logical operations such as union, intersection or difference. In this approach, the modeller must position the primitives to the proper place, then invoke the required logical operator to obtain the desired shape. For example, two primitives can be added together at some point with a union operation to form a part. A hole in a part can be created by intersecting the part with a 'negative' cylinder. Figure 6.5 illustrates how a simple model can be constructed from primitives by logical operations.

Primitive modelling systems work best on parts that do not have complex surfaces. Although primitive modelling can be used to synthesise complex sculptured surfaces, it may be achieved only with some assistance from the user. The main problem is that if sculptured surfaces are to be modelled using primitives, the computer will have difficulty in finding the points where primitives intersect to generate these surfaces because the potential number of possible intersection curves for such surfaces is very large. Thus, a great deal of computer time is needed to produce the required surface model.

As already mentioned, primitive modelling systems are designed to enable users to construct parts with several standard building-blocks. Most modelling systems usually provide a dozen or so building-blocks although only four of them, such as plane, cylinder, cone and sphere, are really necessary to describe the majority of parts. These four primitives are sometimes called natural quadrics. Indeed, according to a survey carried out by one university, most engineering parts can be easily and are actually modelled with natural quadrics. This result of the survey can be explained to a certain degree by a claim from experts that natural quadrics seem to perform better than parts with more complex shapes even when stress/weight ratios are critical. Another crucial reason for the popularity in industry of natural quadrics employed to model parts is that they are easier, thus more economical, to form and machine than complex shapes, yet their performance characteristics are just as good if not more superior. For example, planar surfaces are generated by rolling, chamfering and milling. Cylindrical surfaces can be produced by turning or filleting. Spherical surfaces result from cutting with a ball-nose cutter, whereas conical surfaces are created

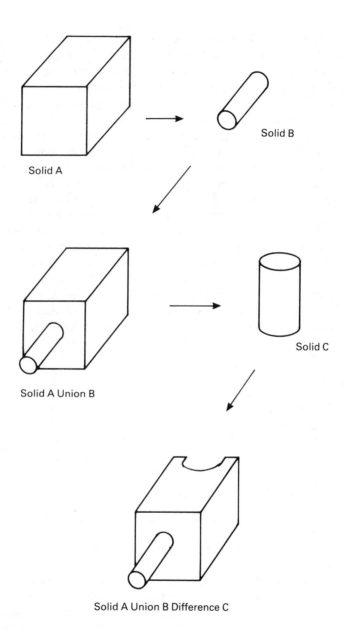

Solid B

Solid A

Solid A Union B

Solid C

Solid A Union B Difference C

Fig. 6.5 — Primitive modelling of solids.

from drill tips and counter sinks as well as by turning. In addition to natural quadrics, the torus is quite commonly used as a primitive for solid modelling because it can be frequently found in surfaces around the edge of a cylindrical hole or boss.

Primitive modelling is certainly a powerful technique to model solid parts, but not all parts can be conveniently or best modelled with primitives. In some cases such as automobile exhaust manifolds which have complicated dimensions, it can become difficult if not impossible to model the part using only primitives. Complex shapes of this type can be modelled much more easily and efficiently by boundary or perimeter modelling.

The principle behind boundary modelling is that part geometry is different from part topology and that they can be defined separately. The topology of an object is the property that describes how its vertices, edges and faces are connected. For example, which two vertices or points are joined up to form an edge in the object, and which edges are connected to form a face. The topology of an object thus provides such information as to what faces share what edges, how many faces meet at a given point, and so on. The geometry of an object defines the dimensions and positions of its vertices, edges and faces, and therefore fixes it in space. In general terms, topology merely describes the connectivity of vertices, edges and faces in the object. Thus, cubes and other parallelepipeds are identical topologically as shown in Fig. 6.6. Introducing dimensions and positions of vertices, edges and faces provides the object with geometry which, together with its topology, completely and uniquely defines it in space.

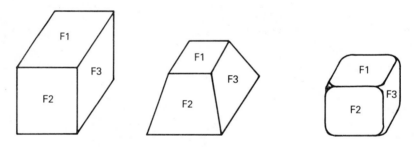

Topologically identical, but geometrically different

Fig. 6.6 – Topology and geometry of solid models.

In boundary modelling, once a particular topology has been defined, many different operations can be performed on the part to adjust geometry without changing basic topology. Effectively, geometry may be considered as a set of properties associated with the topology. Boundary modelling starts with the construction of an outline drawing of the part, similar to the outline drawing commonly used in orthographic views. This view then undergoes a linear sweep, that is, the boundary is raised to produce a part with thickness. This procedure is illustrated in Fig. 6.7. Various degrees of linear sweep may be performed on different portions of the part in order to create different parts of the modelled component. Holes can be produced by defining a circle which is then raised through the top surface of the component. These translations are often referred to as lift operations because they lift the contour up to form a new surface.

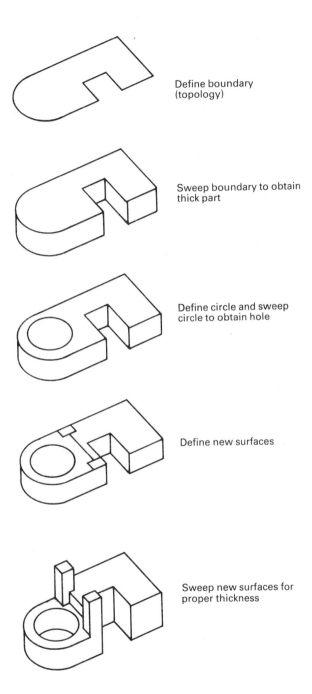

Define boundary
(topology)

Sweep boundary to obtain
thick part

Define circle and sweep
circle to obtain hole

Define new surfaces

Sweep new surfaces for
proper thickness

Fig. 6.7 – Boundary modelling of solids.

Another similar operation, known as a spin, rotates a contour about an axis to generate such parts as lathe-turned components that have axial symmetry. In fact, boundary modelling is particularly suitable for modelling parts that have a great deal of inherent symmetry. The main advantage of treating topology and geometry separately is that it makes it much simpler to incorporate new geometries to the modelling system. The reason is that the definition of new geometries takes place through the addition of subroutines that do not change the main body of the modelling system which defines only topology.

A solid model created using the boundary technique may be stored in the database of a CAD/CAM system as the example shown in Fig. 6.8. In this example, the tetrahedron is composed of four vertices, namely A, B, C and D, which are specified as coordinate data in the record in the database. The network in Fig. 6.8 indicates how the vertices are related to each other to produce edges (a, b, c, d and e) and then how these edges are joined together to form faces (1, 2, 3 and 4) which make up the tetrahedron as a solid.

In boundary modelling, the shapes defined have to follow a set of Euler relationships, named after the Swiss mathematician, which state in polyhedra,

$$\text{vertices} + \text{faces} - \text{edges} = 2$$

When the bodies contain holes and passages, the relationship becomes

$$\text{vertices} + \text{faces} - \text{edges} - \text{holes} - 2 \times \text{bodies} + 2 \times \text{passages} = 0$$

where an edge is the boundary between adjacent faces and a face is a bounded surface. Using the example in Fig. 6.8 of a tetrahedron, it has four faces and six Euler edges. An Euler vertex is at each end of an Euler edge. Several edges joining at one position form one Euler vertex. For example, a tetrahedron has four Euler vertices. The above equations provide a very useful check on the model against the Euler law relationship between the number of faces, edges and vertices to make sure that the model defined remains unambiguous.

Most solid modelling systems include a number of graphics functions to simplify the process of model building. Rotation, translation and scaling are usually provided for the user to manipulate the model. Facilities for duplication and editing often help the user to build up the model. Some powerful solid modelling systems can even generate cutaway views. This function can generally be performed by positioning a large block so that one of its faces forms a sectioning plane. Then this block is subtracted from the solid to produce the sectioned solid. It is also possible to cross-hatch automatically on the sectioned solid surface.

As the two solid modelling approaches have their relative strengths and weaknesses, much work has been carried out to develop a modelling system that combines both primitive and boundary modelling techniques. With these systems, solid models may be created using either approach, whichever is more appropriate to the complexity of the model in question.

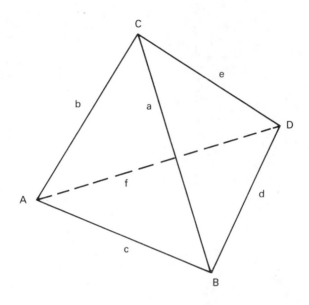

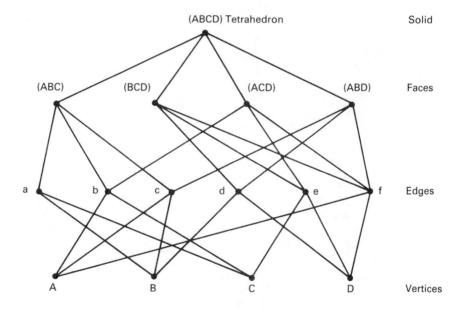

Fig. 6.8 – Data structure for representation of a solid (in this example, a tetrahedron).

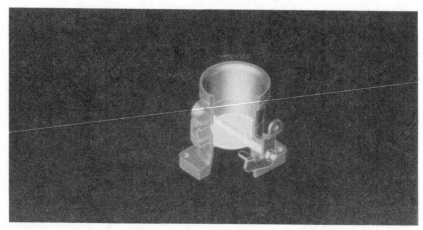

Fig. 6.9 – Solid model of a component with cutaway view.

Fig. 6.10 – Solid model of a product.

Fig. 6.11 – An exploded solid model of a component.

Fig. 6.12 – Solid model of a motorcycle.

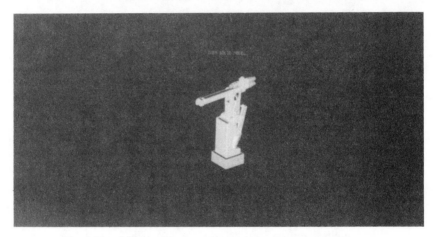

Fig. 6.13 – Solid model of a robot with six-degrees of freedom.

REFERENCES

[1] *CAE Systems and Software Annals,* Penton/IPC Publication, 1982.

[2] Faux, I. D. and Pratt, M. J., *Computational Geometry for Design and Manufacture,* Ellis Horwood, Chichester, 1979.

[3] Groovers, M. P. and Zimmers Jr., E. W., *Computer-Aided Design and Computer-Aided Manufacturing,* Prentice-Hall, 1984.

[4] Ballard, D. H. and Brown, C. M., *Computer Vision,* Prentice-Hall, 1982.

7

Draughting

7.1 INTRODUCTION

Draughting is mainly concerned with the process of producing engineering
drawings and other hard copy documentation. In CAD/CAM, a graphics system
of the kind described in Chapter 4 is generally used to create the drawing
which is then produced on a typical hard copy device such as a plotter. With
computer-assisted draughting, it is not unusual to gain a two to six-fold increase
in productivity for most draughting applications. For example, a large drawing
that requires several days to produce manually may be completed in less than
a day with automated draughting. Indeed, it is possible in some instances to
achieve productivity increases as high as twenty times.

To be able to create engineering drawings, a graphics system will need to
include some additional functions such as dimensioning, text and cross-hatching
etc. The following sections in this Chapter will attempt to explain some of
these essential draughting functions in a graphics system.

For draughting, some CAD/CAM systems use a large digitiser which
resembles a traditional drawing board. Rough sketches can be placed on this
digitiser and traced with a cursor whose position is monitored and stored in
computer memory. When digitising, the cursor is positioned over a point on the
drawing, and a button is pressed to enter the position. In this way, the points
are specified and they are connected usually as lines to form the drawing.
Geometric figures such as arcs, circles or rectangles are generally produced using
special routines. As a result, an idealised form of the sketch is entered into the
computer, which tidies up the drawing by straightening lines, smoothing out
curves and arcs, and orienting lines at proper angles.

7.2 ANNOTATION

Drawings and diagrams are an extremely efficient method of recording and conveying complex information for human comprehension. In a typical engineering drawing, it is very common for it to contain not only graphical data but also textual information such as drawing labels, notes, specifications and dimensions of parts. Thus, it is not surprising to find that most CAD/CAM systems can perform these draughting functions for annotation which will be described in this section.

7.2.1 Arrows and pointers

Arrow heads and pointers figure prominently on most engineering drawings and a quick and simple method of entering them is essential. There are many different styles and sizes of arrow heads and pointers as shown in Fig. 7.1. It is important that the draughting system is flexible enough to allow a large variety of arrows and pointers to be generated for different applications. In Fig. 7.1, it illustrates three of the more common styles of arrow heads, and that the size of arrows may be specified by three parameters, namely, the length and width of the arrow head and the tail of the arrow. Defining arrows and pointers in this way permits the user to easily create a good selection of different arrows and

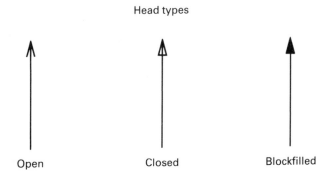

Fig. 7.1 – Arrow formats.

pointers by varying the above parameters to suit the requirements of the drawing being produced. Some CAD/CAM systems provide facilities to enable users to specify their own arrow and pointer styles apart from those already supplied with the system.

7.2.2 Dimensioning

Most systems provide the user with a complete range of dimensioning conventions which cover British, American and German standards. Usually, a user specifies the positions of each of the witness lines, and that of the text or dimension, using the cursor. The system then produces the witness lines to the correct length for the required dimension position, draws the dimension lines and inserts the appropriate dimension. Also, the system will automatically check whether or not there is space to place the dimension between the witness lines and, if not, it is put externally with a split dimension line.

To cater for the many different geometric features commonly found in a drawing, most systems have different methods of dimensioning depending on the geometric feature involved. Figure 7.2 illustrates some of the ways of dimensioning in a drawing. Typically, it is possible for the system to create horizontal, vertical, angular, circular and diameter dimensions. A user can easily perform dimensioning through the use of menu commands which make it very user-friendly. However, there is usually an option available for the user to dimension a drawing manually, in which case, dimension lines can be set out individually. The lettering or numbers are either typed or input from the numeric menu, and a string of text is positioned by means of a cursor.

Despite all the automatic annotation facilities provided by a CAD/CAM system, it is sometimes convenient to run off a plot of a layout or component drawing before annotating it. The plot, which is an accurate drawing, may then be annotated in freehand. The roughly annotated drawing is then placed on the drawing area of the digitiser and annotated in the way described. This technique, though not ideal, is particularly useful for annotating large complex drawings since it allows the draughtsman time to plan carefully the annotation on the drawing. Performing the planning while using the system could be wasteful of computing time. Another advantage in using an accurate plot of the drawing during the annotation stage is that points on the drawing where dimensions are required can easily be located, since moving the cursor on the digitiser to the required point and using the FIND facility will ensure that a desired point is located quickly and precisely.

7.2.3 Text

Text is probably the most common type of annotation and appears in some form on every drawing produced. In addition to outline and dimensional information, most engineering drawings contain textual information such as notes, specifications, comments and titles. Most CAD/CAM systems allow the user flexible control over the format of text, namely the character height and aspect ratio, character set or font, and the direction in which it is written.

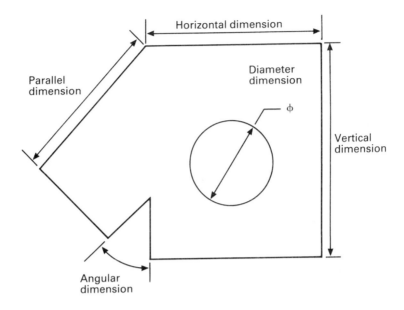

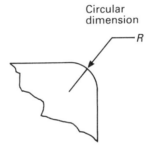

Fig. 7.2 – Dimensioning.

Figure 7.3 shows the different control over the format of text generated. The popular method of implementing text control is to define the above parameters manually and store them in the system common area. The values of these parameters remain the same until specifically changed by the user. To create a block of textual information on a drawing, all a user need do is to digitise a point to indicate the position where the text is to be placed, and then enter the characters, through an alphanumeric keyboard or a keypad on the menu, on the design. It is also possible to justify the text to the position specified as illustrated in Fig. 7.4. Left-justifying is often used to align the bottom left-hand corner of a block of text to the specified position. Similarly, right-justifying and centre-justifying are useful for aligning the bottom right-hand corner and centre respectively of a block of text to the position specified. Some powerful systems can even allow positioning text on an irregular curve.

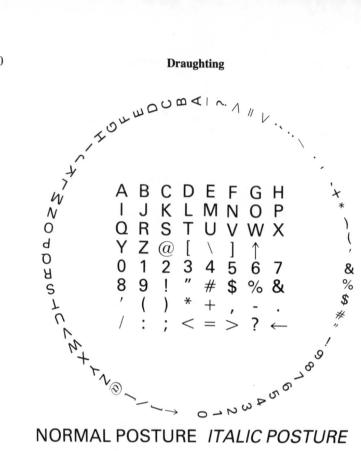

NORMAL POSTURE *ITALIC POSTURE*

Fig. 7.3 – Text control for size, font and angle.

This message
is
left-justified

(a)   left-justified

This message
is
right-justified

(b)   right-justified

This message
is
centre-justified

(c)   centre-justified

Fig. 7.4 – Text justifications.

7.3 CROSS-HATCHING

Very often it is necessary to produce cross-sections in an engineering drawing to reveal the interior details of a certain component. Cross-sections are commonly indicated in a drawing by cross-hatching which is a time-consuming and tedious process if performed by hand. Most CAD/CAM systems can do this task automatically through menu commands. The appearance of cross-hatching varies depending on the hatch angle and the spacing between successive hatch lines as shown in Fig. 7.5. These parameters can be specified and changed by the user

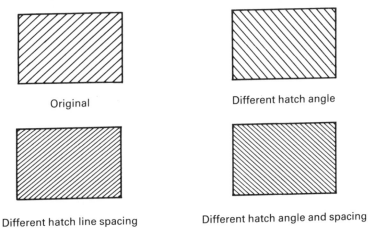

Original   Different hatch angle

Different hatch line spacing   Different hatch angle and spacing

Fig. 7.5 — Cross-hatching.

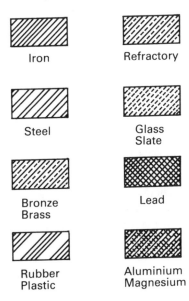

Iron   Refractory

Steel   Glass
                        Slate

Bronze   Lead
Brass

Rubber   Aluminium
Plastic   Magnesium

Fig. 7.6 — Patterns of cross-hatching for different materials.

in a similar manner to those parameters that control the format of text in the previous section. For some applications, it is even necessary to generate different patterns of cross-hatching as illustrated in Fig. 7.6 to indicate the various types of material used.

To create cross-hatching for sections, the user usually needs to specify the points that make up the outline for the boundary of the area to be cross-hatched. If there are holes within the section, it is possible to specify the boundaries of areas, known as islands, within the main boundary that are not to be cross-hatched. An example of this situation is shown in Fig. 7.7. Some sophisticated systems can even cope with this kind of cross-hatching automatically.

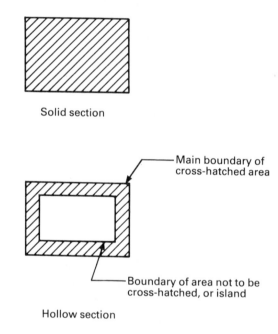

Fig. 7.7 – An example of cross-hatching with hollow section.

7.4 DRAUGHTING EXAMPLES

Here in this chapter, only some of the more common draughting functions in CAD/CAM systems have been described. There are many other different functions that are available on various CAD/CAM systems depending on the particular implementation. It is not unusual to find functions that help the user to generate centrelines, balloons and other annotation features. All the functions often included in CAD/CAM systems to make the lives of the users much easier are far too numerous to list here. More detailed descriptions of these functions can usually be found in the manuals of the particular CAD/CAM system used.

An example of construction of layouts is given here to show the benefits of computer-aided draughting. To make CAD techniques really cost effective when applied to draughting, maximum use must be made of the macro facilities. Any standard items such as valves, filters, and pipe fittings will be stored in the macro library. However, any part of the layout which is subject to repetition may be digitised first into the macro library for subsequent use in the construction of the layout.

The procedure for construction of a pipe network layout is to sketch the layout first with dimensions set down from a set of grid lines or datum. The rough drawing can then be placed on the drawing area of the digitiser and turned into an accurate drawing. As an example, consider the drawing of a simple pipe run shown in Fig. 7.8. We will show how the drawing can be accurately constructed and how documentation such as drawings of plans, elevations, and isometrics can be produced.

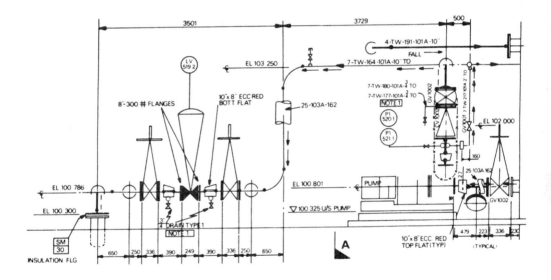

Fig. 7.8 – An example of a pipe run.

The system set-up procedure is similar to that described for the construction of macros except that the drawing data is filed under a code and not a macro library square. Each file may be subdivided into a number of levels which are identified by number. This permits a complex drawing to be digitised in a number of discrete parts, and each part may be overlaid with another to make up the complete drawing. In the present example, the various different pipelines have been assigned to different levels. All annotation is assigned to a certain level, and a flag masked into the data indicates whether it applies to orthogonal or isometric views. Thus a drawing may have two completely different sets of annotation without them interfering with one another.

The first digitising stage is to set up the grid or datum lines and other relevant system parameters. Additionally the user must specify the current pipeline type to enable correct recall of macros and a 'principal pipeline direction' to aid their positioning.

The next stage is to enter the pipe runs together with the standard components such as valves. The most common procedure is to pick macros from the library and place them in the correct position on the drawing. Complete macros can easily be manipulated by using the macro menu facilities. For example, if a macro has been placed in an incorrect position, it can be deleted and then repositioned by the macro handler.

Once all the components have been located properly, the pipe centrelines may be digitised by using the menu and FIND button function. If there are a number of pipe runs of differing diameters, then pipe runs of each diameter can be stored on a different level so that cumulative lengths can easily be determined at the material take-off stage.

When pipe runs are entered, the straight runs are digitised first. The CONTROL 90 and FIND facilities are used to position each run. Arcs and fillets are then added, using the ARC and FILLET modes from the symbol menu.

Each level of digitised data can be displayed in turn for final checking, and levels superimposed to obtain the complete layout. If, during the digitising stage, any area of the drawing becomes complex and difficult to visualise, then that area may be windowed and digitising performed at a large scale. The final layout of the example is shown in Fig. 7.9. It is possible to produce an isometric of the layout by simply allocating one portion of the screen for isometric viewing. An isometric of the layout is shown in Fig. 7.10. An annotated version of the isometric layout is shown in Fig. 7.11.

Figures 7.12–14 show some of the drawings produced on the TIGER system for a robotics project carried out at Imperial College.

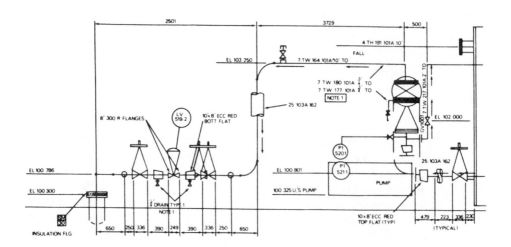

Fig. 7.9 – Final layout.

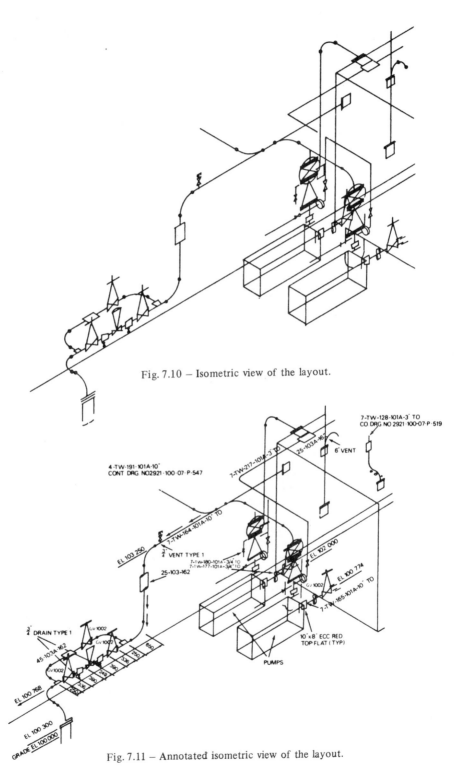

Fig. 7.10 – Isometric view of the layout.

Fig. 7.11 – Annotated isometric view of the layout.

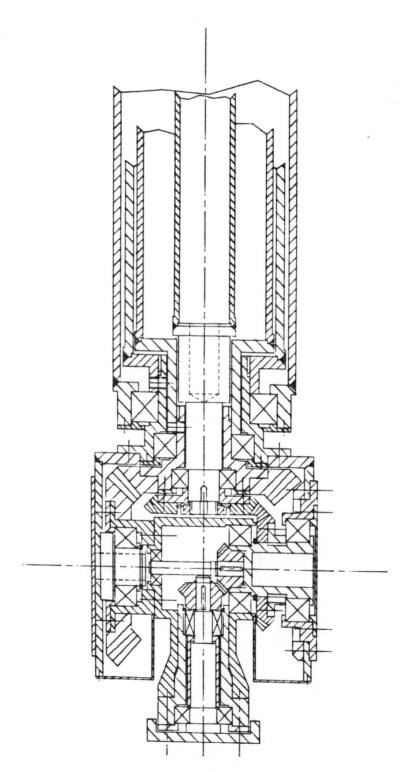

Fig. 7.12 – Cross-section of robot wrist assembly.

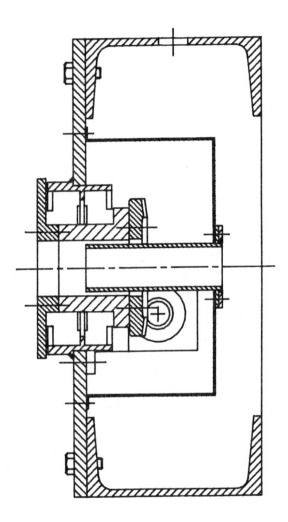

Fig. 7.13 — Cross-section of robot base assembly.

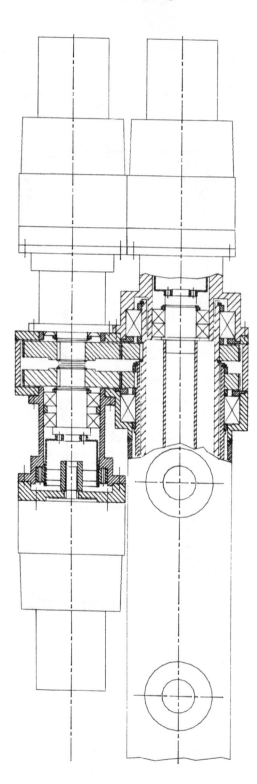

Fig. 7.14 – Cross-section of robot gearbox assembly.

REFERENCES

[1] Besant, C. B., *Computer-Aided Design and Manufacture,* 2nd edn, Ellis Horwood, Chichester, 1983.

[2] *CAE Systems and Software Annals,* Penton/IPC Publication, 1982.

[3] Thompson, D. T., *The Application of Computer-Aided Draughting to Mechanical Engineering Design,* Ph.D. thesis, University of London, 1982.

[4] *CD/2000–170 Reference Manual,* Control Data Corporation, 1983.

[5] Lui, C., *The Use of Computer-Aided Design Techniques for Hypoid Gear Design,* Ph.D. thesis, University of London, 1983.

8

Application of CAD techniques to finite element data preparation

The finite element method is now widely used for the analysis of many engineering problems involving static, dynamic, and thermal stressing of structures. The technique is also used in other branches of engineering as well as reactor physics. In this chapter we shall deal with the application of finite elements to the stress analysis of structures. The technique of finite elements is not often easy to use because of the problem of formulating the necessary input data for a finite element analysis program. The area of greater difficulty lies in representing the geometry of the structure by suitable elements of regular shape. This is called the idealisation.

The input data for a finite element analysis program consists of the geometric idealisation, the material properties, and the loading and boundary conditions. The major problem, and a large portion of the input, is in the geometric representation of the structure by a suitable mesh. When this task is performed manually, it is time-consuming and subject to considerable errors since hundreds of punched cards are often required to describe an average-sized problem. The human errors lead to abortive runs of the finite element analysis program which can prove to be expensive as well as time consuming.

The data preparation stage is extremely tedious and time-consuming and consequently very expensive. Automating this process or mechanising it can greatly enhance the use of the technique.

8.1 AUTOMATIC MESH GENERATION

In many cases the user of a finite element program will reach a point where it becomes desirable to use an existing mesh generating system [1] or to write to one if necessary. Often it is difficult to write a program to generate a mesh for complex structures and the user has to spend many hours drawing meshes in a drawing office and measuring the coordinates of each nodal point. So, in some problems where the number of nodal points is very large, the user frequently has to spend many hours mostly involving data checking and rechecking by tedious graphical methods.

Using a CAD system in the data generation process has considerable advantages over most other methods because the user is able to see the element connections and position of each element directly on a display, as element generation is in progress. Also, if the system is interactive, the user is able to change a mesh instantaneously to arrive at the best mesh arrangement to suit a particular problem by adding or deleting elements. The CAD system can also be used after the finite element analysis to present data graphically so that the results can quickly be assessed.

If a CAD system is to be used in the generation and presentation of data in finite element analysis, it should ideally be capable of providing the following facilities:

Generation of mesh for two- or three-dimensional structures.
Ability to represent curved edges and surfaces.
Ability to control element density and generation of non-uniform meshes.
Facility for concentrating and grading the mesh over any region.
Speedy node and element numbering system which will lead to computational efficiency.
Facility to display idealisation.
Ability to rotate the idealisation from any desired angle.
Option to display any portion of the model.
On-line interactive modification of data for alternative idealisations.
Level of automatic mesh generation with manual override.
Preparation of input data for analysis programs.
User-orientated and easy to use with minimum of input.
Economical with respect to both computer time and manual effort.

8.2 THE FINITE ELEMENT METHOD

The finite element method is now well established, and one of the best books on the subject is by Zienkiewicz [2]. The theory is based on an elastic structure or continuum being represented by many discrete components or elements interconnected at a finite number of nodal points situated on the element boundaries. The displacements of these nodal points are the basic unknown parameters of the problem.

A set of functions is chosen to uniquely define the state of displacement within each finite element in terms of its nodal displacements. Thus the displacement functions uniquely define the state of strain within an element in terms of the nodal displacements. These strains, together with any initial strains and the constituent properties of the material, define the state of stress throughout the element.

A system of forces concentrated at the nodes and equilibrating the boundary stresses and any distributed loads is determined, resulting in a stiffness relationship of the form:

$$\{F\}^A = \{K\}^A \{U\}^A + \{F\}_P^A + \{F\}_{\epsilon 0}^A$$

where

$\{F\}^A$ represents the nodal forces on element A;

$\{K\}^A$A is the element stiffness matrix;

$\{U\}^A$ represents the nodal displacements of element A;

$\{F\}_P^A$ represents the nodal forces required to balance any distributed loads acting on the element;

$\{F\}_{\epsilon 0}^A$ represents the nodal forces required to balance any initial strains such as may be caused by temperature change, if the nodes are not subject to any displacement.

It is not always easy to ensure that the chosen displacement functions will satisfy the requirement of displacement continuity between adjacent elements. Thus the compatibility condition on such lines may not be valid. Concentrating equivalent forces at the nodes ensures that equilibrium conditions are satisfied in the overall sense. Local violation of equilibrium conditions within each element will usually occur.

The choice of element shape and the form of the displacement functions will determine the accuracy of the finite element model.

It is also possible to define the stresses or internal reactions at any specified point or points of the element in terms of the nodal displacements:

$$\{\sigma\}^A = \{S\}^A \{U\}^A + \{\sigma\}_P^A + \{\sigma\}_{\epsilon 0}^A,$$

where

$\{S\}^A$ is the element stress matrix;

and $\{\sigma\}^A$ represents the nodal stresses for element A.

The last two terms are the stresses due to the distributed element loads and initial stresses when no nodal displacement occurs.

The stiffness matrix is related to the nodal forces, nodal displacements, and the type of element. Elements can take a wide variety of forms, the simplest being beam, triangular, and quadrilateral.

There are a variety of computer programs for solving the stiffness matrix, and hence deflections and stresses throughout a structure, for a wide range of elements. Some of the most widely used programs are NASTRAN [3] developed in the USA for the aerospace industry, and ASKA [4] developed in Stuttgart.

8.3 A GENERAL FINITE ELEMENT MESH GENERATING SYSTEM GFEMGS

A data preparation and results presentation system was developed by Ghassemi [5] at Imperial College for use with a variety of finite element programs. The system named GFEMGS is intended to work in conjunction with many of the modules of the draughting system GCADS described in Chapter 5. A user menu (Fig. 8.1) is introduced specifically to cope with functions required in the finite element data preparation or presentation of results that are not in the GCADS facilities.

The GFEMGS system has two different modes of operation, one being automatic mesh generation and the other manual. Automatic mesh generation is used for two-dimensional and axisymmetric shapes. Meshes can be generated for a model using a series of quadrilaterals with sides of different lengths. Each quadrilateral is automatically divided into a number of particular elements, such as triangular elements with three nodes or quadrilaterals with four nodes. The user may subsequently transfer these elements to higher order elements such as triangular with six nodes or isoparametric traingular elements with parabolic curved edges or quadrilateral elements with eight or nine nodes and isoparametric quadrilateral elements with curved edges.

The user may generate a mesh within a quadrilateral such that it may be concentrated on any side or at any point within the quadrilateral. The system can also automatically generate meshes for circular discs, with or without holes with meshes of varying concentrations.

The GFEMGS system has a graphical checking system to check the coordinates and element connections so that any error in the data will be shown to the user via the storage tube display or keyboard. For example, if after generating a mesh of a particular type of element a mistake has been made, then the check facility will identify any missing element on the display, or display any elements with a suspicious shape such as obtuse-angled elements. The user is always guided through the data preparation and actions are recommended at various stages during mesh generation.

The manual data preparation system makes extensive use of the draughting system. Each element is added to the model by using the FIND routine and the display. Thus a triangular mesh can be added by digitising two fixed points using FIND mode, and the third point of the triangle is moved by moving the cursor on the display until the desired position is obtained. An element can be deleted by digitising its nodes using FIND followed by an answer to a request to delete or ignore. The element is always displayed in bright-up form as soon as it has been identified so that it is clear to the user that he has selected the desired element.

GFEMGS has an automatic nodal point numbering system which minimises the stiffness matrix bandwith. If the user has manually numbered the nodes then he may display the stiffness matrix on the display. If the bandwith is greater than a permitted value then it may be renumbered using the automatic system.

The user may identify the boundary conditions for a mesh directly from the digitiser. Nodes can be suppressed in any direction, or the loading specified. The

GENERATE MESH FOR ANY QUADRANGLE	GENERATE CONCEN MESH QUADRANGLE	GENERATE MESH FOR PART DISC	GENERATE TRIANGLE QUADR-ANGLE	GENERATE HIGHER ORDER ELEMENT	CONNECT ISOPARA-METRIC ELEMENT
ADD ELEMENT TO MESH	DELETE ELEMENT FROM MESH	ADJUST NODES OF ELEMENT		DISPLAY INDIVI-DUAL NODES	DISPLAY ALL NODES
PROJECT ANY PLANE ADD 3D	DOUBLE X-Y PLANE PROJECT	3D COOR-DINATE FILE	3D ELEMENT CONNEC-TIONS	CHECK ANGLES & ELEMENT NUMBERS	PRINT ANGLES OF ANY TRIANGLE
INPUT BOUND-ARY CONDIT	ELEMENT WITH EXTERNAL FORCES		DISPLAY STIFF MATRIX	MINIMISE BAND-WIDTH	
DISPLAY RADIAL & AXIAL STRESS	DISPLAY CIRCUMF & SHEAR STRESS	CALCUL MAX SHEAR STRESS	DISPLAY STRESS IN 3D PICTURE	DISPLAY DEFORM-ATION	

Fig. 8.1 – USER menu for finite element data preparation and presentation system GFEMGS.

load conditions are applied at the appropriate nodes as soon as the mesh has been generated and the elements which take the external forces have been identified.

Three-dimensional meshes are generated by a 2½-D technique. This involves the generation of a mesh on a plane, followed by a projection of this mesh onto a second plane parallel to the first. The projection can be smaller or larger than the original mesh. The process can be repeated many times until the desired size is obtained.

GFEMGS may be used to generate three-dimensional meshes using any of the following elements:

Pentahedronal elements with six nodes or pentahedronal macro elements built of three tetrahedrons;

Hexahedronal elements with eight nodes or hexahedronal macro elements built of six tetrahedrons;

Triangular membrane elements in 3-space (three or six nodes); or quadrilateral plane membrane elements in 3-space (four, eight, or nine nodes).

Up to the present, GFEMGS has been run in conjunction with the finite element program ASKA which is available on the Imperial College CDC 6400 computer. Once the finite element data preparation stage is complete, the data is dumped to tape and transferred to the CDC 6400 for input to ASKA. On completion of the finite element calculation the results are transferred back to the CAD system for the presentation of results. GFEMGS allows the user to display graphically deflections of the mesh. Stresses are also shown graphically by a vector notation where the magnitude of the stress is proportional to the length of the vector and the direction of the stress indicated by the direction of the vector. Arrowheads at each end of the vector are used to indicate whether the stress is tensile or compressive. A second method may be used to display stresses. This takes the form of an oblique stress diagram which shows the stress projected away from a datum plane. Examples of results presentation are discussed in the next section.

8.4 AN APPLICATION OF GFEMGS

The use of the complete system is best illustrated by applying it to a practical example. Such an example is a standpipe from a nuclear reactor system which is shown in section in Fig. 8.2. The standpipe is subjected to an internal gas pressure of 39.8 bar when in use. The outside is subject to a water jacket at a pressure of 6.8 bar. The standpipe is also subjected to axial loads producing a pressure of 38.7 bar at the top and 234.9 bar at the base. The loading boundary conditions are shown in Fig. 8.3. Since the problem is axisymmetric, it was decided to use an axisymmetric ring element with a triangular cross-section (TRIAX3 in ASKA format) for the finite element mesh. The standpipe cross-section was divided into ten different sections as shown in Fig. 8.4. The mesh for each section was generated automatically by digitising the four boundary points of each section followed by specifying the number of nodes one each side of the sections together with a mesh concentration factor.

The nodes in each section were numbered after the mesh was generated in a section. The mesh is generated automatically starting from a specified corner and the sequence of node numbering following the sequence of the generation of the mesh.

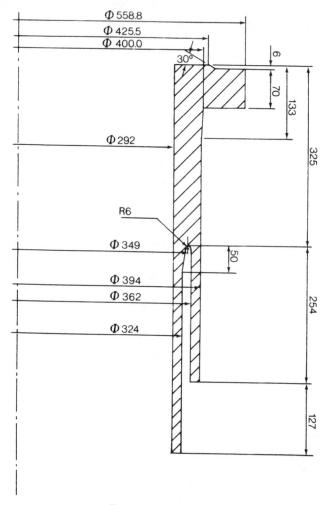

Fig. 8.2 – Standpipe.

One section was joined to another via a common set of nodes, and the numbering of a particular section was taken from the preceding section so that the common nodes are only numbered once. Each section is generated in a separate file to permit easy modifications to that section.

After the completion of mesh generation in all sections, the nodal points of each section were combined into a single file. A similar operation takes place for the element connections.

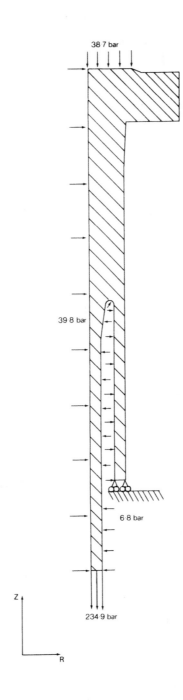

Fig. 8.3 — Boundary conditions.

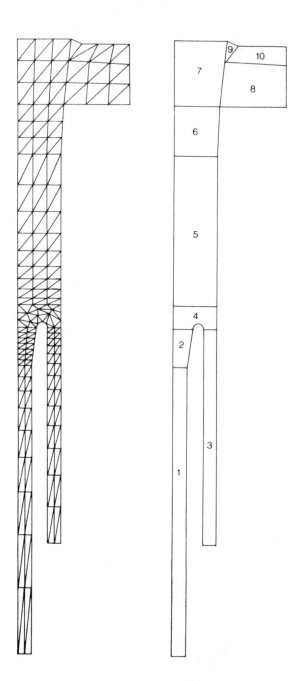

Fig. 8.4 – Mesh generation for the standpipe.

The next stage was to check the nodal points and element connections using the checking facilities which display the complete mesh. Errors are identified on the screen and any obtuse angles between elements are indicated. The order of nodal and element numbers was examined and any serious errors were rectified.

The type of freedom was selected for each node, and the appropriate boundary conditions were applied at nodes where external forces were present.

The stiffness matrix was then displayed to examine its general pattern. If the bandwidth is large due to poor numbering of the matrix where the difference between any two adjacent node numbers is large, then the whole mesh may be renumbered using an automatic renumbering facility which minimises the bandwidth.

Finally the mesh coordinates, node numbers, element connections and numbers were transferred to magnetic tape for transference to the CDC 6400 main frame for input to ASKA. A record was also output to the line printer.

The complete process of generating the mesh shown in Fig. 8.4 was completed in less than forty minutes. A total of 438 statements was generated.

The finite element program ASKA was then run and the results output to tape for transference back to the CAD system. The ASKA run time for the present case was 120.5 seconds.

8.5 RESULTS PRESENTATION

The results consist of nodal point displacements, radial, axial, circumferential and shear stresses.

The results may be displayed graphically on the CRT display or plotter or in numerical form on the line printer. The nodal displacements for the standpipe are shown in Fig. 8.5 and the displacements may be magnified using any desired multiplying factor. The stresses are shown in Fig. 8.6 to 8.13. Each set of stresses is displayed in two forms, one in vector form and the other as an oblique diagram. In the case of the vector form, the directions of the arrowheads indicate whether the stresses are compressive or tensile and, in the case of the oblique diagram, the nature of the stress in indicated by the direction of the projection away from a datum plane.

Finally, two further examples of meshes generated using the system are shown in Fig. 8.14.

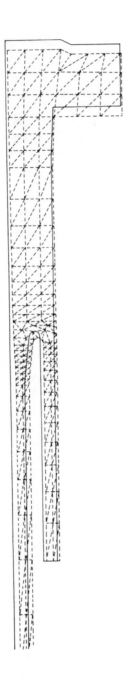

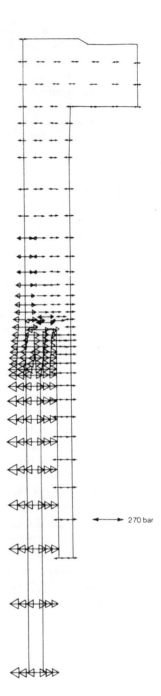

Fig. 8.5 – Nodal point displacement.

Fig. 8.6 – Circumferential stress

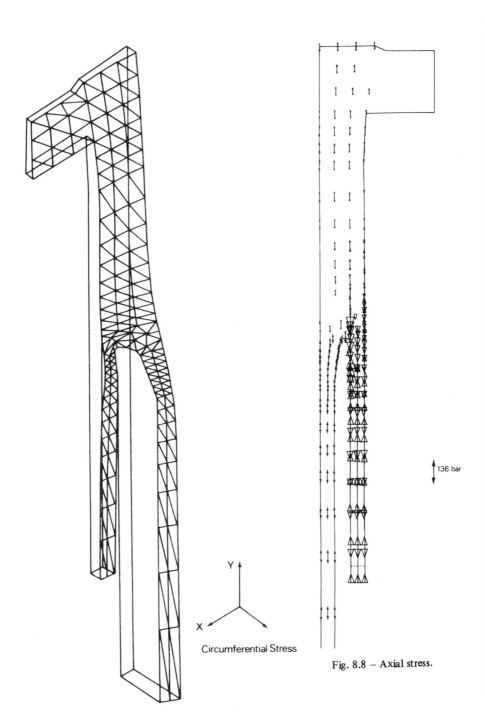

Fig. 8.7 – Circumferential stress.

Fig. 8.8 – Axial stress.

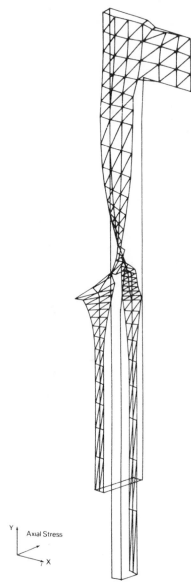

Y
Axial Stress

X

Fig. 8.9 – Axial stress.

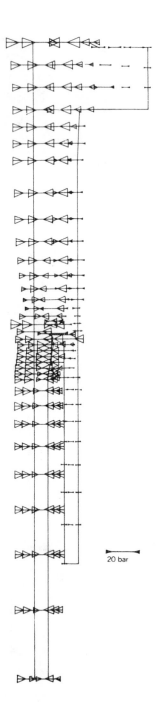

20 bar

Fig. 8.10 – Radial stress.

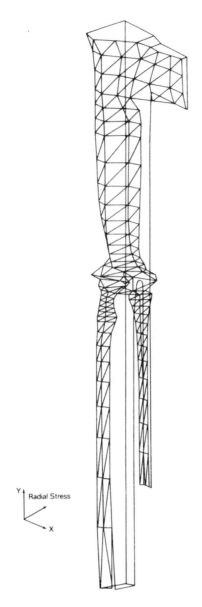

Fig. 8.11 — Radial stress.

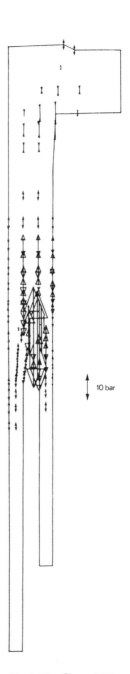

Fig. 8.12 — Shear stress.

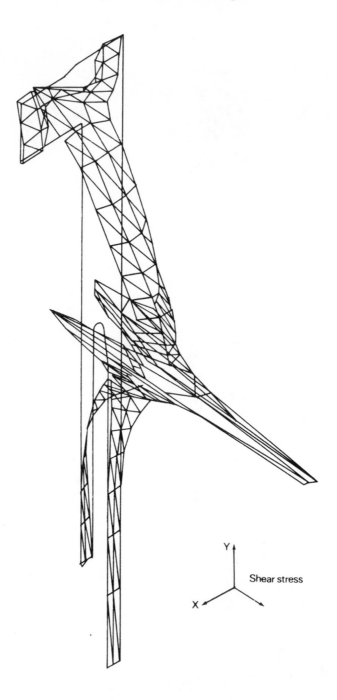

Fig. 8.13 – Shear stress.

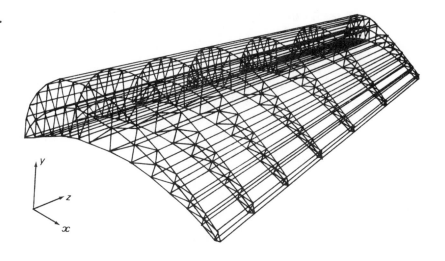

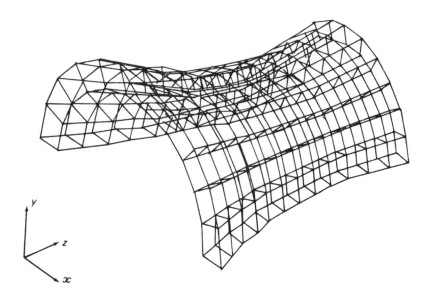

Fig. 8.14 — Meshes generated using the CAD system
for an aircraft wing and a cooling tower.

8.6 THREE-DIMENSIONAL SHAPE DESCRIPTION AND MESH GENERATION

In the previous sections the data preparation system described dealt with 2-D problems. Three-dimensional problems were described by utilising a $2\frac{1}{2}$-D technique which is based on describing a three-dimensional mesh by a mesh digitised in two dimensions and then giving the mesh a uniform depth. With the increasing power of computers, the finite element technique is becoming increasingly used, not only for stress analaysis but also in areas such as heat transfer and reactor physics. This increase in utilisation of the finite element method has increased the requirement for true three-dimensional shape definition and mesh generation.

The graphical definition in three dimensions has been thoroughly described in earlier chapters. It was shown that there are a number of ways of entering three-dimensional data into a computer. The first method is based on digitising two or more orthogonal views which are subsequently assembled within the computer to give the required three-dimensional shape. The second method consists of digitising contours in two dimensions with the third dimension normal to the surface, being added numerically. A surface patch technique is then used to interpolate the contour data in order to produce data of sufficient quality for the graphical representation. Finally, a third method may be used which is based on photogrammetry where two perspective drawings or photographs are digitised and the data converted into a three-dimensional format.

Of these three methods of graphical data description, the first method was found to be more suited to complicated shapes. The method was varied slightly in that a complex shape is always divided into a number of basic building blocks which are assembled at a later stage. The use of the building block technique may seem a tedious method but it is the most efficient one for computers for small core sizes. Each building block can be described in core and the mesh generated in the building block without large numbers of disc transfers. The data generated can be stored in a series of files which can easily be edited and finally assembled into a single file. Unlike 2-D mesh generation, different types of building blocks are designed to produce different types of elements and the method of input for each building block is specifically defined. The extension of the original 2-D Imperial College finite element data preparation system to 3-D was performed by Etela [6].

8.7 INPUT METHODS AND THE MACROBLOCKS

There are six different types of input shapes designed to generate four different types of building blocks called 'macroblocks'. All inputs are handled and serviced by an interactive overlay called 3D MESH. When this overlay is loaded into core by digitising above the appropriate square on the menu table, the system requests the number of input points, since the different types of input shapes are recognised by the difference in the number of input points. Therefore it is important for the user to know the type of input required and the number of input points for the desired input shape. The order in which the points of the input shape are digitised is also rigidly defined and, on providing the number of input points, '3D MESH' will proceed to guide the user to digitise one numbered input point after the other in either the x-y plane or the z-x plane. Owing to the difficulty in representing a three-dimensional object on a plane surface, each 3-D point is defined by two points on the digitising surface: the first point in the x-y plane and the second in either the z-x plane or the y-z plane. From these points the cartesian x, y and z coordinates of the three-dimensional point are then derived.

Since the points on any input shape are specifically numbered, the program using overlay 3D MESH is capable of directing the user to input point number 'N' in the x-y plane and z-x plane, 'N' starting from unity to the specified number of input points, in steps of one. After accepting the last input point, the program enters an editor mode when the user is permitted to change the coordinates of any of the input points. The user is requested to specify the number of the input points he wants to change and is directed to redigitise the point in the appropriate planes. The rigid definition of the order of entry of the points in any of the six input shapes, compels us to consider the six input shapes individually:

a) A straight-edged block with uniform thickness, as shown in Fig. 8.15. In this case only five input points are required to fully define the shape although there are eight corners. The problem of digitising two planar points to represent one 3-D point encourages cuts in the number of input points where the cuts will not affect the accuracy in the overall

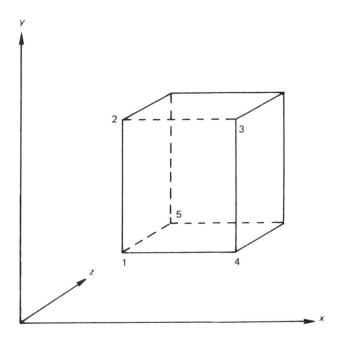

Fig. 8.15 – Straight-edged block of uniform thickness.

definition of the shape. This input shape is used by the first two types of building blocks: MACROBLOCK 1 and MACROBLOCK 2. These building blocks are used in generating 8-node brick elements and 20-node isoparametric elements respectively. The first parameter supplied to the overlay 3D MESH is 5 for this input shape, and the second parameter, the macroblock number must be 1 or 2, for the program to load the appropriate overlay into core.

b) The second type of input shape is one in which a planar view has curved edges, but uniform thickness (Fig. 8.16). In this case the orthographic views of the structure will be taken such that the surface with the curved edge must be digitised in the x-y plane. Only nine input points are necessary for full definition of the shape; they must be digitised in the order shown in Fig. 8.16. Again this type of input shape is only useful for the first two types of building block, but in practice it is better utilised by the second type of building block which maintains a better approximation to the curve along the edges.

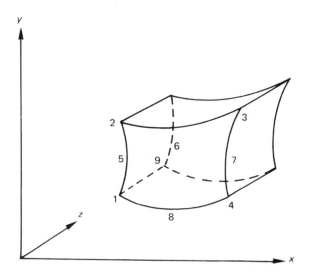

Fig. 8.16 – Block with curved edges and uniform thickness.

c) The third type of input is similar to the first. The difference is only in the number of input points. Instead of five, eight input points are required for this irregular block with straight edges (Fig. 8.17). Theoretically, input shape (a) is a special case of this type of input where only one depth or thickness is required by the program to compute the co-ordinates of the sixth, seventh and eighth corners. The node numbering routines are responsible for obtaining and using the coordinates of the box.

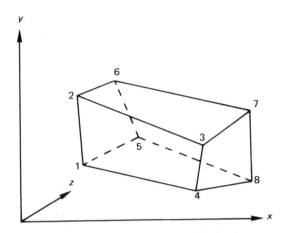

Fig. 8.17 – Irregular block with straight edges.

d)   For the fourth type of input shape (Fig. 8.18), twenty points are used as in the second type of input shape (b), the latter being a special case of this type of input. '3D MESH' will guide the user by requesting for a specific input number, first in the x-y plane and then the z-x plane. It must be emphasised again, that, owing to the complexity of the routines involved in the generation of the node numbering of the element nodes, preparing the shape function parameters of the node, and finally generating the nodal point coordinates, the order of input

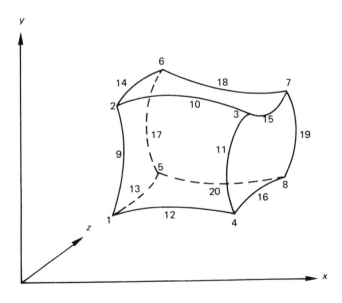

Fig. 8.18 – Irregular block with curved edges.

should be strictly followed to avoid spurious results. This input shape can be used by the first two building blocks but is more useful for MACROBLOCK 2.

e) The fifth input shape is for solids of revolution whose cross-sections have straight edges (Fig. 8.19). Six input points are necessary to give a complete definition of such figures. The first two points define the centre line and the last four points define the section to be revolved about the centre line. The relative positions of the first two points are not critical; as long as they lie on the path of the centre line, as in Fig. 8.19, the second point can be at 2 or 2'. The centre line can have a three-dimensional skew, and the three direction cosines are computed internally by the mesh generating routine. This type of input shape is used by the third and fourth building blocks: MACROBLOCK 3 and MACROBLOCK 4. These building blocks are used in generating 8-node and 20-node cylindrical elements respectively.

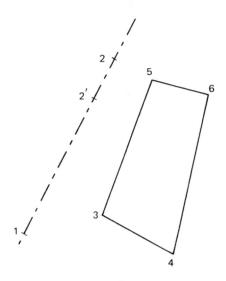

Fig. 8.19 – Solids of revolution with cross-sections of straight sides.

f) The last input shape is also designed for solids of revolution, and only differs from the last one by having ten 3-D input points (Fig. 8.20). The first six points are like those defined in the last input shape, which take care of the definition of the centre line and the four corners of the shape to be revolved. The last four points define the midpoints on the shape of revolution. Thus a parametric curve is introduced on the lines of the shape to be revolved. As with the last input shape, the revolution is done in accordance with the right-hand corkscrew rule with the direction of the centre line from point 1 to point 2. This input block is most suitable for MACROBLOCK 4 although it can be used by MACROBLOCK 3.

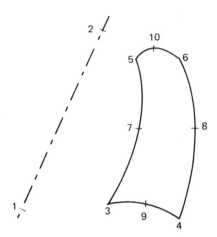

Fig. 8.20 – Solids of revolution with cross-sections of curved sides.

So far only two groups of input have been discussed, first the numerical input of the number of the input points to be digitised, and secondly the actual points digitised on the table. Once the number of 3-D points digitised on the table is equal to that specified by the user, the program jumps out of the digitising input loop and requests the user to certify the acceptability of these stored points. If any of the points are not acceptable to the user, for instance if the program requested point number three in the x-y plane and the user digitised what should be point number five in the particular input shape, the program jumps into an editor phase, guiding the user to correct the point which has been digitised in error. At this stage all the input points can be changed and, if the initial number of input points specified is less than that required, more points could be added. The program can only get out of the editor mode when the user requests for a null point to be edited. Once the user is satisfied with the input points, the program requests for the macroblock number, to call the right overlay into core for the building block required by the user.

Before loading the appropriate service overlay for the required building block into core, 3D MESH stacks its third entry point on the overlay stack, so that control of the system is returned to it after the mesh is developed by the service routine. All the routines used for the generation of meshes within the separate building blocks are written in such a way that, if the number of input points specified at first entry to 3D MESH is not compatible with the routine, control is automatically handed over to 3D MESH at the second entry point where the user can correctly input the required macroblock. The user can exit from 3D MESH by supplying a macroblock number of six — a dummy macroblock number used for exit purposes. When a mesh generating routine is successfully executed and control handed over to 3D MESH, the element connection generating routine is loaded only if a coordinate file is stored on the system disc. After the element connections are generated and a file created, or if there is no coordinate file, 3D MESH will ask the user for more macroblocks or an exit from the program. If the addition of more building blocks to the workspace is requested, 3D MESH places its first entry point on the overlay stack and calls the overlay responsible for displaying the three orthogonal views. After displaying the three orthognal views, '3D MESH' is loaded into core and the first request from the program 'NUMBER OF INPUT POINTS' appears on the VDU.

The use of the overlay 3D MESH as a coordinating program for the whole 3-D mesh generation cannot be overemphasised. It is responsible for loading and coordinating four separate mesh generation overlays, an element connection data generating overlay, and, last but most important, itself.

8.8 FORMULATION OF THE MESH

By discretsation of a structure into macroblocks, a complete mesh can be generated by generating meshes in each macroblock such that there would be no discontinuity of the mesh across the boundary between macroblocks. The basic shape of the macroblocks has previously been shown to be a quadrilateral in two dimensions and a hexahedral in three dimensions. For more realistic modelling of structures with moderate curves, the edges on the quadrilateral and hexahedral can each be given a midpoint for a quadratic curve.

Isoparametric transformations are used to relate the actual geometry of the macroblock to a unit size or scaled block. From the unique mapping of the actual cartesian coordinates to the scaled coordinates obtained from these transformations and the number of elements required in the macroblock, an interpolation is performed on the scaled block and related to the actual macroblock to determine the nodal geometry.

To illustrate the method, consider the three-dimensional mesh of Fig. 8.21a where the cartesian coordinates of all the eight corners are given. The transformation of the (x, y, z) coordinate to the scaled (ξ, η, ζ) coordinates of Fig. 8.21b can be written in general form:

$$U = \alpha_1 + \alpha_2\xi + \alpha_3\eta + \alpha_4\zeta + \alpha_5\xi\eta + \alpha_6\xi\zeta + \alpha_7\eta\zeta + \alpha_8\xi\eta\zeta$$

where U is the displacement component $(x, y$ or $z)$.

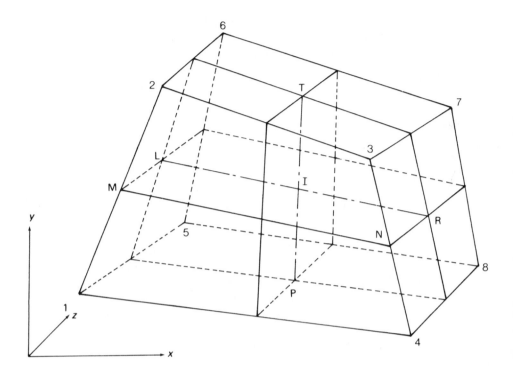

Fig. 21a – Actual block.

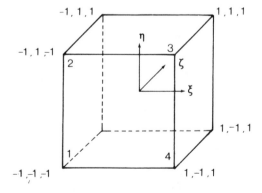

Fig. 21b – Scaled block.

By substituting the actual value of U and the corresponding scaled coordinates at the eight corners, into the above equation, the coefficients (α_1, α_2 etc.) can be found, and the equation written in another general form:

$$U = [N_1, N_2, N_3, N_4, \ldots, N_8] \begin{bmatrix} U_1 \\ U_2 \\ U_3 \\ U_4 \\ \vdots \\ U_8 \end{bmatrix} \tag{8.1}$$

or

$$U = \sum_{i=1}^{8} N_i U_i \tag{8.2}$$

where N_i is called a shape function and is associated with each node and defined in terms of the curvilinear coordinate system $\xi \eta \zeta$, each of which has a value ranging from -1 to 1 on the opposite faces. Typical shape functions for a variety of elements are given in a general form by Zienkiewicz *et al.* [7].

The fourth and fifth are the inverse of the second and first respectively, performed to bring the object to its original coordinate system.

The complete transformation representing the five steps is:

$$T R_1 R_2 R_\theta R_2^{-1} R_1^{-1} T^{-1} \tag{8.3}$$

Let a, b, c be the direction cosines of the axis of rotation. Let $v = (b^2 + c^2)^{1/2}$ and (x, y, z) the coordinates of a point through the axis of rotation.

Then the matrices in equation (8.3) are:

$$T = \begin{bmatrix} 1 & 0 & 0 & 0 \\ 0 & 1 & 0 & 0 \\ 0 & 0 & 1 & 0 \\ -x & -y & -z & 1 \end{bmatrix} \qquad R_1 = \begin{bmatrix} 1 & 0 & 0 & 0 \\ 0 & c/v & b/v & 0 \\ 0 & -b/v & c/v & 0 \\ 0 & 0 & 0 & 1 \end{bmatrix}$$

$$R_2 = \begin{bmatrix} v & 0 & a & 0 \\ 0 & 1 & 0 & 0 \\ -a & 0 & v & 0 \\ 0 & 0 & 0 & 0 \end{bmatrix} \qquad R_\theta = \begin{bmatrix} \cos\theta & -\sin\theta & 0 & 0 \\ \sin\theta & \cos\theta & 0 & 0 \\ 0 & 0 & 0 & 0 \\ 0 & 0 & 0 & 1 \end{bmatrix}$$

The combined use of isoparametric mapping and matrix transformation techniques make it possible to generate the mesh to fit a range of building blocks. The production of a unique mesh by joining one or more surfaces of a building block to that of an existing model is achieved by a simple temporary numbering technique. The faces of each hexahedron are identified by certain numbers and an algorithm was developed for numbering the nodes of the mesh developed in the current macroblock. By specifying any face to be joined to an existing model, the temporary numbers of the nodal points on the common faces are recognised and masked. In our examples there are eigth shape functions

$$N_1 = (1-\xi)(1-\eta)(1-\zeta)/8$$
$$N_2 = (1-\xi)(1-\eta)(1-\zeta)/8 \text{ etc.}$$

Rewriting equation (8.1) only for the x-coordinates the general form will be:

$$[x = N_1, N_2, \ldots, N_8] \begin{bmatrix} x_1 \\ x_2 \\ \vdots \\ x_8 \end{bmatrix} \tag{8.4}$$

where x_n is the x-coordinate at corner n, and $N_i \equiv N_i(\xi, \eta, \zeta)$.

Although the use of isoparametric mapping produces negligible errors in the majority of cases, the exact modelling of circles and small solids of revolution require the use of matrix transformations of which there are three primitive types: scaling, translation and rotation. Complex transformations can be expressed as a sequence of primitive transformations or a concatenated expression yielding a single transformation matrix that has the same effect as the sequence of primitives. Three-dimensional rotation transformations are complex ones, especially when the axis of rotation is not aligned with any of the coordinate axis. For such cases we must concatenate five transformations:

The first translates any fixed point on the axis to the origin (T).

The second performs the appropriate rotation about the x and y axes of the coordinate system such that the unit vector of the direction cosines is mapped into the unit vector along the z axis (R_1 and R_2).

The third transformation performs the desired rotation (R_θ) about the z axis of the new coordinate system.

The geometric coordinates of each common node are checked with those on the existing model and corrected if there is any shift. Next the nodes that are not masked are then given permanent node numbers from the first node number that will be supplied by the user. Finally, the node numbers given to the common nodes are extracted from the previous data file and written into a new data file for the current macroblock. Initially, two data files of coordinate and topology are created. Boundary conditions and loading data can then be created by using the appropriate overlays. The data generated for each macroblock can be checked and modified if necessary before concatenating all the data files into a single file for each set of data.

8.9 USING THE 3-D SYSTEM

The 3-D System is used in a similar manner to that described for 2-D. However, some additional requirements must be met on the input of data because it is in full three dimensions. The following information must be supplied by the user in order to generate the required meshes and data files for each macroblock. The data input is divided into three main groups:

1) Mesh generation parameters
   a) The number of input points (5, 6, 9, 10 or 20);
   b) The coordinates of the numbered input points in the x–y and x–z planes;
   c) The number of elements along defined x, y and z directions of the macroblock.

2) Coordinate modification parameters for a unique mesh description
   a) Common faces on the current macroblock.

3) File creation parameters
   a) First node number;
   b) File number of current macroblock;
   c) File number of previous macroblock to which the current building block is connected;
   d) First element number.

A typical problem which can be handled by the system is shown in Fig. 8.22, this being a slice from a rotor disc of a jet engine. The slice represents 1/26 of the complete rotor and the fir tree slot, for containment of the blades, is cut at an angle of 10° to the axial centre line of the engine.

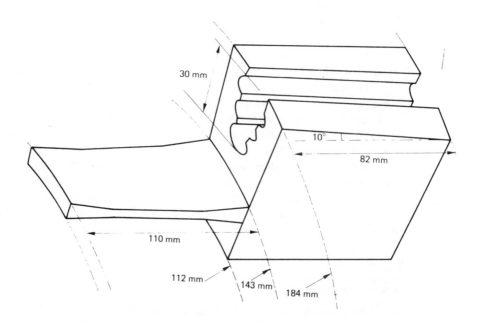

Fig. 8.22 – Slice from rotor disc and shaft.

The mesh was generated from two drawings consisting of a cross-section in the axial–radial plane and a true view looking along the slot as shown in Figs. 8.23 and 8.24 respectively. The customer requiring the analysis specified that the mesh should comprise 20-node isoparametric elements. The finite element stress analysis program used for this application was FINEL [8].

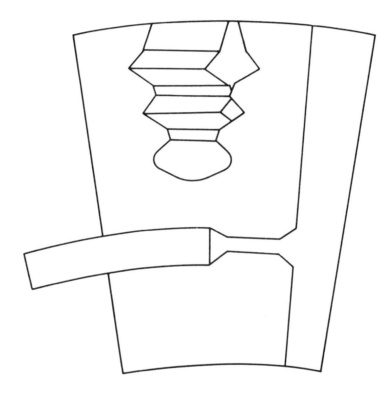

Fig. 8.23 – Axial view on engine.

The whole structure was divided into 22 macroblocks, the first 19 on the disc and the rest on the shaft. The macroblocks allocated to columns with edges of rapidly changing curvature, such as around the fir tree root, were described with 20 input points while the remainder were described by eight input points. The coordinates and topology data were arranged into a format acceptable to the program FINEL.

The boundary conditions are input via the menu. However, in the present case 'the loading on the specimen is centrifugal and cannot easily be described as point loads or transformed into equivalent point loads manually. A small program was therefore written to produce the equivalent point loads at the nodes on the surfaces that are loaded centrifugally and this program was added to the menu. In fact this new program can now be used to generate loading data for uniformly

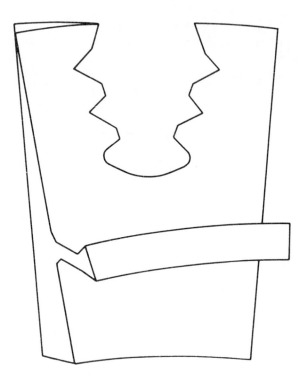

Fig. 8.24 – View along slot 10°.

distributed loads. Adding an extra program to the menu is an easy task, owing to the modularity of the system.

The loading of the specimen is shown in Fig. 8.25 and Table 8.1. The figure

TABLE 8.1 TOOTH NORMAL LOADS (N/mm)

			Top tooth	Middle tooth	Bottom tooth
Blade		h_1	609.2	630.0	654.6
CF		h_2	799.3	778.5	753.9
Torque			90.9	90.9	90.9
Total	Front	L	708.4	687.6	663.0
		R	700.1	721.0	745.5
	Rear	L	700.1	721.0	745.5
		R	708.4	687.6	633.0

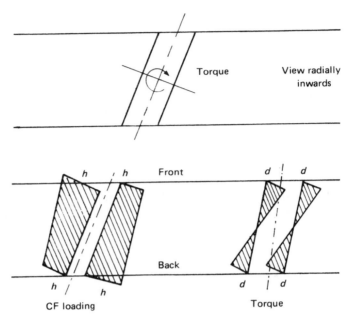

CF loading   Torque

Fig. 8.25 – Blade loads for disc model.
Blade centrifugal force, 250,000 N.
Blade torque, 450,000 N.mm.

and table show a distributed load changing from one value on the front to another
at the rear. Since the surfaces of the teeth that are loaded are plane and not curved,
the element surfaces that are loaded are rectangles. The proportion of the total
force acting at the nodes on the element surface given in Figs. 8.26 and 8.27 are

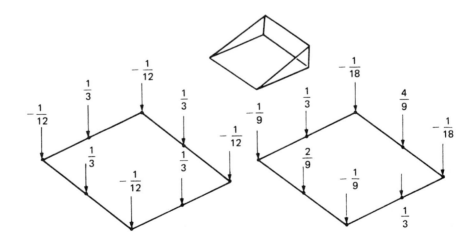

Fig. 8.26 and 8.27 – Edge and surface loading for quadratic elements under distributed load.

superimposed to give equivalent point loads. This information together with the direction cosine of the pressure load on each surface was used to produce the load data for FINEL.

The overall mesh initially generated by the system is shown in Fig. 8.28.

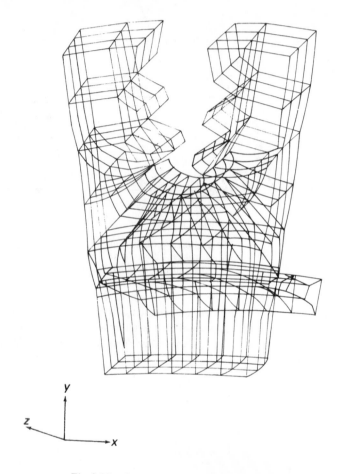

Fig. 8.28 – Overall mesh for case study specimen.

Figures 8.29 and 8.30 show the mesh at various angles to test for any discontinuity and were produced by the plot subroutine in FINEL.

REFERENCES

[1] Zienkiewicz, O. C. & Phillips. D. V. An automatic mesh generation scheme for plane and curved surface bky isparametric co-ordinates. *Int. J. Numer. Meth. Engn.*, **3**, 519-528, 1971.

[2] Zienkiewica, O. C. *The finite element method in engineering science.* McGraw-Hill, London, 1971.

[3] McCormick, C. W. (ed) *The NASTRAN User's Manual.* National Aeronautic and Space Administration.

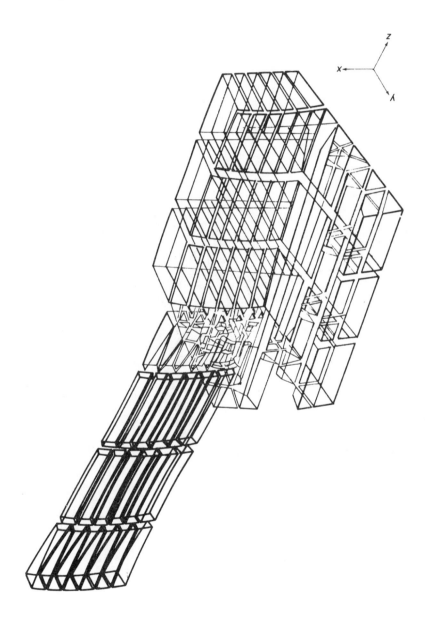

Fig. 8.29 — Discontinuity test plot.

[4] *ASKA linear static analysis user's reference manual.* ISD — report no. 73. Stuttgart, 1971 (ASKA UM. 202).

[5] Ghassemi, F. *Data presentation for finite element analysis.* PhD thesis, University of London, 1978.

[6] Etela, D. T. *Data preparation system for finite element stress analysis in three dimensions.* Ph.D. thesis, University of London, 1982.

[7] Zienkiewicz, O. C., Iron, B. H., Ergatoudis, J., Ahmad, S. & Scott, F. C. *Isoparametric and associated element families for two and three-dimensional analysis.* Tapir Press, Trondheim, 1969.

[8] Hitchings, D. & Mathews, F. L. *FINEL — Finite elements in theory and practice.* Aeronautical Engineering Department, Imperial College, London, 1978.

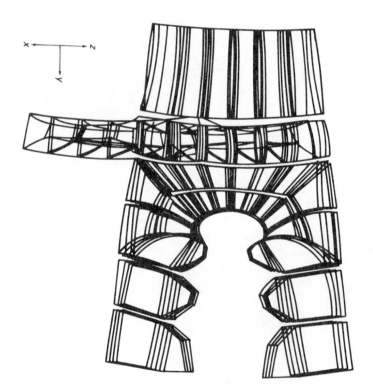

Fig. 8.30 — Discontinuity test plot.

9
Computer-aided manufacture

In the previous chapters the term computer-aided design (CAD) has been defined as almost any design activity involving the use of a computer. More specifically CAD is more associated with the creation of graphics and the interaction of the graphics with engineering analysis involving arithmetic operations such as in stress analysis, heat transfer or fluid dynamics. Thus a shape or physical structure is defined by a list of suitably coded numbers and these numbers can be used as data input to an engineering analysis program.

The use of CAD is becoming more widespread as smaller computers and intelligent computer peripherals such as graphic displays, digitisers and plotters become lower in cost. Furthermore, it has been shown that smaller computers can assist in the non-computer dialogue by their real-time interactive accessibility to the user, thus permitting a greater flow of information between the design and computer. This advance in computer technology is resulting in dedicated CAD systems in various fields of engineering design such as the use of CAD techniques in large-scale integrated (LSI) circuit design.

Computers have also been widely applied to manufacture and the term computer-aided manufacture (CAM) is used for manufacturing procedures that use computers to assist in the planning and production of manufacturing processes from inventory control to the programming of machine tools. More recently the gap between computer-aided design and computer-aided manufacture has been closed and integrated CAD/CAM systems are now commercially available. An integrated CAD/CAM system can therefore be defined as a system where the link between design and manufacture is accomplished by the use of a computer. The concept of such systems is based on the use of information and data from the CAD process directly in the CAM procedures thus avoiding the independent generation of data for computer programs in the manufacturing area. Thus the information which is common to a larger number of programs is usually stored in the mass storage units of the computer and it is often referred to as a database. Such a database for an integrated CAD/CAM system is shown in Fig. 9.1.

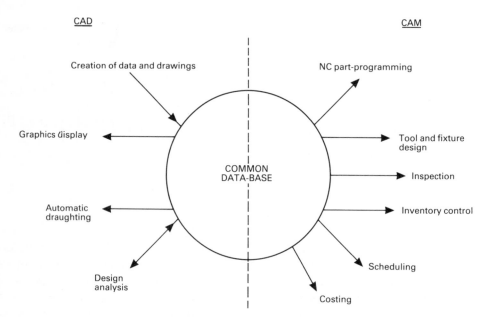

Fig. 9.1 – General division of tasks in an integrated CAD/CAM system in mechanical engineering.

CAD/CAM techniques have been applied to mechanised engineering for many years in specialised industries such as the aircraft industry. In this industry the design and manufacture of complex parts is common and such applications have been well suited to the use of computer techniques. For example, aircraft wing sections can readily be designed with the aid of computer graphics and the ensuing data used as input to a manufacturing program which will produce the input data for a numerically controlled (NC) machine tool so that manufacture can take place.

The high cost and unavailability of skilled labour together with the need to increase efficiency has resulted in medium and small companies using CAD/CAM techniques. The falling cost of computers and in particular the availability of microcomputers has now brought the capital cost of such systems within the range of many firms. The growth in CAD/CAM applications is likely to be very large in the next decade.

A variety of CAD/CAM systems will be discussed in this chapter and proceeding chapters. These systems will range from those based on mainframe computers to those using minicomputers and to future systems which will make extensive use of microcomputers.

9.1 EXISTING CAM SYSTEMS – THE APT SYSTEM

The use of computers in design and manufacture in the mechanical engineering field started as two separate activities with no apparent link between them. This may be due in part to the separation of activities in a firm into departments, such as the drawing office, planning office and production control. Each department tended to be rigidly responsible for its activity and if computers were introduced each department was responsible for applying the new technology to its activity. This tended to result in each department having to duplicate information during the data preparation stage before running a computer program which was applicable to its activity. A good example of this duplication of effort arose when numerically controlled (NC) machine tools were first introduced. Computer programs were specifically produced for the manufacturing departments for the preparation of control tapes for these N/C machine tools. The first such program was Automatically Programmed Tools (APT).

APT is a programming language which allows geometrical data to be specified together with tool motion statements for any NC material processing machine. The development of APT started in the Electronic Systems Laboratory of Massachusetts Institute of Technology in collaboration with the United States aircraft industry [1]. Conceptually APT is a computer program and a programming language. It is capable of allowing a user to produce complex three-dimensional geometrical contours of tool motion with respect to a complex three-dimensional design. The user of APT defines the geometry of the workpiece and the requisite tool motions using simple English-like statements. Its objective is to relieve the numerical control part-programmer from the time-consuming task of calculating relative motions between the tool and workpiece to produce a certain shape such as a complex contour. It can be used effectively to produce control tapes for simple turned parts as well as the more complex shapes.

The APT systems consists of three parts, namely the APT part-program, the APT program processor and the APT program post-processor. The part-program is first written in APT language to specify the geometry of a component or workpiece together with a certain tool and the direction of relative motion between the two. There are provisions in the language to allow for the specification of machine tool dependent data such as feedrates and spindle speeds. This information is not processed by the APT processor but is used later by the APT post-processor.

The APT processor is a very large program normally resident on a large storage unit which is run mostly on a mainframe computer. It accepts APT post-programs and produces an output which is called the Cutter Location Data

(CL DATA). This data is then normally passed onto an APT post-processor which produces a numerical control tape for a specific machine tool. Thus the post processor is machine-tool-dependent and converts the generalised CL DATA into a specific numerical code. The flow chart for an APT system is shown in Fig. 9.2. The post-processing activity can be considerable for older types of NC systems with limited built-in intelligence. This is necessary since the program would have to provide information on accelerations, decelerations and velocities of the tool or cutter in order to maintain the correct path. In modern systems, machine control parameters and characteristics are built into a minicomputer or microcomputer controller attached to the machine tool.

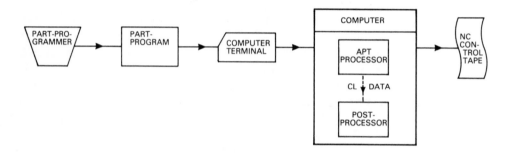

Fig. 9.2 – Macro flow chart for APT system.

A disadvantage in the use of the APT system for many of the applications in manufacturing industry is the large overhead in terms of computing capacity and power, for relatively simple components. APT can be used via a time-sharing system but the system response can prove to be slow when used in this mode. Finally, use of the APT system is dependent on the availability of a post-processor for a machine tool, otherwise the control tapes for that machine tool cannot be produced.

One of APT's biggest advantages is that it has become a world wide standard for NC machines. Variations of APT have been developed outside the United States; for example, NELAPT was developed at the National Engineering Laboratory in the UK, EXAPT in Germany and IFAPT in France and all these systems produce APT CLDATA formatted outputs to a specific CLDATA standard [2]. Some of the derivations of APT have special features not found in APT. For example, NELAPT has a useful feature for pocket milling or rough turning. This allows the programmer to mill out a complex geometrical pocket by the use of a few English-like statements that specify the contour of the pocket and the diameter of the tool. The NELAPT processor will then produce a series of parallel cuts so that the specified contour is cleared [3]. The minimum computer hardware configuration on which the NELAPT system can be implemented is as follows:

— 28K of 24-bit words of primary memory;
— 4 magnetic tapes and disk storage;
— 1 input medium such as a computer terminal;
— 1 line printer;
— 1 output medium such as a paper punch for NC tapes.

It can be seen from these requirements that the NELAPT system can run on a large minicomputer as well as a mainframe.

Details of the APT language will now be discussed in the following sections.

9.1.1 APT symbols and words

The APT language is made of symbols and words. A certain number of words, based on the English language, are designated APT words and are contained in a dictionary. There are many hundreds of words in the APT vocabulary to which definite meanings have been assigned. The programmer must learn these words. The vocabulary is divided into main words and auxiliary words or modifiers.

Examples of main words include:

CALL, CIRCLE, CUTTER, GO, GOTO, LINE, POINT, TOOL

and examples of auxiliary words or modifiers are:

ARC, CENTRE, LEFT, PARLEL, TANTO

Main and auxiliary words are separated by an oblique stroke, as for example

CIRCLE/CENTRE, PI, TANTO, LI

where PI is the centre of the circle and its tangent is LI. Both PI and LI must be defined somewhere within the program.

A substantial part of the APT language is devoted to enabling the programmer to provide geometric descriptions of a part to be machined, to the computer, and to define the way in which the tool should move over the surface of the defined geometry in order to machine it. The remainder of the language is devoted to producing computer instructions such as for calling subprograms.

An example of a geometric statement is

PTI = POINT/7, 8

PTI is the symbolic representation of a point whose X coordinate is 7 and whose Y coordinate is 8.

An example of a motion statement is

GOLF/LI, PAST, L2

Move to the left and then move the tool along a line called LI until it passes a line called L2.

In this last statement GOLF is an APT main word and PAST a modifier. PTI, LI are identifying names and are given to geometric expressions such as points and lines and cannot be APT words. Other types of statements not related to the geometric or motion statement are those related to computational aspects. For example CLPRNT is an instruction to the APT system to produce a printed list of all cutter location coordinates that have been computed. The computation results are those of the APT system, before post-processing.

9.1.2 Geometric definitions in APT

The APT language enables the definition and machining of 3D surfaces as well as 2D parts. A geometric expression is used to define a shape or form. The APT language contains a number of basic definitions or primitives for different geometric forms, such as POINT, LINE, PLANE, CIRCLE, CYLINDR, ELLIPS, CONE and SPHERE. Complex shapes can be defined by building-up from a number of primitives.

Points. In APT a point can be described in a number of different ways and a few are now illustrated:

(a) By coordinates

          POINT/X coordinate, Y coordinate, Z coordinate

    Example: PTI = POINT/3, 2, 1
             or PT2 = POINT/4, 5

In the latter case no Z coordinate is given and it is therefore assumed to be zero.

(b) By the intersection of two lines

          POINT/INT OF, symbol for a line, symbol for a line

    Example: PT2 = POINT/INF OF, LIN I, LIN 2

(c) By a centre of a circle

          POINT/CENTRE, symbol for a circle

    Example: PT3 = POINT/CENTER, CI

Lines. A line can be expressed in many different ways and a few examples are now presented:

(a) Through two points

          LINE/symbol for a point, symbol for a point

    Example: LI = LINE/PTI, PT2

(b) By a point and a tangent to a circle

          LINE/symbol for a point, LEFT, TANTO, RIGHT symbol for a
          circle

where the modifiers LEFT or RIGHT are applied looking from the point towards the circle.

    Example: LI = LINE/PI, LEFT, TANTO, CIRI

Planes. Similar to points and lines and can be defined in a number of different ways, such as:

(a) By three points that are not in the same straight line

> PLANE/symbol for a point, symbol for a point, symbol for a point

Example: PLI = PLANE/PTI, PT2, PT3

(b) By a parallel plane and the perpendicular distance between the two planes

$$
\text{PLANE/PARLEL, symbol for a plane,} \quad
\begin{matrix}
\text{XLARGE}\\
\text{XSMALL}\\
\text{YLARGE}\\
\text{YSMALL}\\
\text{ZLARGE}\\
\text{ZSMALL}
\end{matrix}
\quad \text{, offset}
$$

distance between the two planes.

Example: PL2 = PLANE/PARLEL, PLI, ZSMALL, 5

Here, PL2 is parallel to the plane PLI, and is lower 5 units in Z. See Fig. 9.3. If the planes are not parallel to the main planes, two modifiers are appropriate, and either of them could be used in the statement.

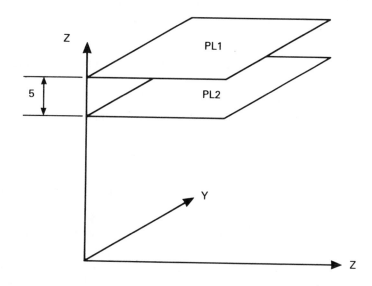

Fig. 9.3 – PL2 = PLANE/PARLEL, PL1, ZSMALL, 5.

Circles. The circle can be expressed in 10 different ways, and three of these are now shown:

(a) By using three points on the circumference.

> CIRCLE/symbol for a point, symbol for a point, symbol for a point

> Example: CI = CIRCLE/PN2, (POINT/5, 7, 4), PNTI

(b) By the centre and a point on the circumference

> CIRCLE/CENTER, symbol for a circle centre point, symbol for a point on the circumference

> Example: C2 = CIRCLE/CENTRE, (POINT/3, 6, 2), PTI

(c) By the centre and the radius

> CIRCLE/CENTRE, symbol for a circle centre point, RADIUS, radius of circle

> Example: C3 = CIRCLE/CENTER, PTI, RADIUS, 4

The CIRCLE/statement defines a circular cylinder perpendicular to the XY plane. The CYLNDR/statement is used to define a cylindrical surface that could not be defined with a circle statement.

9.1.3 Motion statements

The first stage of APT programming is the definition of the part using geometric expressions. This is then followed by a specification of tool movement using motion statements. Each motion statement will move a tool to a new location or along a specified surface. Two groups of motion statements are available to the APT programmer, these being for point-to-point and contouring operations.

Point-to-point motion statements

Motion statements are available for positioning the tool at a desired point and their format is as follows:

> FROM/symbol for a defined point
> — indicates the initial position of the cutter centre.

> GOTO/symbol for a defined point
> — position of the tool centre at a specified point

> GODLTA/$\Delta X, \Delta Y, \Delta Z$
> — positions the cutter in the specified increment from its current position.

Comments relevant to these statements are as follows:

(a) Instead of 'symbol for a defined point', a statement (POINT/X,Y, Z) or X,Y, Z coordinates may be written.

(b) FROM provides the initial location from which a motion is to start and is placed as the first motion statement in the program. The operator adjusts the machine to make the location of the tool coincide with this programmed location

(c) The GOTO/statement will move the tool along a path from the present location to the specified point. The GODLTA/statement will move the tool the specified incremental distance from its present location.

(d) In a drilling operation the GOTO/statement is used to position the drill above the required hole. A GODLTA/statement is then used to move the drill into the workpiece. A second GODLTA/statement is used to remove the drill.

Contouring motion statements

In APT programming, it is assumed that the part remains stationary and the tool moves. Three surfaces control the tool motion in contouring. These are shown in Fig. 9.4 and are as follows: the tool end moves on the 'part surface', the tool slides along the 'drive surface' and the motion continues until the tool meets the 'check surface'. Initially the tool must be brought to the controlling surfaces by an initial motion statement which has the following format:

> GO/cutter specifier, drive surface,
> cutter specifier, part surface,
> cutter specifier, check surface.

Three cutter specifiers can be used which are **TO, ON** and **PAST**. Their use is illustrated in Fig. 9.5.

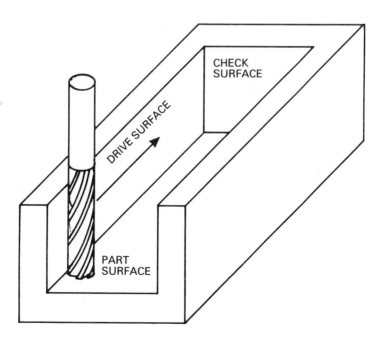

Fig. 9.4 – The surfaces which control the cutter in APT.

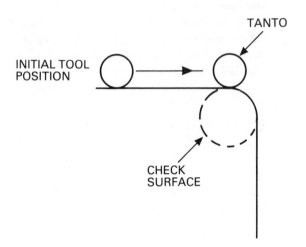

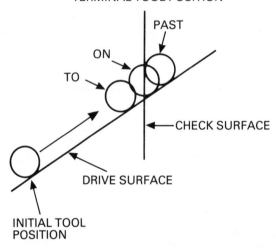

Fig. 9.5 — Cutter specifiers.

Examples   GO/TO, CIRCI, ON, PLI, TO LIN I

The drive surface of a GO/statement will be the surface cut along the next motion statement. The part surface is established for all the subsequent motion statements. The initial motion statement only appears once in a part program and it brings the cutter from the set or start point to the workpiece. The actual cutting is controlled by another type of statement called the intermediate motion statements.

There are four variants of intermediate contouring motion statements used in APT. The most useful one has the following format:

motion word/drive surface, cutter specifier, check surface

Example   GOLFT/DRS, TO, CKS

The drive surface is the surface where the cutting takes place and the check surface defines the end of the cutter motion.

There are six different motion words, illustrated in Fig. 9.6, these being:

GOLFT GORGT GOFWD GOBACK GOUP GODOWN

In intermediate motion statements, four different types of cutter specifiers, illustrated in Fig. 9.5, can be used:

TO ON PAST TANTO

Comments relevant to these words and statements are as follows:

(a) Motion statements are programmed from the viewpoint of the tool.
(b) Each motion statement is dependent on the preceding statement for establishing the direction of motion.
(c) The check surface of the current motion usually becomes the drive surface of the next statement.

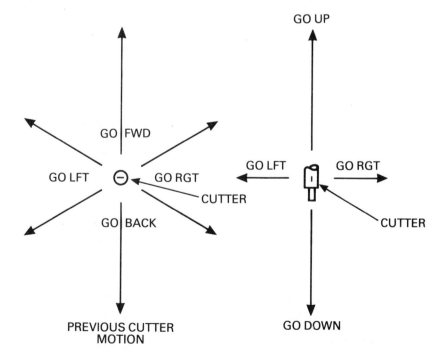

Fig. 9.6 – Directions of contouring motion words.

9.1.4 Additional APT statements
Approximately 70% of an APT program consists of geometric and motion statements. There are additional statements necessary for an APT program to perform. Some of these additional statements are now discussed.

Post-processor statements
The post-processor allows the transformation of a post-processor control statement into an appropriate code of the control. Typical examples of post-processor control statements are:

COOLNT/ON	Turn coolant on (m08 code)
FEDRAT/250	Tool feedrate is 250 mm/min
SPINDL/1250, CCLW	Spindle speed is 1250 rpm in counter-clockwise direction
TOOLNO/4500, 10	Tool number 4500, with 10 units in length
SPINDL/ON	Turn on spindle (m03 code)
END	End of program (m02 code)

Cutter specifications
The tool can be specified by the statement CUTTER/ followed by up to seven arithmetic parameters where the first is the cutter diameter. For example

CUTTER/20
– indicates a cutter 20 units in diameter

CUTTER/20, 5
– indicates a cutter 20 units in diameter with an end radius of 5 units.

Tolerance specifications
All containing motions are reduced to a series of straight-line motions required to approximate to a given curve, the straight lines departing from the given curve or surface by no more than a specified tolerance. The tolerance words are followed by an arithmetic parameter.

Examples	OUTTOL/0.001	Outer tolerance – affects overcutting that is the outside of the curve, departing by no more than 0.001 units from the contour.
	INTOL/0.001	Inner tolerance – affects undercutting.
	TOLER/0.01	Outer and inner tolerances are equal.

Initial and termination statements
The first statement in an APT program begins with the word PARTNO. Any information may follow on the same line after PARTNO.

Example   PARTNO CAM PLATE

The PARTNO statement identifies the name of the part.

The last statement in any APT program will be FINI. It defines no successor and has the effect of terminating the program.

9.1.5 Conclusions

The APT vocabulary presented here is only a small part of the overall vocabulary. The statements outlined permit a level of 2D part programs for point-to-point and contouring. The complete APT system permits a variety of different methods for constructing the same geometry and covers 3D as well as 2D geometry. APT programming covers turned and milled components.

The following rules are suggested for the APT part programmer:

(a) Spelling of APT words must be exactly as specified and only capital letters are permitted.

(b) Generally there is a punctuation mark, such as a comma, equal sign or slash, between every two words or numbers.

(c) There is no punctuation mark at the end of a statement.

(d) A period is permissible only as a decimal point and a period at the end of a statement is an error.

(e) Spaces can be inserted or omitted in most APT words such as GO RGT or GORGT. There are a few exceptions such as PARTNO.

(f) It is recommended that identifying words of geometric expressions are used with appropriate names if possible, i.e. LFTSID.

(g) An identifying word or name cannot be an APT word.

(h) Identifying words can be any combination of letters or numbers but the first must be a letter.

(i) Spaces can be inserted in identifying words.

(j) Geometric expressions of shapes must be defined and given an identifying word prior to using that identifying word in a statement. It is therefore recommended that definitions of the part geometry be placed before the motion statements.

(k) Motion instructions are written from the viewpoint of the cutter. This means that the cutter moves left or right as if the programmer was on the cutter.

(l) Use of the $ sign indicates that a statement is continued on the next line.

It is recommended that APT programs are structured in the following way:

```
PARTNO Part name and number
MACHIN/Post processor name
Description and definition of the part geometry
Cutter and tolerance specifications
Machining conditions
Motion instructions
Spindle coolant off
FINI
```

An example APT program now follows and is related to the part shown in Fig. 9.7.

```
PARTNO CAM PLATE NO 1                                    10
MACHIN/UNI                                               20
SETPT  = POINT/−10, −10, 2.5   $$ CLEAR OF PART          30
P1     = POINT/15, 5, 0                                  40
P2     = POINT/4, 40, 0                                  50
P2HIGH = POINT/4, 40, 12       $$ ABOVE POINT 2          60
P3     = POINT/76, 40, 0                                 70
P4     = POINT/65, 5, 0                                  80
P4HIGH = POINT/65, 5, 12       $$ ABOVE POINT 4          90
PL1    = PLANE/P1, P2, P2HIGH                           100
PL2    = PLANE/P2, P2HIGH, P3                           110
PL3    = PLANE/P3, P4, P4HIGH                           120
PL4    = PLANE/P1, P4, P4HIGH                           130
PRT    = PLANE/P1, P2, P3      $$ PART SURFACE          140
CUTTER/10                                               150
TOLER/.01                                               160
SPINDL/2000, CLW                                        170
FRDRAT/2500                                             180
FROM/SETPT                                              190
GO/TO, PL1, TO, PRT, TO, PL4                            200
GOLFT/PL1, PAST, PL2                                    210
GORGT/PL2, PAST, PL3                                    220
GORGT/PL3, PAST, PL4                                    230
GORGT/PL4, PAST, PL1                                    240
GOTO/SETPT                                              250
STOP                                                   260
FINI                                                   270
```

Remarks

10 Identifies the part name.

20 Indicates the post-processor name.

30 SETPT is the starting point of the tool. A part is replacd when the tool is at SETPT.

40 Definition of point P1.

50 Definition of point P2.

60 Definition of point P2HIGH.

70 Definition of point P3.

80 Definition of point P4.

90 Definition of point P4HIGH.

100 Definition of plane PL1.

110 Definition of plane PL2.

120 Definition of plane PL3.

130 Definition of plane PL4.

140 Definition of part surface PRT.

150 A cutter with a diameter of 10 mm is used.

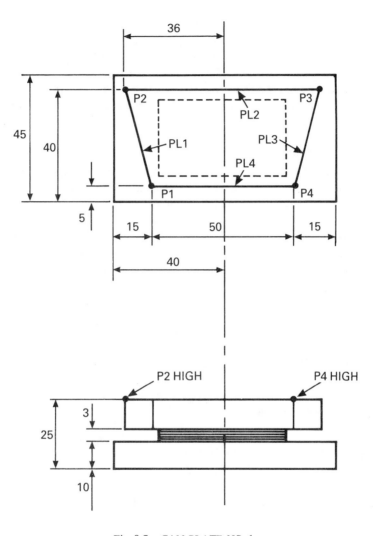

Fig. 9.7 – CAM PLATE NO. 1.

160 The cutter end will always be within 0.01 mm of the true mathematical surface along which it moves.

170 The spindle rotates clockwise at 2000 rpm.

180 The desired feedrate is 2500 mm/min. The distance required for acceleration and retardation, to hold overshoot within tolerances, is automatically computed when post-processed.

190 Defines initial location of cutter.

200 Initial motion statement which moves the cutter to the surfaces named. Since PRT is the part surface, the cutter will stay on it throughout the succeeding motions. PL1 is the drive surface and has to be the surface cut along in the next motion statement.

210 The cutter moves along PL1 until PL2 is past.
220 The cutter moves along PL2 until passing PL3.
230 The cutter moves along PL3 until passing PL4.
240 The cutter moves along PL3 until passing PL1.
250 The cutter moves to SETPT which is out of the part.
260 The STOP statement stops the machine and turns off the spindle.
270 Indicates that the program is terminated.

9.2 CAD/CAM systems

Over the past ten years many manufacturing companies have become increasingly aware of the importance in linking design and manufacture. This can result in a reduction of lead-time, reduce drawing office effort, simplify production planning and reduce scrapped components due to faulty manufacture. The availability of powerful minicomputers and associated graphics facilities has resulted in a number of relatively low cost CAD/CAM systems which utilise a common database for design and manufacture, thus eliminating the regeneration of graphical data for the manufacturing programs. A number of firms now offer such CAD/CAM systems, typical firms are Computervision, Counting House and Ferranti Infographics. The Computervision system CADDS3 (Mechanical Design/ Numerical Control) [4] is based on the use of interactive graphics facilities for inputting and editing geometrical data into a three-dimensional database from which all outputs and post-programs compatible with the APT CLDATA can be generated. The system supports a set of geometrical elements similar in their format to the APT language. The hardware of the system is based around the following configuration:

 — 1, 16-bit minicomputer with 196 K 16-bit primary memory;
 — graphic input terminal with tablet;
 — graphic output plotter;
 — printer and paper tape punch for NC tapes;
 — disk back-up storage.

A flow chart showing the comparison between the Computervision CAD/CAM system and the manual APT system is given in Figs. 9.8(a) and (b). Most of the other available systems work on a similar principle to that of Computervision.

Most integrated CAD/CAM systems use geometric contours as a link between design and manufacture. This is due to NC languages having the requirement for precise definitions of geometry in order to calculate tool paths. Geometrical contours are either defined by a string of straight line segments and arcs in two dimensions [5] or by a series of straight line segments in three dimensions. In the latter case spline-blending techniques are often used to produce smooth contours [6]. Two-dimensional contours are more widely used since the majority of NC machines are capable of continuous control of only two axes. Some typical applications of a CAD/CAM system are given in the next chapter. In this case the system hardware is based on a 16-bit minicomputer and is similar to that of some of the commercially available systems.

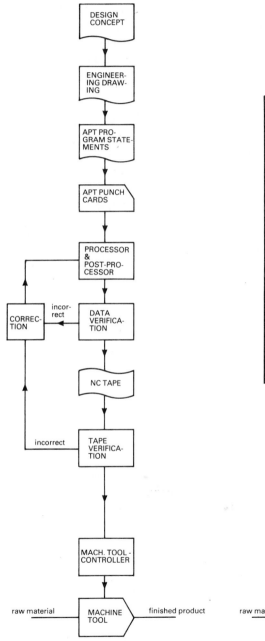

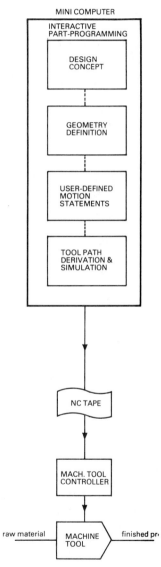

Fig. 9.8(a) – Standard APT flow chart.

Fig. 9.8(b) – Computervision's part-programming flow chart.

CAD/CAM systems are not just systems for generating NC data from a graphics database but have a much wider concept in terms of design and manufacture. Some of the most comprehensive systems now available permit considerable integration between design and manufacture. These systems provide tools for 2D draughting and 3D modelling with access to filing systems based on IGES (International Graphics Exchange Standard) and comprehensive library facilities. Integrated links are provided to analysis systems such as FEM and CAM. Also facilities are available for management such as scheduling, reporting and cataloguing. One of the most well known of these comprehensive systems is MEDUSA [7] which was produced by Cambridge Interactive Systems Ltd.

MEDUSA is a system of highly interlinked computer programs which carry out a wide range of functions for CAD/CAM. MEDUSA can be used in a very wide range of applications, from mechanical engineering such as shipbuilding, heavy industrial machinery and consumer products, to civil and structural engineering, electrical schematic design, PCB layout and architectural design. The overall structure of the MEDUSA system is shown in Fig. 9.9 and Fig. 9.10 shows a diagram of all the possible installed options which are available to users. In practice a particular installation would only contain some of these options.

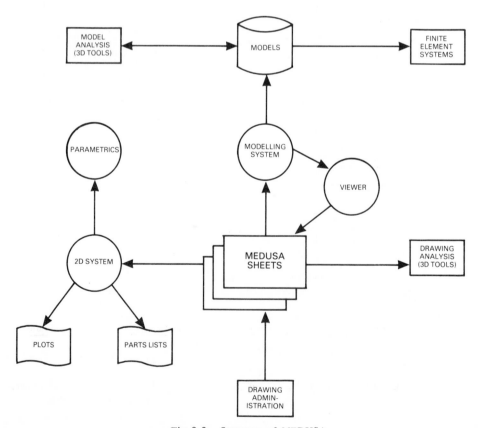

Fig. 9.9 – Structure of MEDUSA.

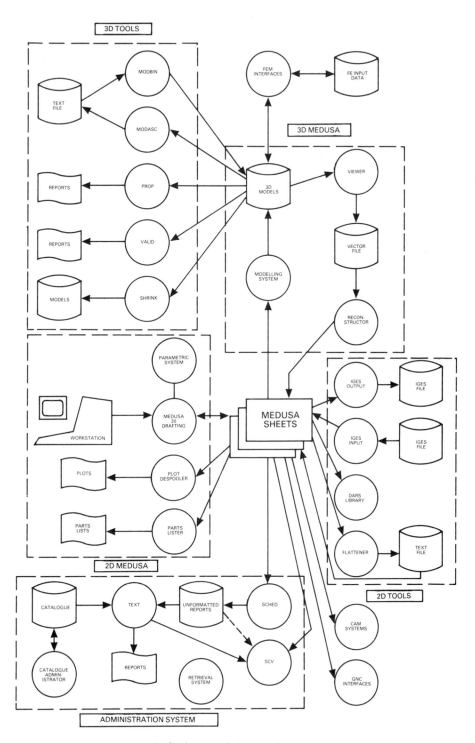

Fig. 9.10 – Detailed MEDUSA structure.

The MEDUSA system is designed to run on a wide range of computers such as those in the PRIME or DEC–VAX ranges. A small installation with two work-stations may be supported by a computer with 1M bytes of memory while a large installation of twenty workstations will require a computer with a fast CPU and 8M bytes of memory. Disk capacities range from 35M to 600M bytes, depending on the size of the installation.

The workstations consists of a graphics display supported by a tablet and joystick. An alphanumeric screen is used to display error messages, prompts and other textual information. Communication by the user is via the tablet or a keyboard. Extensive use is made of menu facilities during data input.

The MEDUSA system is entered via the workstation and in most cases using the 2D draughting system. All the necessary facilities are provided to allow the designer to create, edit and manipulate engineering graphics ready for inter-pretation by other programs within MEDUSA. The data fed into the computer is stored on a file known as the MEDUSA sheet and each sheet is made up from a number of elements such as lines, texts, prims (commonly used symbols) and clumps (dummy elements used to associate other elements). The data can be visualised through view boxes for examining orthogonal views for example.

The 2D graphics data can be passed via the MEDUSA sheets to the 3D Modeller where various orthogonal views are combined to give a 3D model.

The 2D and 3D systems are linked in with a number of optimal CAD/CAM packages. For example, CAM programs exist for allowing the direct production of NC tapes for a wide range of applications and machines.

9.3 THE USE OF MICROCOMPUTERS IN CAD/CAM SYSTEMS

Over the past five years there has been increasing use of microcomputers both in CAD and NC machine tools. In the case of CAD systems, microcomputers have been used extensively to provide local intelligence to computer peripherals such as graphics displays, plotters and disk and magnetic tape controllers. This has resulted in offloading some of the functions originally performed by the minicomputer in the CAD system. The result is that more workstations can be attached to one minicomputer with each workstation providing a good response to the user.

The role of microcomputers in NC machine tools has been to perform NC functions normally executed by hard-wire logic circuits or minicomputers and to further improve the performance of the NC control system. Standard NC control functions for machine tools in which microcomputers can play a useful role can be summarized as follows:

– part-program and data interpretation and distribution;
– control of information flow;
– vector interpolation;
– feedrate control;
– sequencing logic;
– positioning servomechanism interface;

— compensation for mechanical inaccuracies such as lead screw pitch error, backlash in gears;

— tool compensation;

— maintenance and fault detection.

New and powerful control systems are now becoming common on machine tools and these controllers are easing the problems of programming the machine tool. In some cases the machine tool can even be programmed at the machine tool itself using powerful built-in graphics facilities for geometrical data input and simulation of matching processes. Such systems are manufactured by Yamazaki [8] and Fanuc [9] and a recent system by R D Projects is described below.

9.4 A CNC SYSTEM WITH INTEGRATED INTERACTIVE GRAPHICS

Modern machine tools with CNC control systems are often offered with integrated graphics systems to aid programming. These integrated systems are of particular interest to small companies that do not have large computing capacity for running systems such as APT. Furthermore they are interested in systems which are very user-friendly so that complex tasks can be undertaken by semi-skilled staff. One of these comprehensive systems is produced by R D Projects Ltd and is available for the Batchmatic range of turning machines, being known as the RDP Batchmatic CNC. The Batchmatic machine tools consist of a range of turning machines from a three-axis bar feed type to a chucker.

9.4.1 The hardware

A block diagram of the RDP/CNC system is shown in Fig. 9.11. It is based around an executive 16/32-bit microcomputer with 515k of DRAM memory with several communication ports. The executive computer is connected to the axis computers via an IEEE-488 communication bus. Each axis computer contains an 8-bit microprocessor and is interfaced to the servomotors that power the slides. The axis computer performs the servo loop closure, provides velocity control and performs interpolation.

The operator communicates to the system via a front panel of membrane construction containing keys with tactile feedback. A colour graphics unit is an integral part of the system to provide visual information on data fed to the system or data such as cutter paths that are generated by the executive computer. The graphics controller provides for 8 colours and displays 640×288 pixels. The front panel with its graphics display is shown in Fig. 9.12.

Part-programs can be input or output from the system via a paper tape reader/punch unit. Also the system contains a rigid disk which is connected via a parallel interface to an executive computer which is used for storing part-programs and data. The executive computer also contains communication ports for linking the machine tool with other systems such as a robot or factory production control computer.

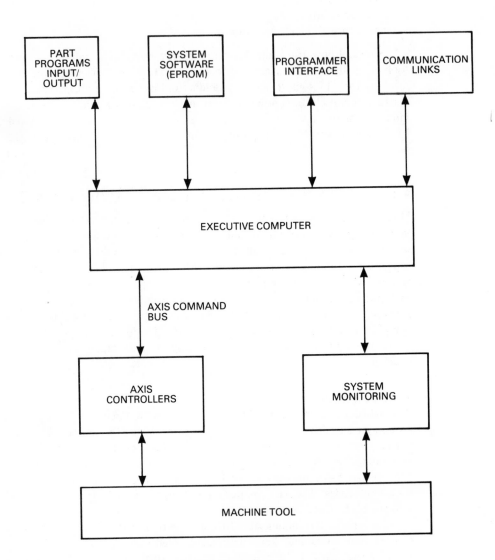

Fig. 9.11 – RDP–CNC system structure.

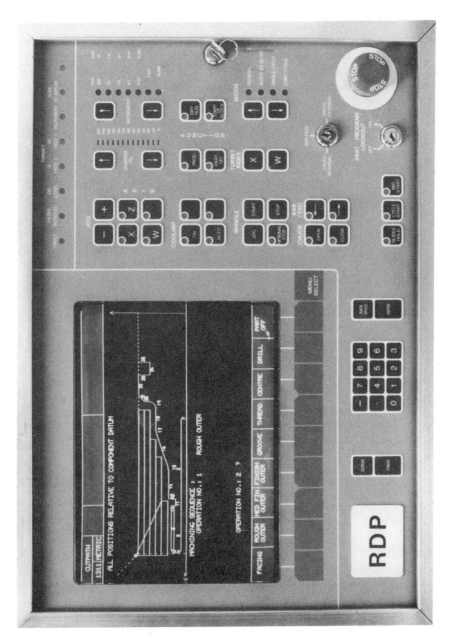

Fig. 9.12 – The front panel of the RDP/CNC system. (Courtesy of R D Projects Ltd.)

9.4.2 The software

The software executive is divided into a number of different tasks. Each task performs a logically separate function in relation to the controller, such as computer-aided part-programming (CAPPS), diagnostics and condition monitoring. An operating system kernel shares the main processor amongst the different tasks to provide concurrent operation of all the tasks. Any hardware input or output is done directly from the task requiring it. The operating system ensures that there is no contention between more than one task. The disk-based filing system allows for files in relation to part-data, part-programs and system programs.

9.4.3 The front panel

The front panel shown in Fig. 9.12 is divided into two principal areas. The right-hand side is for the manual control of the machine tool and incorporates keys machine functions – such as for switching the coolant on or off, spindle on or off, jogging the various sides, indexing the turrets etc.

    The left-hand side of the panel comprises the interactive graphics display and the keys for using the graphics features. The display is situated in the upper left-hand side of the panel and nine 'soft' keys are situated directly beneath the display to allow options to be selected from those displayed on the menu above the keys. The numeric keypad, backspace and enter keys are used for entering part data, tool offsets, etc.

    A screen button switches the left-hand side of the panel between NC operation when machining and CAPPS operation when part programming.

9.4.4 NC machining

The RDP CNC system accepts part-programs on paper tapes which have been written for the machine tool by a part-programmer in the conventional way or via APT. The system contains all the essential features required of a three-axis turning machine and some of these are now itemised.

(a) Axes control. Three axes are controlled simultaneously (X, Y, W). Circular interpolation is available in the Z axes. The minimum programmable increment on any axis is 0.001 mm or 0.0001 inch

    Jog: A range of jog increments are available from 100 to 0.01 mm or 10 to 0.001 inch.

    Feedrate. The feedrate is programmable in mm/rev, mm/min, inch/rev, inch/min.

    Interpolation rate: The linear interpolation rate is 100 Hz and the circular interpolation rate is 50 Hz.

    Slide at datum: an LED is indicated when a slide is at its datum position.

    Dwell: Dwells up to 99.999 s may be programmed.

    Thread cutting. Parallel or taper threads may be cut in plunge or compound modes.

    Absolute/incremental programming: The X, Y, W axes may be programmed using a combination of absolute and incremental dimensions.

9.4.5 NC display

The NC display has five pages, the units may be in either metric or imperial. The display pages are Coordinates and Feeds, Absolute Coordinates, Relative Coordinates, Run Setup Parameters and Run Modify. The run modify refers to changes that may be required during a run such as insertion of a block, tool offset adjustments, batch quantities or datum shifts.

9.4.6 Diagnostics

The system contains a wide range of diagnostics from machine to part-program diagnostics. Messages are displayed on the screen along with the sounding of an alarm when a fault occurs.

9.4.7 CAPPS programming

The real power of the RDP/CNC system lies in its CAPPS system which incorporates a level of expert system technology. The CAPPS system permits an inexperienced machine tool user to define a part shape for the CNC system, given an engineering drawing of the part, select the correct tools for machining the part and to generate the correct tool paths for the various machining processes necessary to create the actual part. The system contains a library of information which can be brought into use by the programmer in order to aid various tasks in the programming process.

9.4.8 Geometric definition

The first task of the programmer is to define the shape of the part and the blank from which the part is to be machined. A drawing is entered with the aid of the menu and soft keys together with the numeric keypad. A drawing may be entered in a variety of formats, the two most important being by radius or diameter. The shape of the part is defined by the aid of menus of arrows. The programmer simply starts at the intersection of the part drawing with the centre-line and inputs each line of the drawing. For example a vertical line is input by pressing a vertical pointing arrow followed by the dimension on the keypad. The technique is simple and very fast to use. When the drawing has been input to the system, a surface finish can be specified for each surface.

The programmer has numerous aids in creating a part, for example, a calculator facility is available on the soft keys so that dimensions can be entered as a mathematical expression. Operators such as $+, -, \times, \div$, square, sin, cos, tan, etc. are available. The interactive graphics system also contains full editing facilities for inserting, deleting or changing lines. The dimensions of line segments can also be displayed at any time.

9.4.9 Tool path generation

Much of the tool path generation is automatically performed by the system software. An option is available which permits the operator to make specific choices if desired. For example the system will select certain tools from its library to perform specific machining tasks. The operator may decide to change the tool selected by the system.

The programmer sets various tools on the turrets in line with an operation sequence also chosen by the programmer. The operations may be simulated to show the interactions between the turrets. Once the operations have been selected, the executive software generates a set of default parameters that include the area to be machined, cutting data and other information specific to the type of operation.

The tool path is then simulated on the colour graphics display showing the tool path superimposed on the shapes of the blank and part. The programmer can always make changes if desired.

When all the operations have been specified, the part program is automatically generated and stored on the disk. The tool path file or even the point data can be edited at any time without having to completely regenerate all the data. For example the tool paths can be read off the disk into the DRAM of the computer and changed by the programmer using the editing facilities.

9.4.10 Tool library
The system contains data for up to 100 tools, each tool being specified by tool number and tool name type (turning, drilling, roughing, grooving etc.). Tools may be created in metric or imperial and the geometric data is input in response to a series of questions. The tool geometry is stored in detail so that the tool path generation software can create the part-program with no further input from the programmer.

9.4.11 Material library
The material data file contains space for cutting data on 20 materials. Five materials, namely ENS, EN16, aluminium, stainless steel and cast iron have recommended data stored on file, the remainder being entered as required. The cutting data includes the appropriate feeds and speed for each material. The feeds and speeds are optimised by the tool path generation software taking into account the depth of cut and power of the machine.

9.4.12 Files
The file module is for the input and output of part data and part program files. Nine utilities are provided from Paper Tape Reader, Paper Tape Punch, DNC to Disk Copy, to name a few.

The overall system provides assistance when required, via help statements, so that even the inexperienced user can program the machine tool. If the programmer inputs bad data or information the system diagnostics will put out an error message along with suitable advice.

REFERENCES
[1] *APT part-programming*. ITT Research Institute, McGraw-Hill, 1967.
[2] Leslie, W. H. P., *Numerical control user's handbook*. McGraw-Hill, 1970.
[3] Leslie, W. H. P. Numerical Control Programming Languages. PROLAMAT '69, North-Holland, 1970.

[4] *Design/Engineering/Manufacturing Automation.* Computerisation Corporation, January 1977.
[5] Wilkinson, D. G. The use of contours as an interface between CAD and CAM. Int. Conf. on computers in engineering and building design CAD 74. London, IPC Press Ltd., 1974.
[6] Besant, C. B., and Craig, D. P. The use of interactive CAD techniques and machining in mould design and manufacture. Proceedings of 18th Int. MTDR Conf., 1977.
[7] *Reference manual. MEDUSA.* CIS Ltd. 1985.
[8] *Reference manual.* Yamazaki UK Ltd, 1982.
[9] *Reference manual.* Fanuc, 1982.

10

The use of microcomputers in CAD/CAM systems

Microcomputer systems are now available in large numbers and are finding many applications in industry from management type activities to engineering analysis. They are also finding their way into many areas of the CAD/CAM field and it is therefore appropriate to discuss the microcomputer within this context. There are now many engineers who are enquiring if or how the microcomputer might be applied to his or her application. Microcomputer systems will now be considered in the context of possible use in CAD/CAM and their limitations discussed. Suggested ways of using microcomputers will be presented, and in the following chapter a CAD/CAM system based on microcomputer technology will be discussed.

10.1 MICROCOMPUTER SYSTEMS

Most engineers are familiar with the conventional microcomputer system which is used for developing programs and running them. Such systems commonly consists of an 8-bit or 16-bit microprocessor together with ROM and RAM to form the basic computer unit. Mass storage is usually in the form of dual floppy disk units and cassette magnetic tape. The input/output unit is normally a keyboard with an alphanumeric display. Hardcopy output is in the form of a low-cost serial printer. Recent systems such as the FUJITSU and APPLE offer a colour graphics display with colour graphics capability and a printer that can output graphics as well as alphanumerics all at a relatively modest price.

Some microcomputers have had limitations when applied to CAD/CAM tasks. One such limitation has been the lack of processing power in standard 8-bit MOS microprocessors for the execution of arithmetical operations required for the computer manipulation of design and manufacturing data, at acceptable speeds. A further limitation has been the lack of high-level scientific programming languages on microcomputers which has resulted in making the programmers task difficult for applications software development involving any analytical calculations and data processing. Finally, there have been two further limitations connected with the microcomputers hardware. One is the limited size of direct or primary memory addressable by a standard 8-bit microprocessor (typically 64K bytes). This has meant that large programs must be overlayed into the primary memory from a back-up memory such as a disk. In order to use this technique effectively, the rate of data transfer between the primary memory and the back-up memory must be high. This leads to a further problem in that older microcomputer systems rely on floppy disk units for back-up memory. Not only do these units have limited storage capacities compared to hard disk units found on minicomputers but they are also much slower in data access and transfer.

Many of the limitations mentioned are rapidly being overcome although resulting in a modest cost penalty in some cases. High-level scientific languages are now becoming common on many microcomputer systems. These are available on microcomputers with disk operating systems (DOS). BASIC was probably the first interpretive computer language implemented on microcomputers. The difficulty with all interpretive programming languages is that the interpreter must be resident in memory in order to run an application program. This causes a further restriction in the size of applications programs and this could be an important factor in CAD/CAM where the programs tend to be extensive. Interpretive languages have a further restriction in that they produce a slower execution speed in applications programs when compared to a compiler. However, there are now microcomputer based compilers for high level programming languages such as FORTRAN and PASCAL. The 8-bit microcomputers are unable to support full versions of these compilers owing to lack of processing power. Thus simplified versions of these compilers are used on many microcomputer systems and this shows up in the lack of program statements for more complex arithmetic functions and data manipulation.

Some manufacturers have been increasing the power of their 8-bit microcomputer systems in order to overcome some of the limitations but still keep the cost low. This has been achieved by using more modern 8-bit microprocessors such as the Motorola 68B09 which is a 2 MHz processor with some 16-bit instructions. Two of these microprocessors are incorporated into the new FUJITSU system allowing this system to support an extensive PASCAL compiler. However, the main thrust in microcomputer development has been in the 16-bit microcomputers whose processing power is in some cases comparable with that of minicomputers. These 16-bit microprocessors include in their instruction sets multiplication and division of 16-bit signed and unsigned binary numbers. This results in a far greater capability in performing arithmetic operations which is crucial in CAD/CAM applications. In fact a Motorola 68000 16-bit microprocessor can match most 16-bit minicomputers for speed in adding, subtracting, multiplying and dividing.

A further development in microcomputer technology has been the introduction of inexpensive hard disk back-up storage based on Winchester technology. These disks are capable of storing up to 10M bytes of programs and data with transfer rates of up to 8M bytes/s and an average access time of 30 ms [1]. The Winchester disk units are approximately two to three times the cost of a floppy disk but their performance is at least an order of magnitude better and therefore represents a much more cost-effective and practical disk system for CAD/CAM applications.

10.2 CAD/CAM SYSTEMS BASED ON MICROCOMPUTERS

It has been suggested in the previous section that the 16-bit microcomputers are forming the basis of new CAD/CAM systems which is almost certainly true. However, there are many instances where a powerful 8-bit microcomputer system would suffice and this would result in the minimum cost system. It is the possibility of utilising these low cost systems in CAD/CAM that is so attractive to engineers.

One mthod of utilising these systems is to create workstations which are located at the most convenient place and to give the user an adequate response. Thus each task such as design, manufacture, production control, materials scheduling etc., would have its own workstation, each one working to a common database. The workstations would all be interconnected by a network system in order to provide intercommunication between the workstations and to provide each workstation access to more computing power should it be necessary. Anderson and Jenson [2] have suggested a variety of methods for interconnecting microcomputers (Fig. 10.1). One of the most practical systems for the present application is the centralised routing network with a star-like structure. This system utilises a single microcomputer for communication purposes and controls the flow of information between the various workstations. Such a system is shown in Fig. 10.2 and a typical workstation in Fig. 10.3. This type of system has the advantage that it provides a degree of local intelligence which allows the user to prepare data and perform simple calculations with a good response time. In CAD applications this phase is often time-consuming and so the computer system during this period must provide the necessary facilities on a very cost-effective basis. Access to more computing power or large data storage files can always be achieved via the network. The number crunching facility can either be an array processor consisting of a number of microcomputers working in parallel or a powerful minicomputer. Such a computer would perform design analysis calculations such as stress analysis, heat transfer, or manufacturing analysis such as machine optimisation calculations. Parallel processing microcomputer systems are already beginning to appear in such applications and offer an extremely cost-effective solution to activities where fast processing power is required, such as is found in CAD simulation studies. The beauty of the system described is that it can be used effectively by both large and small organisations. In the case

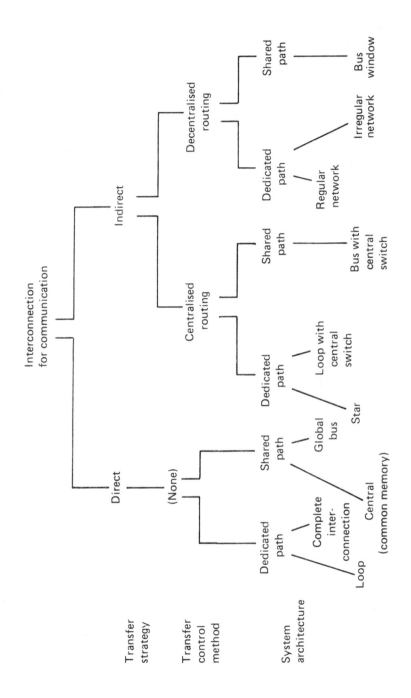

Fig. 10.1 – Multicomputer structures taxonomy (reproduced from Anderson and Jenson [2].

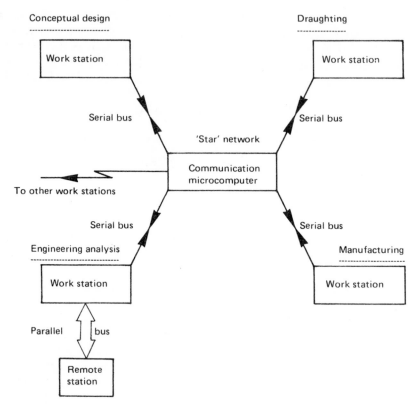

Fig. 10.2 – Overall hardware configuration for a distributed processing CAD/CAM system.

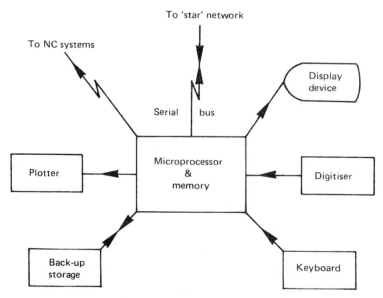

Fig. 10.3 – Hardware configuration for a typical work station.

of the very small companies, the CAD/CAM system would be a single micro-computer. The important point is that, whatever system is used, provision must be made for expansion and this expansion should be done with the minimum of software changes. In other words the software required to drive a single CAD workstation should be similar if the workstation is integrated into a network system.

10.3 HOW TO CHOOSE THE MICROCOMPUTER SYSTEM

There are many parts to a microcomputer and a large number of manufacturers from which to choose. A discussion on the availability of microcomputer tech-nology will now be presented dealing with the actual microprocessors right through to the system peripherals.

10.3.1 The microprocessor

The microprocessor is the CPU of the microcomputer and as such performs the arithmetic functions, logical operations and sequence control of programs. In deciding on a microprocessor for a particular application it is necessary to consider the processing power and the manufacturer's support for both com-plementary hardware and software. The processing power of a microprocessor is determined by its architecture. That is the number of data and address registers, stacks and the interrupt handling. Also important are the addressing modes, instruction sets and the type of semiconductor technology used in the construction of the chip. Most microprocessor manufacturers produce a range of hardware components such as ROM and RAM which are compatible with their microprocessor and with those of other manufacturers.

Currently there are nearly thirty microprocessor manufacturers producing 8-bit devices of comparable processing power. The larger manufacturers such as Motorola, Intel and Zilog offer the widest range of compatible components. It is unusual for standard 8-bit microprocessors to have Multiply and Divide in their instruction sets, the exceptions being the Motorola 6809 microprocessor. Standard 8-bit microprocessor instructions are normally executed within 1 μs, depending on the clock frequency of the microprocessor. These instructions include a comprehensive set of program control codes for conditional branch, subroutine jumps and interrupts with priority.

A workstation not requiring to perform excessive data manipulation or arithmetic functions could utilise a microcomputer with an 8-bit microprocessor. The performance of the standard 8-bit microprocessor can be, in some cases, enhanced by the addition of arithmetic LSI chips. For example, Advance Micro Devices manufacture the AM 9511 arithmetic chip which can be incorporated into a microcomputer system with a minimum requirement for external hard-ware [3]. The AM9511 APU (arithmetic processing unit) uses 16-bit and 32-bit numbers in either fixed or floating point format and performs arithmetic opera-tions including multiplication and division. Instruction execution times range from 8 μs for 16-bit fixed point operations to 6 ms for exponential operations using a 2 MHz check frequency.

The choice between 16-bit microprocessors is now extensive. One of the earliest 16-bit microprocessors was the TMS 9900 manufactured by Texas Instruments. The TMS 9900 has a comprehensive instruction set including 16-bit Multiply and Divide but suffers from the lack of hardware general registers. It should be remembered that the TMS 9900 was an early development in 16-bit technology and at the time LSI technology did not permit large size chips which could provide all the necessary facilities included in present day 16-bit microprocessors. This meant that the TMS 9900 could only accommodate an index register which could be programmed to point at any memory location. The memory location pointed at what would be the first 16 sequential 16-bit memory resident registers. This means that even for register operations there is a need to access the memory and consequently access time is longer, resulting in a reduction of program execution speed. Almost all modern 16-bit microprocessors have a general 16-bit register located on the chip. The Intel 8086 was the first standard 16-bit microprocessor, with processing power comparable to a standard minicomputers CPU. The Intel 8086 has eight standard general registers. The first four are general-purpose 16-bit arithmetic registers which are used for arithmetic or logic operations. In addition there is a set of four pointers and index registers used for program control. The Intel 8086 has a comprehensive instruction set and addressing modes including indirect addressing which is useful for implementing high-level language compilers. The Intel 8086 can address up to 1M byte of memory in 64K bytes memory segments. This is a clear advantage over standard 16-bit mini computers CPUs which can only address 128K bytes of memory without extra support in the form of memory management hardware.

The Zilog Z8001 is another 16-bit microprocessor and its addressing capability exceeds that of the Intel 8086. The Z8001 can address up to 8M bytes of memory directly in memory segments of 64K bytes. The number of hardware general-purpose registers of which eight are general-purpose. These may be used as data registers for byte (8 bits), word (16 bits) and long words (32 bits) data operations. The other seven registers are for program control as stack pointers and index registers. The 68000 also has a 24-bit program counter and can therefore address up to 16M bytes of memory directly.

From the point of view of processing power in CAD/CAM local workstations, any of the more recent generation of 16-bit microprocessors should be adequate. The microcomputers based on these microprocessors could even handle the more complex processing and calculations. Their widespread use will depend on their ability to provide high level language support in the form of good operating systems. One such operating system is UNIX which supports languages such as PASCAL, FORTRAN 77 and 'C'.

10.3.2 Primary memory

The cost of semiconductor memories are falling each year. This is mainly due to the ability of semiconductor manufacturers to pack more memory onto one chip. This is well illustrated in Fig. 10.4. Memory costs used to be a significant factor in a computer system but with the increasing size of memory on a single

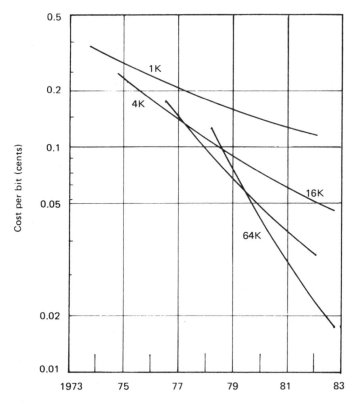

Fig. 10.4 – Reduction in cost of semiconductor memories (extract from *Microelectronics* a *Scientific American* book, September 1977).

chip and the consequent reduction in unit memory cost, memory is no longer a major cost consideration.

The type of primary memory needed in CAD/CAM workstations is mostly random access read/write memory (RAM). This type of memory is used for the temporary storage of programs and data which can be randomly accessed by the microprocessor at speeds only limited by the access time of the memory chips. This access time can be as short at 125 ns for some memory devices. In CAD/CAM applications a large and fast primary memory is desirable since programs can be seen without overlaying thus reducing the processing time for large programs. Furthermore, large amounts of data can be processed quickly which is particularly useful in many CAD areas, such as in three-dimensional data transformations, i.e. rotating a picture on a graphics display. Even though some microcomputers can access 16M bytes of primary memory directly, it is not really necessary to have such a large memory for most CAD/CAM applications.

10.3.3 Back-up storage
The most common type of mass storage found on microcomputer systems is the floppy disk. The biggest disadvantage of the floppy disk system is the maximum

storage capacity being limited to 1M bytes which is certainly nor adequate for many CAD/CAM applications. Furthermore they are limited when it comes to data transfer or access being about 100K bytes/s. Hard disks based on Winchester technology are generally thought to be the best solution to mass storage back-up although they have a more expensive drive than that for a floppy disk. However, the storage capacity can be higher than 70M bytes which is more practical for CAD/CAM work and data transfer rates of up to 8M bytes/s are possible.

The Winchester disks are not removable and are therefore not suitable for permanent storage of data files or programs. The most suitable storage medium for permanent records is the magnetic cassette of the 3M type. A type of drive unit for a 3M type cassette is the Perex.

All of the different types of input and output devices used in CAD have been discussed in Chapter 3. These CAD input and output devices were devices mainly used on minicomputer systems. Many of these devices could equally well be used on microcomputer-based systems although there have been trends in recent years to use low-cost CAD peripherals, more in keeping with the cost of micro-computer-based systems. There has also been a trend towards using micro-computers more effectively so that CAD peripherals could be simplified and therefore reduced in cost.

The principal input device for the microcomputer-based systems has been the keyboard. Graphics are input from a keyboard by utilising a range of primitives such as lines, cylinders, cubes, etc. to build up complex shapes. A primitive can be called from memory via a keyboard instruction and then given a precise dimension. The primitive is the positioned using a graphics display, using a cursor facility which is driven by a set of keys, a joystick or a mouse. This method of inputting data is more popular for CAD/CAM applications than using the digitiser or tablet because the keyboard gives high precision, of geometrical data required, directly and is a much cheaper solution to the inputting of data. Furthermore, the modern generation of young engineers, many of whom have small microcomputer systems with graphics at home, are familiar with this method of inputting information into a computer.

The main output device is the graphics display. In recent years there has been a move towards the refreshed graphics display since microcomputers are taking over the picture-refreshing task and thus reducing the cost of display. Furthermore the refreshed display does offer the advantage of local picture editing or picture movement over the storage tube. This is important in CAD/CAM applications where the simulation of machine operations, such as in cutter path derivations, is required. A typical refreshed display which could be used in CAD/CAM work is the Hewlett Packard HP 264 BA [4]. The combined alpha-numerics/graphics unit is an intelligent computer peripheral with its own microcomputer system in charge of the execution of display control functions. The unit has a single screen which is used for both alphanumeric and graphics display. It uses the raster scan deflection method for the display of both alpha-numeric and graphics. There are 720×360 pixels (dots) on the screen. The alphanumerics display has its own memory and can hold up to 6000 ASCII characters with each character occupying a 7×9 dot matrix. There is a separate graphics memory which allows access to individual pixels. The graphics controller has a hardware vector generator capable of displaying straight line vectors on the screen under remote or local control. The unit has a comprehensive set of input keys in its keyboard including eight programmable keys. This peripheral can be linked to a microcomputer system via an RS232 compatible serial asynchronous communications bus.

Hardcopy output for alphanumerics is normally performed on a serial printer. Some of the modern printers such as the Centronix can also give a hardcopy for graphics and this is a useful combined alphanumerics/graphics hardcopy unit. This type of device is comparatively low in cost and can satisfy many of the CAD/CAM graphics requirements.

The more conventional plotters described in Chapter 3 are available to the microcomputer systems to cover the applications involving the more complex graphics.

10.3.4 Communications system

Communication between computer peripherals and between computer systems is becoming a subject of ever-increasing importance. In the case of CAD/CAM workstations and of general factory monitoring systems an effective communications network is of paramount importance. There are now a number of international standards for communications bus systems. These bus systems currently fall into two categories, the serial bus and the parallel bus. The main difference between these two bus systems is that in serial transfers discrete bits, constituting characters data are sent serially while in parallel transfers these items of information are sent simultaneously.

The most common serial bus standard is the EIA-RS 232 asynchronous communications standard [5]. This standard covers the electrical specification for bit-serial transmission as well as its physical specification. The bus uses plus and minus 12 volt pulses to perform information transfers. This voltage level is easily obtained from microcomputer systems power supply units. The maximum rate of data transfer is 2×10^4 bits/s over a wire length of 30 metres. The main advantage of this standard is in its international acceptability. Most microcomputer manufacturers produce a variety of special-purpose programmable chips for communication interfaces using this standard.

The RS 232 has the disadvantage that it is slow for some applications and is susceptible to interference by electrical noise over long distances. The RS 422 standard [5] is another asynchronous serial communication standard but capable of much high data transfer rates than the RS 232 standard. Typical data transmission speeds are 1M bits/s and much better noise immunity is achieved over long distances. The RS 422 has yet to achieve the same hardware support as for the RS 232 and this is its present disadvantage.

In order to improve transmission rates some manufacturers have resorted to synchronous serial communication bus systems. The synchronous method of data communication has the advantage over the asynchronous method in that fewer items of information are sent for a given data. For example in the RS 232 asynchronous standard the transmission format contains at least two extra bits per character (data) and these bits are the start and stop bits used by the receiver for the detection of individual characters. In synchronous transmission data are sent as a continuous stream of bits with no start or stop bits. This means that with the synchronous system the receiving logic must be compatible with the transmission system in order that the data is re-synchronised and correctly decoded. As yet there is no recognised international standard for synchronous data transmission and manufacturers therefore adopt their own systems which results in some incompatibility between various manufacturers' equipment.

For data communication between local workstations in a CAD/CAM applications area, high data transfer rates are not usually required since most of the work is performed at local stations. The data that is transferred from one workstation to another is one of transferring data from one mass storage unit to another which usually means slow data transfer rates will suffice. The RS 232 standard is therefore an adequate choice for such purposes.

The parallel communication systems are much more specialised than serial systems but their use is becoming significant as multi-microcomputer systems appear in CAD/CAM systems. In such systems several microcomputers could be working in parallel and the data transfers between the microcomputers could be 1M bytes/s on an 8-bit parallel bus system. The most widely adopted parallel bus standard for communication between microcomputers is the IEEE 488 [6] system.

This standard covers the handshaking protocol and electrical specifications of a bus used for 8-bit parallel data transfers at transfer rates of up to 1M bytes/s. This system is only suitable for transmission over short distances owing to its noise sensitivity.

10.3.5 Software for CAD/CAM systems

The software for a CAD/CAM system entails the implementation of a database structure together with a number of utility programs for the creation, storage, display and management of design and manufacturing data. Also required is a set of application programs for the analytical manipulation of the data.

Also, in a multiple microcomputer system with local workstations, a set of communication programs are necessary in order to permit data transfer between the various workstations.

REFERENCES

[1] Snigier, P. System designers guide to 8-inch hard disk drives. *Digital Design* **9**, 8, 1979.

[2] Anderson, C. A. and Jenson, E. D. Computer interconnection structure: Taxonomy, characteristics and examples. *Computing Surveys* **7**, 4, 1975.

[3] Gupta, B. K. Arithmetic processor chips enhance microprocessor system performance. *Computer Design* **19**, 7, 1980.

[4] Reference Manual, Graphics Terminal 2648A. Hewlett Packard Co., 19400 Homestead Road, Cupertino, California 95014, USA.

[5] Zaks, R. and Lesea, A. *Microprocessor interfacing techniques.* Sybex Inc., 1978.
[6] Loughry, D. C. What makes a good interface? *IEEE spectrum,* November, 1974.

11

A microcomputer-based CAD/CAM system

The main purpose of this chapter is to indicate how a CAD/CAM system may be developed, based on the ideas outlined previously. Work has been in progress for a number of years at Imperial College which is aimed at providing an integrated inexpensive CAD/CAM system, suitable for both large- and small-sized firms. The work described here is based on the research of Pak [1] and Khurmi [2] into low-cost CAD/CAM systems. The objective of this work was to take a commercially available microcomputer system, with a good operating system able to support a high-level language such as PASCAL as well as BASIC and assembly languages, and integrate it with various computer peripherals to form a work station suitable for CAD/CAM tasks. Software was then written for this system to perform the following functions:

Utility programs for data creation, data management and interactive graphics.
Application programs for NC part-programming, including cutter path derivation and simulation.
Communication programs for serial data transfer between the CAD/CAM work station and a CNC machine tool.

11.1 THE HARDWARE FOR THE CAD/CAM WORK STATION

The configuration of the hardware for the CAD/CAM work station is shown in Fig. 11.1. The system consists of a Southwest Technical Product (SWTP) 6809 microcomputer system, a Hewlett-Packard HP 264A combined graphics and

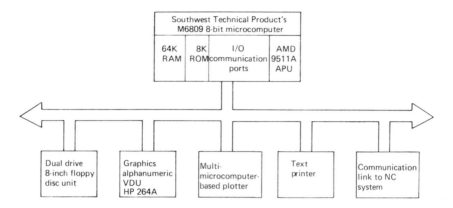

Fig. 11.1 — The basic hardware of the CAD/CAM work station.

alphanumeric display unit and an A0 size flat-bed plotter designed and constructed at Imperial College.

The microcomputer system is based on a Motorola 6809 8-bit MOS microprocessor with 40 K bytes of RAM and a dual 8-inch flexible disc storage unit with the total storage capacity of 1.2 M bytes [3].

The system has eight programmable input/output ports which can be used for serial and parallel communication with peripheral units and other computer systems.

The Hewlett-Packard HP 264A display is of the raster scan variety with 720 X 360 pixels on the screen. It is described in Chapter 10.

The plotter is an A0 size flat-bed machine designed and manufactured at Imperial College. It is a stepper motor-driven plotter with a distributed processing multiple-microcomputer controller. The plotter is capable of straight line vector generation at speeds up to 250 mm per second. This peripheral is also linked to the microcomputer CAD/CAM system via an RS232 serial asynchronous communication bus.

A character printer is also linked to the microcomputer system to provide hardcopy listings of part-programs and data.

11.2 SOFTWARE DEVELOPMENT AIDS

The SWTP 6809 microcomputer system has a basic computer operating system called Flex [4]. This operating system allows file-structured disc storage of various programs and data. It consists of the following; the file management system (FMS), the disc operating system (DOS) and the utility command set (UCS).

The programs contained in the UCS perform such tasks as the saving, loading, copying, renaming, editing, deleting and listing of disc files. These programs normally reside on the system disc and are only loaded into memory when required. The basic set of commands supplied by the manufacturer may be expanded without the necessity of modifying the entire operating system.

The DOS forms the communications link between the user and the FMS via a computer terminal. All standard commands, including those in the UCS, are accepted through DOS. Functions such as file specification, computer terminal input/output and error reporting are also handled by DOS. The FMS forms the communications link between the DOS and the actual disc controller. It performs all file allocation and removal on the disc. All file space is allocated dynamically and the space used by a file is immediately re-usable as a file is deleted. The FMS also permits any non-standard commands developed by the user to communicate directly with the disc storage unit by circumventing the DOS. This feature allows the storage and retrieval of special-purpose disc files whose data formats are not compatible to those supported by the basic Flex system. The SWTP 6809 microcomputer also supports the following programming language development systems: Motorola 6809 assembly, BASIC and PASCAL.

When the DOS and FMS have been loaded into primary memory, the DOS and FMS software only occupies 8 K bytes of RAM leaving the bottom contiguous 32 K bytes for user programs and the UCS programs. At the top of the memory map a small memory space is allocated for 1 K byte ROM resident executive programs which allows the initialisation of the Flex operating system.

11.3 THE SOFTWARE FOR THE WORK STATION

The utility software of the work station will be discussed in this section. It is made up from the data structure, the creation of data and its editing, and finally the display of graphical data.

11.3.1 The database

The two main considerations when implementing a database for a CAD/CAM system are its contents and structure. The database discussed here is intended for firms manufacturing some of their own products, including manufacture by CNC machines. Therefore the database must include a digital representation of the geometry and technology of products. The precision with which the geometric data is stored must be adequate for the manufacturing processes.

The structure of a database is defined by the way in which individual items of data are internally stored in the computer. This in turn determines the ease with which various application programs can interface with the database in order to insert, retrieve or modify data.

The way in which the database is structured in the present system bears a resemblance to that for the minicomputer-based CAD system described in Chapter 4. Once again, the main problem with data structuring is that computer memories are one dimensional in access, the memory locations are numbered sequentially and the logic of the CPU is designed to fetch data and instructions from memory in a sequential manner unless programmed to do otherwise. If records were stored sequentially, the process of inserting new records or deleting an old record would be slow, since the whole file of records would need to be processed and updated each time a change was made. For this reason, sequential organisation by itself is not suitable for a database structure for CAD applications.

Another possibility is to store records with associated names or labels in a table on a randomly organised basis. The table is then used for each record reference as shown in Fig. 11.2. The table is normally called a 'symbol table' or a 'dictionary'. The table can become large for large files and searching it can take a long time. Giloi [5] has proposed various methods for increasing the rate of record access in a random organisation, but all of these methods require some kind of computation which is at a premium in 8-bit microcomputer-based systems.

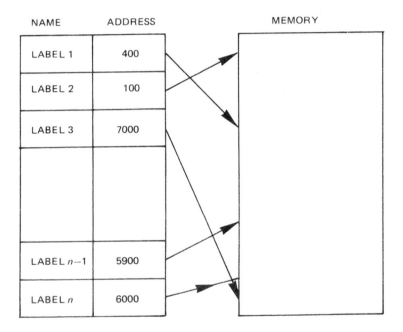

Fig. 11.2 – Table method for random data organisation.

A better method is to use a list organisation in which records are chained together by pointers. Thus records may be scattered throughout the memory but linked together by pointers which may be words containing addresses or records.

Records may be chained together in such a way as to form a 'ring'. One record is often designated the 'head record' of a ring and pointers from other records to the head record or from other rings can be useful. The ring system is easy to update. A new record can easily be created and the pointers rearranged to include the new record in the ring. To delete a record the pointers to that record are made to point to the next record beyond the one to be moved. In the case of a complete ring both the forward and reverse pointers must be changed.

More complex database structures can be created from the ring format by structuring the ring hierarchically. This type of structure (see Fig. 11.3) is often used in CAD applications. It permits access to any record from another record via the rings. The structure is easy to update or modify, although processing can become significant if the structure is complex. The pointers go horizontally around the rings in both directions and vertically between rings. If the structure becomes complex, the overhead in terms of storage can become unacceptable for a small microcomputer system.

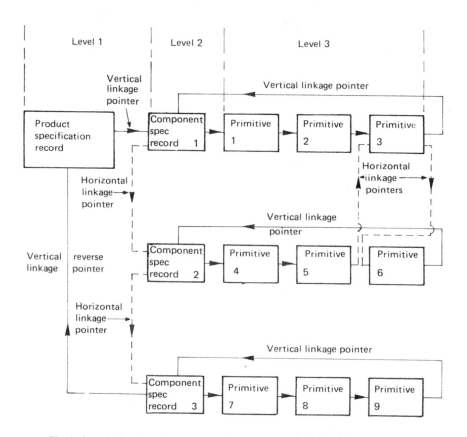

Fig. 11.3 – A three-level hierarchical structure. Note: horizontal linkage pointers are connectors in a given level, vertical linkage pointers connect records of different levels. For clarity, backward pointers and reverse pointers are omitted.

Pak [1] devised a compact data structure for small microcomputers based on the hierarchical system (see Fig. 11.4). In this structure Level 1 has a unique record called the 'product specification record'. This record is connected via a vertical ring to Level 2 of the hierarchy called the 'component level'. In the component level there is one horizontal forward ring which connects various 'component specification records'. Each component record is also connected through a vertical ring to its associated 'primitive specification record' residing in Level 3 of the hierarchy. In Level 3 there are no horizontal linkages and all records are stored sequentially.

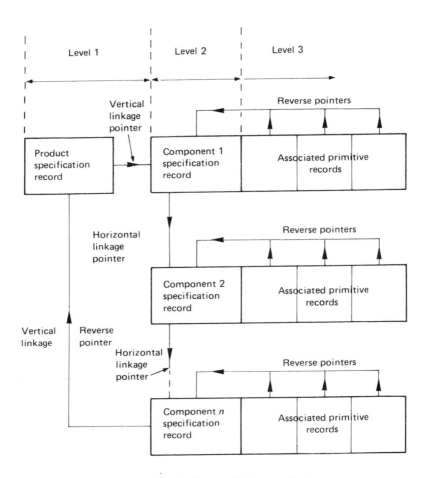

Fig. 11.4 – CAD/CAM system's database structure.

11.3.2 The creator/editor program

The creator/editor program CREATE is responsible for the interactive creation, modification and storage of components within a product in the system database. The CREATE program was written in Motorola 6809 assembly language in order to achieve the necessary speed and ease of data record decoding. Writing programs in assembly language does have the disadvantage that programs are not easily portable to different computer systems. However, the software has been structured in a modular fashion to enhance clarity and consequently to allow for easy updating to more advanced computers as they become available.

The database, which is fundamental to the CREATE program, is described in the previous section. It is made up of three main elements, namely, product specification, component specification and primitives. The product specification contains information pertinent to the managerial aspects associated with the product such as the material quantities required, contract number, name of customer and delivery dates. The product is then subdivided into several components, each representing information required to manufacture it from a single machining process to a complicated sequence of processes. Finally, the components are subdivided into primitives which are a collection of geometrical attributes such as lines, cylinders or taps which define the geometry of the component. Thus, by using the standard primitives or by defining new primitives where necessary, geometries for turned or milled parts can be quickly and easily entered into the computer.

The CREATE or EDIT commands are entered via the alpha numeric keyboard. The record input repertoire is simple but effective. Product or component input is performed by inputting the PRODUCT name or COMPONENT name and material code. These are then stored in the appropriate data fields, together with the correct linkage pointers. There are a wide range of primitives from LINES, CIRCLES, FILLETS, CYLINDERS to THREADS and TAPS. Two examples or primitives are as follows:

TAPERED CYLINDER

Format: TCYLINDER attributes 1, . . . , attribute 4.

Description: A Primitive Specification Record for a tapered cylinder is stored in the workspace. The primitive is then displayed.

Attribute 1 — orthogonal plane perpendicular to the longitudinal axis of the cylinder.
Attribute 2 — initial cylinder diameter.
Attribute 3 — cylinder length.
Attribute 4 — final cylinder diameter.

Example: TCY, ZY, 1000, 2000, 1500 (Fig. 11.5).

THREAD

Format: THREAD attribute 1, . . . , attribute 5.

Description: A Primitive Specification Record for a threaded cylinder is stored in the workspace. The primitive is then displayed.
Attribute 1 — orthogonal plane perpendicular to the longitudinal axis of the cylinder.
Attribute 2 — outer cylinder diameter.
Attribute 3 — cylinder length.
Attribute 4 — inner cylinder diameter.
Attribute 5 — pitch.

Example: THR, ZY, 2000, 3000, 1800, 200 (Fig. 11.6).

The CREATE software also contains a command repertoire for display control, editing and filing. These commands bear a resemblance to those used in the minicomputer-based system described in Chapter 4. These commands include SCALE, STATUS, VIEW, LINE TYPE, DIMENSION, SAVE, LOAD, DELETE, LIST, EDIT etc.

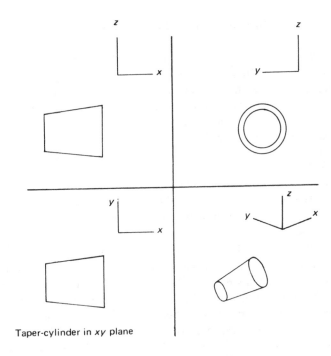

Taper-cylinder in *xy* plane

Fig. 11.5 – Tapered cylinder primitive.

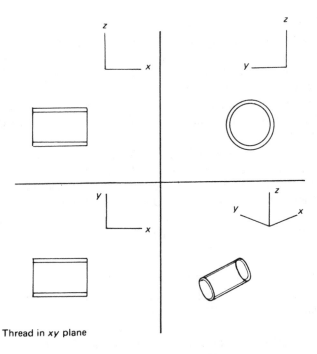

Thread in *xy* plane

Fig. 11.6 – Thread primitive.

Two examples of commands are as follows:

VIEW
Format: VIEW plane.
Description: View command allows the display of a single orthogonal view of
                all the components or a particular component stored in the
                workspace.
Example: VIEW, XY.

SAVE
Format: SAVE file name.
Description: The current product in the workspace is transferred to the disc
                storage unit under the file name specified.

Special note should be made of the EDIT module as it contains a number of submodules which are fundamantal to the whole system. The EDIT software allows components to be recalled and modified interactively. Primitives may be analysed graphically as well as numerically in a similar manner to a text editor. By sequentially stepping through the primitives which define a component geometrical errors can quickly be and easily detected. Once an error has been detected, two powerful commands DELETE and INSERT allow entire primitives to be deleted and replaced.

11.3.3 The display programs

The DISPLAY package is an interactive, modular graphics system responsible for the comprehensive display of components formulated using the CREATE package. DISPLAY has been developed in the high-level language PASCAL. PASCAL allows formulation of algorithms and data in a form which clearly exhibits their natural structure, thus resulting in a suitable and economical data representation which is a desirable feature in CAD/CAM software programming.

DISPLAY allows interactive display and draughting of an entire product or its individual components in first-angle orthographic projection. The various software modules which make up the DISPLAY package are shown in Fig. 11.7. Display features such as ISOMETRIC VIEWS, CLIPPING, WINDOWING, ROTATION, DIMENSIONING, LABELLING, SCALING and SECTIONING form the basis of the DISPLAY package.

11.3.4 Cutter path derivation and simulation

There are two distinct phases that occur before producing the final commands which are given to a CNC machine tool. They are:

— The processor which takes information from the system data base on the
  geometry and technology of a component to be machined and produces

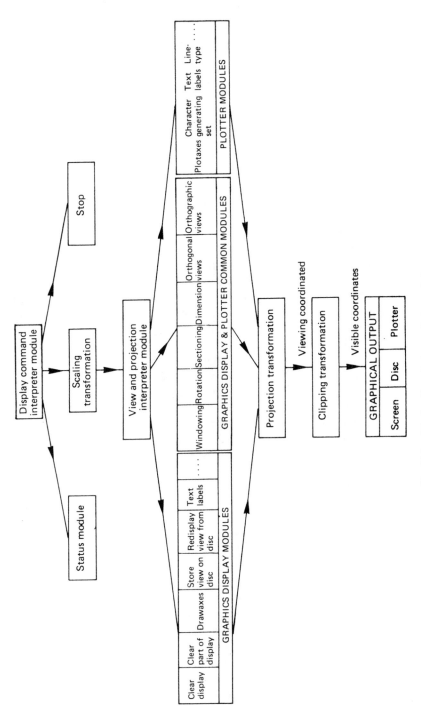

Fig. 11.7 – Display/draughting package modules.

the cutter location data (CLDATA). The CLDATA is independent of the actual machine tool control system.
— The post-processor which adapts the general data produced by the processor and converts it into the specific format required by the machine tool control system.

The input to the processor phase is the database of the system which contains all the specifications required to completely define the geometry of a part. Records, in the form of primitives, are obtained from the database and recreated in graphical form. The CAD/CAM system also needs information on the characteristics of the machine tool to be used, the geometry of the blank workpiece and its material and geometry of the cutting tools before it can produce the CLDATA.

The characteristics of a machine tool can be separated into the mechanical limitations of the machine tool and limitations of the CNC system. Mechanical constraints may include axes strokes, maximum power available at spindle(s) and feed drives, maximum feed rate and spindle speed, automatic tool changing or turret indexing mechanisms and incremental resolution of feed drives. Constraints on the CNC system may include type and speed of continuous contour interpolation, automatic tool changing logic and constant cutting speed algorithm.

The CAD/CAM system has been developed for both turning and milling applications. As much information as possible is stored in library form to permit as much automation as possible in the production of the CLDATA. For example, the user has access to a tooling library for the selection of tools for various cutting operations.

Machining parameters such as spindle speeds and feed rates are determined automatically on the basis of the material that is being machined and the machining operation required. A simulation of the cutting process is displayed on the screen as the cutter path derivation is being performed. The output of the processor phase is a CLDATA file which contains information such as the number of rough cuts, number of fine cuts, tool changes, threading parameters, spindle speeds and feed rates.

The post-processor has the function of generating a part-program compatible with the CNC machine tool. The program has an additional task of direct transmission of part-programs to the CNC system via a RS232 serial communication bus. The CNC system used for the present work has a resident high-level programming language called the Machine Tool Interpreter Language (MTIL) and resembling the BASIC programming language. The presence of a high-level programming language with the CNC system allows the construction of a collection of generalised instructions, or subroutines, within the system. The MTIL also permits easy debugging of programs. With this arrangement only the data required to activate these subroutines have to be generated prior to program execution. Thus the post-processing has been simplified to the conversion of CLDATA records from integer binary format to a format accepted by MTIL.

11.4 THE MACHINE TOOL CONTROL SYSTEM

Modern machine tools incorporate a control system which is based on either a minicomputer or a microcomputer. Most controllers feature a single central processor which is linked to a variety of peripherals as shown in Fig. 11.8. Such systems tend to contain complex software which is costly to develop. Furthermore, changes in the operating system, to take account of a new variable for example, cannot easily be made.

New control systems have recently been developed which utilise a distributed processing technique. This allows multi-axis variable control systems to be constructed without the cost overheads associated with a single minicomputer CNC system. These new systems are based on multiple microcomputers and their basic architecture is a hierarchical one of a master and slaves configuration. Employing such an architecture maintains system modularity, flexibility and expandability. The concept of the general control system is that it can be customised to the specification of a wide range of machine tools.

There are a number of variations of the distributed processor technique for machine tool control systems. One technique is to allocate hardware with the appropriate software to each control task such as interpolation, feed rate, loop closure and machine I/O. This approach was adopted by Stute [6] at Stuggart University and was later used by Gildemeister on some of their machine tools. A block diagram of a typical implementation of this system is shown in Fig. 11.9. Another approach to distributed processing was adopted by Dalzell [7] at Imperial College, and this concept is shown in block diagram form in Fig. 11.10. This system has three levels in its hierarchical structure, Level 1 being used for controlling the variable such as d.c. or stepper motors drives, Level 2 for the definition of variables such as feed rate, threading, axis or spindle and Level 3 is the delegator or master unit. All microcomputer units performing each function are connected to an IEEE-488 bus. This system utilises a feed rate controller which is common to all the variable functions such as axis controllers. Such a concept overcomes any problems of synchronisation between the various axes and spindle. Furthermore, the addition of a machine variable such as an additional axis has only minimal impact on other elements of the system. The majority of the additional computing power required is added with the new variable controller, except for a small addition to the software in the delegator.

The system developed at Imperial College has been applied to two axis turning machines using stepper motor axis drives and to a three axis mill with d.c. motor axis drives. The system is also being used on a variety of machining centres using d.c. motor axis drives.

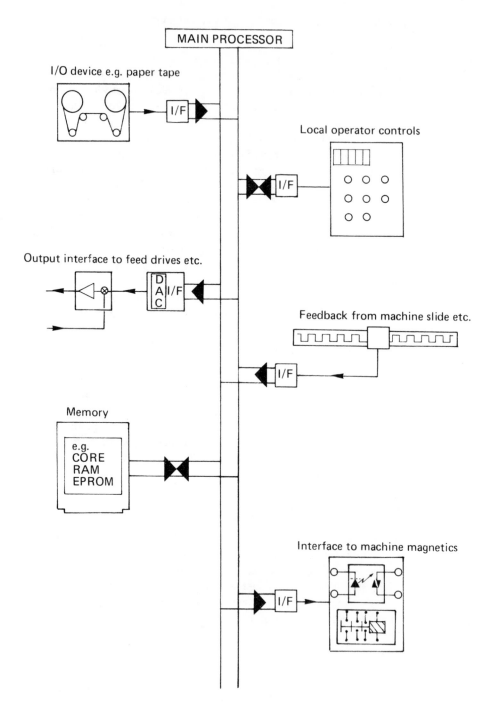

Fig. 11.8 – Single central processor CNC block diagram.

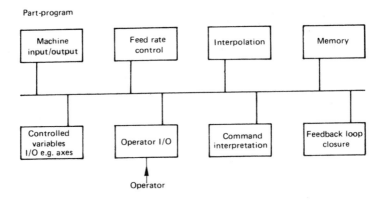

Fig. 11.9 – Multi-processor control task-orientated CNC system block diagram.

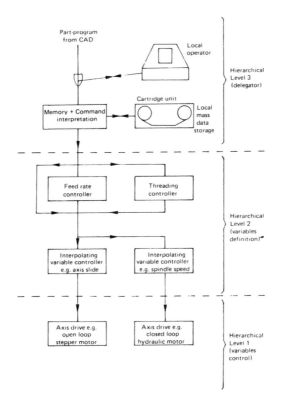

Fig. 11.10 – Practical machine function-orientated CNC system block diagram.

11.5 USING THE CAD/CAM SYSTEM

The CAD/CAM system described has been used with the distributed multi-microcomputer CNC system, both on turning and milling applications. The overall software system integration is shown in communication diagram format in Fig. 11.11.

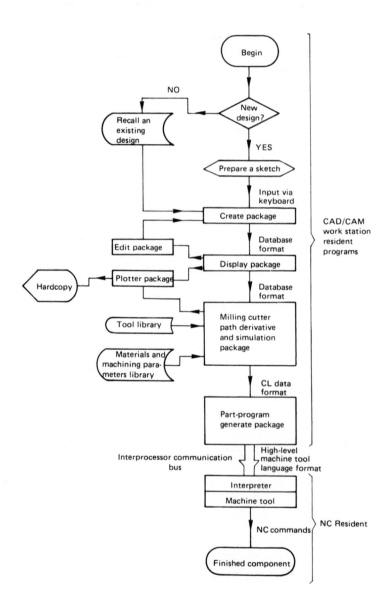

Fig. 11.11 – Macro interpackage communication diagram.

Consider a simple example of pocket milling of a shape shown in Fig. 11.12 which was entered into the computer using the CREATE software. The procedure for determining the various cutter paths and generating the required CLDATA is shown in flow chart form in Fig. 11.13. This procedure is easy to follow since the user is guided through the operation via a series of questions displayed on

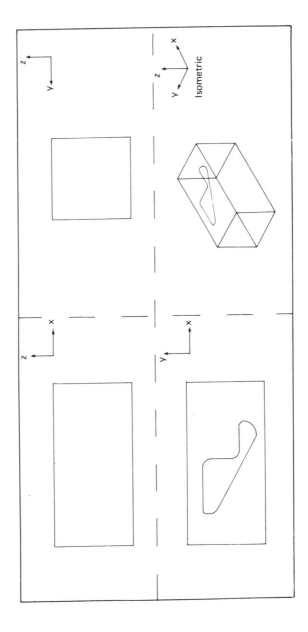

Fig. 11.12 – A typical view of a component displayed in four views.

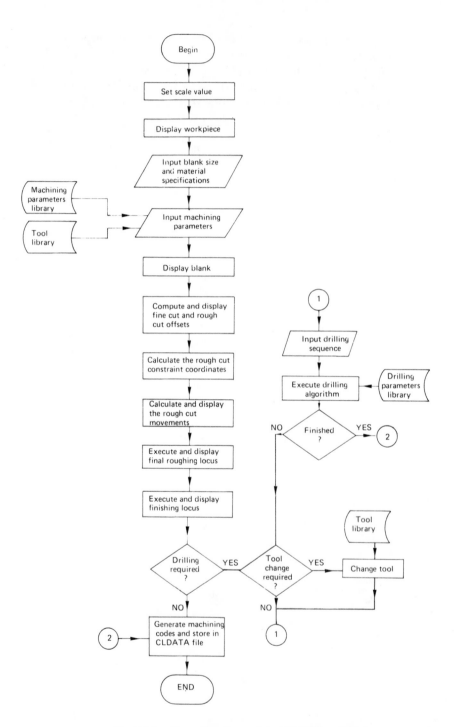

Fig. 11.13 – Macro flow chart of the MCPDRS package.

the screen. This procedure consists of first checking the geometric data entered by looking at the various views shown in Fig. 11.12. The blank size is then entered while tools and machining parameters are selected from tooling and machining data libraries. The libraries contain commonly used tools (cutters), materials and their corresponding machining speeds and feeds. From this information the blank, rough cut and fine cut offsets are displayed as shown in Fig. 11.14. When extensive machining is required the rough cut locus is displayed

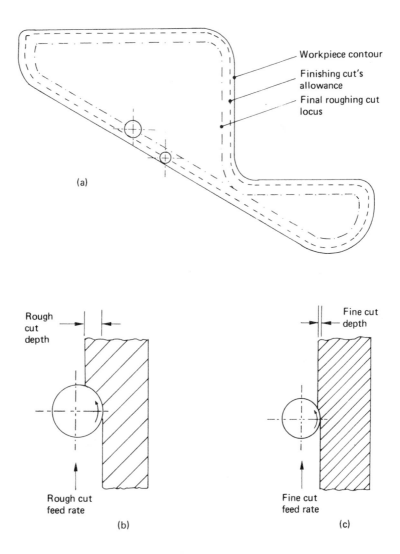

Fig. 11.14 – Visualisation of the workpiece's geometry and contouring rough cut and fine cut loci. (a) Removal of material during (b) rough cut and (c) fine cut processes.

along with the final rough cut and fine cut. A typical example is shown in Fig. 11.15.

The tool changes, spindle speeds, feed rates and cutter paths are stored in an integer binary format in a cutter location data file (CLDATA) for subsequent

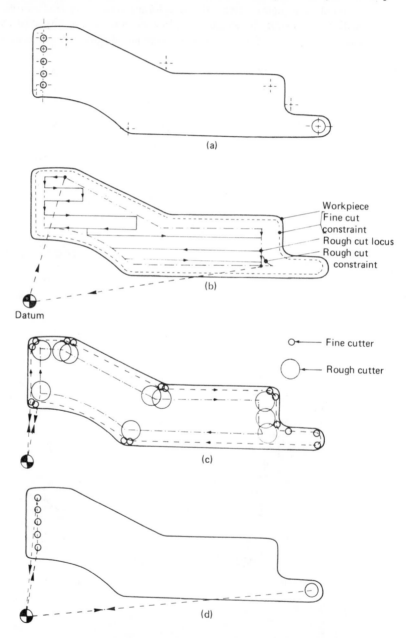

Fig. 11.15 – (a) Geometry of the workpiece. (b) Fine cut and rough cut constraints and rough cut locus. (c) Final rough cut and fine cut loci. (d) Drilling operation locus.

access by the part-program generation program. This program takes the CLDATA, converts it from binary to ASCII format and transmits it, via a standard RS232 serial communication bus, to the CNC system.

The overall approach of this type of CAD/CAM system is to bring this technology to a much wider range of companies. Its easy use by unspecialised labour resulting in fast machine tool programming times makes the technology very attractive to industry. Some machine tools incorporate a graphics input station containing cutter path derivation as part of the control system. Typical systems offered for turning applications are the FANUC 3T and MAZATROL systems. It is probable that in the longer term that such input devices will not stay on machine tools but be away from the machine tools, in either the drawing or planning office, since, as automation gathers momentum with the introduction of powerful flexible manufacturing systems (FMS), the shopfloor will not be a suitable place for personnel.

REFERENCES

[1] Pak, H. A. *Microprocessors in computer aided design and computer aided manufacture.* Ph.D. thesis, University of London, 1981.

[2] Khurmi, S. K. *Computer aided design and manufacture and the microprocessor.* Ph.D. thesis, University of London, 1982.

[3] *System Note Book – 6800 Computer System.* Southwest Technical Product Company, 12, Tresham Road, Peterborough, England.

[4] *Reference Manual Flex Operating System.* Technical Systems Consultants Inc., PO Box 2574, West Lafayette, Indiana 47906, USA.

[5] Giloi, W. K. *Interactive Computer Graphics.* Prentice-Hall, 1978.

[6] Shute, G. & Wom, H. A modular function oriented multiprocessor NC system. *Annals of CIRP*, **27**, 1978.

[7] Dalzell, D. T. *Intelligent machine tools and the microprocessor.* Ph.D. thesis, University of London, 1981.

12

The application of CAD/CAM techniques to the design and manufacture of complex shapes

12.1 INTRODUCTION

The importance of computer-aided design and computer-aided manufacture to the engineering industry has been discussed in previous chapters. Also, the importance of bringing CAD systems together with CAM systems into an integrated CAD/CAM system has been emphasised so as to exploit the data created during the design phase, in the manufacturing phase by using a common database.

In the previous chapter the application of microcomputers in CAD/CAM systems was discussed. It must be recognised, however, that most 8-bit microcomputers have a limited application in CAD/CAM, particularly if complex three-dimensional shapes are to be described and manufactured. Modern 16-bit microcomputers with floating-point hardware or the latest 32-bit microcomputers are much more suitable for the complex tasks, owing to their more powerful performance characteristics. Currently CAD/CAM systems that can deal with the most complex tasks such as CAM-X [1] utilise powerful minicomputers such as the Digital Equipment Corporation VAX range of computers. It must be stressed that the boundary between the microcomputer and minicomputer is becoming blurred as VLSI technology increases.

At present integrated CAD/CAM systems are mainly used in problems involving surface definition, particularly in the aircraft industry. These problems are often so complex and produce difficulties in handling very large quantities of data for both design and manufacture that computerised methods are essential.

The increasing use of computer graphics facilities and the availability of supporting software has given designers the tools they need to produce computerised geometrics and several successful packages have been developed using interactive graphics. One of these is GPP [2] (Graphic Post-Programmer) – used extensively by the British Aircraft Corporation's Military Aircraft Division. Similar interactive graphics software tailored to the needs of the motor industry have been developed by Leyland at Cowley. IBM have also marketed the FMILL and APTLOFT systems which are non-interactive surface-machining packages.

Less sophisticated industries than the aircraft industries are now using NC and CNC technology and are rapidly becoming interested in CAD/CAM systems. Many research institutes have investigated methods for integrating CAD and CAM by linking the geometry-descriptive part of CAD with the manufacturing technology of CAM and CNC.

We have already seen that when an item has to be manufactured the first decision to be made is whether to use numerical control technology or more conventional methods. If the object is to be produced in mid-volume quantities (> 100) or is complex in geometry involving the machining of curved surfaces, then numerical control methods are applicable.

In the case of NC methods a part-programmer would produce a part-program manuscript from an engineering drawing. This involves the coding of the geometry of the component, a repetition of work carried out during the design process, followed by the coding of statements to describe the tool motions required to carry out machining. However, if the task involves complex three-dimensional curves, then it would be necessary to resort to the use of APT in order to describe the geometry. Again if APT is used coding of the geometry is a repeat of the design stage even though it has been simplified by the characteristics of the APT statements.

The more modern CAD systems permit a close integration between design and manufacture. They allow design information to be passed directly onto manufacturing programs such as APT. There are now a number of such commercial systems available, such as turnkey systems by Ferranti Infographics (CAM-X) or Computervision. The integrated CAD/CAM system developed at Imperial College by Craig [3] and Olama [4] will be described in order to illustrate the essential features of such systems.

We have already stated that the similarity of output from the design process defining the geometry of the design and the input for the manufacturing process is a prerequisite for an integrated CAD/CAM system. This means that the design and manufacturing software can be completely separate but share a common database. The design process is made as interactive as possible in order to give the designer as much flexibility as possible, but once the design is frozen, the software for the manufacturing process will take the design data, with the minimum of man-interaction, and will produce a part program for submission to the APT system. Alternatively, for some machine tools the system will produce G and M codes directly without recourse to APT, thus saving much computing time.

The CAD/CAM system now to be described was developed for components requiring either turning, drilling or milling operations. It is capable of dealing with complex tasks such as moulds for casting patterns for gas turbine blades or centrifugal water-pump impeller blades.

12.2 THE CAD/CAM SYSTEM

The software has been designed to run on a variety of computer systems ranging from a DEC-PDP-11 or VAX to a wide range of UNIX-based machines. Typical hardware configurations consist of a DEC-PDP-11/45 with 128K bytes of core or RAM, two 10M bytes cartridge disks and a 9-track magnetic tape unit. The CAD workstation consists of a Tektronix 4014 storage tube display, an alphanumeric visual display unit and a Talos solid state digitiser. Alternative systems making use of powerful microcomputers consist of a Cambridge Microcomputer Company (CMC) system with a Motorola 68000 based processor, floating-point hardware, 1M byte of RAM, 80M byte Winchester disk and a 9-track magnetic tape unit. The workstation consists of a high resolution (1024 × 1024) refreshed graphics colour monitor driven by a Matrox graphics engine giving bit-mapped graphics, with 4-bit planes and 16 colours. The workstation also contains an alphanumeric display and a tablet. There are now a number of UNIX-based workstations suitable for this application such as the Tektronix 6000 series shown in Fig. 12.1.

Fig. 12.1 – Tektronix 6205 intelligent graphics workstation.

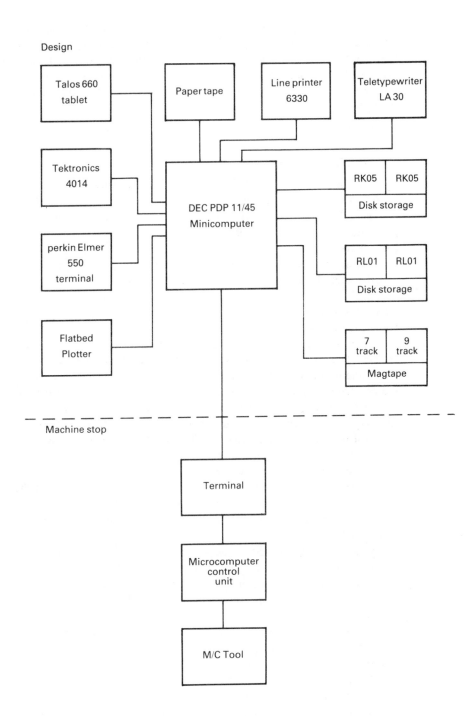

Fig. 12.2 — The CAD/CAM system.

Milling of complex moulds for turbine blade applications was performed on a three-axis CNC Bridgeport milling machine containing an R D Projects controller which was connected on-line to both the DEC and CMC based CAD systems. Data could also be taken from the CAD systems and input to APT processors on a variety of mainframes. A typical overall CAD/CAM layout is shown in Fig. 12.2.

12.3 SYSTEM SOFTWARE

The system software is based on the general graphics system described in Chapter 4 and the manufacturing programs are called from the USER section of the menu. The interface between design and manufacture is a file known as the workspace file. It is based on fixed length records, each record consists of coordinates (x, y, z) and five numbers which qualify the coordinate. These numbers are packed into one integer word by the computer and where necessary a further record may be used to describe a coordinate.

12.3.1 CAD software

While the CAD software has already been discussed in previous chapters, certain aspects of it which relate to the machine tool application will now be amplified. We are concerned with solids which may be represented graphically in a number of different ways. Orthogonal projects are the well-established method of drawing from its projection in at least two views which are plane areas although more views may make the visualisation simpler. The basic means of input from the system is a relatively simple matter in order to obtain three-dimensional coordinates from such drawings [5]. The input facilities of the system obviate the need for these drawings to be up to standard engineering drawing office practice since rough sketches will suffice.

We have seen previously how the two-dimensional surface of the digitiser is divided into areas corresponding to an X–Y, Y–Z or Z–X plane in order to digitise in three-dimensions. It is possible to use more than three planes but usually only three are needed.

It is necessary to digitise the same coordinate in two planes in order to obtain a three-dimensional coordinate. However, the trailing origin facility obviates the need to do this in many cases since if a series of points lie on one plane, say at a particular z level, then the x, y coordinate may be digitised until a new z coordinate is necessary. This approach makes the digitising of cross-section and contour data a very simple process. All the standard draughting facilities such as FIND, DRIVE and CONTROL 90 give the user the power necessary to handle 3D digitising.

In many cases, it is useful to digitise with respect to a reference grid where the grid may be used as a reference skeleton of the object to be defined. A three-dimensional grid is set up in the following way. A grid origin is defined by digitising a point in a two-dimensional view and then grid increments are set up independently in any of the orthogonal directions. For example, the cursor may be constrained to move on a particular grid in the X and Y directions whereas its movement could be left entirely arbitrary in the Z direction. The grid points can be displayed on the storage tube and ten grid files are available for storage and retrieval of commonly used grid configurations.

Points are entered off the grid using DRIVE mode and it is also possible to disable the grid temprarily to input some arbitrary points and then to enable it again.

When the user wishes to define a component which is clearly two and a half-dimensional, then all that is necessary is to digitise one of the required plane faces of the component and enter DOUBLING mode. The required plane is indicated to the program and the thickness is defined. The program automatically generates the data for the solid and stores it in the workspace file.

Components which have one or more axes of symmetry may be generated using the mirroring facilities of the system which allow data to be mirrored about any plane defined by the user.

Some of the more difficult work in manufacture is the machining of curved surfaces. Consequently the specification of curved surfaces in CAD is important. Curves on a surface may be defined in two ways. The first method produces an approximation to a curved surface and is performed by using a method called continuous digitising. Here the user traces the outline of the curve with the digitising cursor and data points are stored automatically by the computer at short intervals along the curve. This is a speedy way of producing contour and cross-section data but it is only as accurate as the user can digitise. In the situation where data points are known at sparse intervals along a required curve, the second method should be used. Here, the set of points on the curve are defined by the user, either by digitising or using DRIVE mode and a cubic spline is passed through the points using CURVEFIT. This technique ensures that a curve is fitted to the specified points such that continuity of curvature and slope is maintained at all spans [6].

There are surfaces which do not easily lend themselves to definition in mathematical terms, such as moulds for castings. The worst cases could involve sculpted surfaces of an artistic nature. These surfaces tend to be described by a series of sections together with contours where necessary. A set of program modules was developed to cope with such surfaces and will produce from the input data, a surface which is smooth and contains all the data points. The technique is based on the use of the facilities already described. The method of obtaining a surface involves the specification of a number of sections from some given data.

There are three methods available for specifying sections. In the first method the user is asked to specify two points on a line. This line is then used to intersect with the data in the X—Y plane of the drawing and return the x, y and z coordinates of the intersection points in an ordered form.

In the second case two points on a line are defined and a third point is digitised to indicate the bounds on a sweep. After this the program finds the intersection for a series of lines parallel to the one digitised within the bounds of the sweep. In the third case a centre may be defined and a number of lines at varying angles are passed through this centre and the intersections are found and stored as before.

When the user feels that a sufficient number of cross-sections have been defined, the spline blending modules are then initiated. In this routine the lines as defined previously are tested for intersection between each other in the X−Y plane. When intersections are found between two lines a cubic spline is generated by the set of points on each line and the Z ordinate at the point of intersection is found for each line. These ordinates are averaged and are taken as being points on the two lines and are added to the data describing those sections. Unsatisfactory points are eliminated at this stage. When this process has finished the output is the raw data plus a set of cubic splines which intersect in three dimensions with the raw data and themselves. The next stage is to generate a set of splines in the X and Y directions which contain the raw data and the generated splines. The final output is the raw data (as input) plus a family of splines which can readily be used to machine the surface.

Examples of a surface defined by raw data, raw data with blended splines, raw data with a family of orthogonal splines and tool movements are shown in Figs. 12.3 to 12.7 respectively.

Finally, it is possible to edit mistakes in the data. Annotated drawings may be produced and filing facilities are available for archival storage of completed designs.

12.3.2 CAM software

The CAM program modules are entered after the definitions and acceptance of the component designed by the user. In the design section it was deemed essential that the user should be in complete control of the machine and input was made as flexible as possible; the approach in the manufacturing area is somewhat different. In the latter case the NC modules play the leading part and operate on the workspace data. Interaction between the user and machine is still necessary but is minimised since the user mainly supplies information to the modules following a request from them.

The system is capable of generating manufacturing data for NC, or CNC turning, drilling or milling operations. The system can generate APT type output ready for an APT post-processor or output in G-code format ready for a CNC controller.

Entry or exit to or from the CAM programs respectively is made via a menu instruction ENTER/EXIT MANUFACTURE MODE. On entry to manufacturing mode there are two options open to the user, one being to use existing data on a part already in the process of being created. This information may include materials, cutting data, start point and cutters. The second option is reserved for the creation of new part manufacturing procedures.

If a new part is to be manufactured then a large number of decisions have to be made, such as the selection of a machine tool type, cutter definitions, cutting data, surface definition etc. The system has sufficient expertise built into it to assist an inexperienced user to make such decisions. These points are now discussed in some detail.

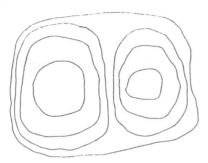

Fig. 12.3 – The original contour data describing the required surface.

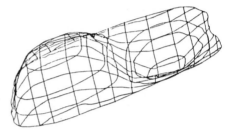

Fig. 12.6 – Here the data from the previous figure has been used to generate the family of orthogonal splines which are used to define the surface.

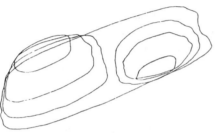

Fig. 12.4 – A perspective view of the above data.

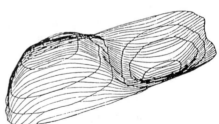

Fig. 12.7 – This is a plot of the computer-simulated tool movement over the surface which has been calculated by the CAM program using the design data from Fig. 12.5.

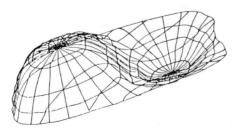

Fig. 12.5 – The set of blended splines defined by the user to describe the surface.

(a) Machine tool selection
The user must select a machine tool, from the existing library, on which the component is to be manufactured. If a suitable machine is not available in the database then an option can be selected for defining the specification of a new or suitable machine tool. Once the machine tool type has been selected only menu instructions relevant to that type are available to the user. For example, if a turning machine is selected then milling instructions are not available for use.

On selecting a machine tool, information such as number of spindles, number of turrets, number of stations per turret, indexing axes, direction of indexing, drilling facilities, maximum slide travel, machine power and spindle speeds are taken from the database and stored in the manufacturing file ready for use. The user also has to define the initial material shape and size, i.e. bar or billet and size.

(b) Material of workpiece
The selection of material can be defined in one of three ways. The first option is to select the material type, its metallurgical structure and heat treatment from the material properties library. The second option is to select the material from standard material tables stored in the database. These are common materials specified in British, German or American standards. The third option is to specify the material properties in detail. When the material is specified recommendations are displayed on the screen to help the user in selecting cutting data and tool geometry.

(c) Cutter definitions
When a cutter has been selected, it is necessary to supply information on cutter geometry and size. Where there are turrets on the machine tool, the user must specify the turret station for the tool together with an offset number.

(d) Cutting data
There are three methods of defining cutting data. The first is by the selection of the cutting operation such as heavy roughing, roughing, light roughing or finishing. The second is by selecting the surface finish. In this case the degree of finish is combined with the geometry of the cutter and the system selects the cutting data using this combination. The third method is based on the depth of cut and feedrate chosen by the user.

(e) Power determination
When a machining operation is specified, the power required for the machining process is calculated. This calculation is based on the combined information derived from the material selected, the cutter geometry, the cutting data and the machine tool specification. The material properties combined with the cutting data gives the cutting speed. The spindle speed is then determined from a knowledge of the cutting speed and machine tool specification. Knowledge of the tool geometry and the cutting data leads to a calculation of the average

cutting chip thickness and with information of the material the specific cutting force can be obtained.

The cutting power can be determined from cutting forces and speeds and compared with the power available of the machine. If the cutting power exceeds the available power, the user is informed and asked to respecify the initial data.

Once the machine specification and the component have been defined, the data required for the manufacture of the component can be selected from the manufacturing file. The component size is compared with the maximum component size that can be machined on the selected machine tool. If the size is greater than the permitted size an error message is displayed with an instruction to choose another machine tool.

(f) Milling

Milling and turning are quite different operations in manufacture and the CAM software for these operations is described separately when appropriate.

(g) Surface definition

Data are selected from the workspace file for defining surfaces which are to be machined. These data are stored in a surface file and checked for surface identification. There are four types of surfaces that can be identified and coded. These codes, together with any additional data required to complete the surface definition, are stored in a file ready for interpretation by the machining software. These surface types are as follows:

(1) Surface with a constant z: the surface code is 1 and the z height is defined.
(2) Surface defined as a plane: the surface code is 2 and four extra records are used to define the direction cosines of the perpendicular to the plane together with three points on the plane.
(3) An extruded surface: the surface code is 3 and extra records are required to define the surface dominance and the extension bounds.
(4) Special surfaces: the surface code is 4 and these refer to sculpted or true 3D surfaces.

(h) Boundary definition

The boundary of the surface to be machined is defined so that the machining limits can be calculated. The boundary can be defined in two ways. The first is to select lines and curves from the surface file to define the boundary. The system then determines how these lines and curves are joined relative to each other to form the boundary. The second method depends on the selection of points on the surface to define the boundary.

The boundary information is then stored in the boundary file and the boundary displayed together with a message on whether it is closed or open.

(i) Machining limits

Once the boundary and cutter geometry have been defined, the machining limits of the surface can be calculated. If the cutter or boundary are not properly defined the system will issue an appropriate error message. If the cutter and boundary are properly defined then a window is set around the $x-y$ projection of the boundary data and displayed to the screen size for checking. At each point on the boundary the user is asked to select the cutter position relative to the vector on the boundary at that point. The cutter position may be selected at any point and the system allowed to repeat this selection for the remaining points. The cutter has nine positions, relative to a vector at any point and these are shown in Fig. 12.8. The cutter position is selected by indexing through the nine positions until the appropriate position is found.

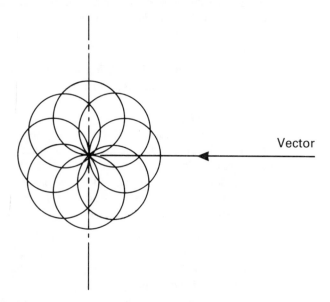

Fig. 12.8 — The nine possible positions at any point relative to the boundary vector.

A cutter position file is created and then checked for cutter interference. If there is cutter interference and it can be corrected, the system will modify the file to avoid this interference. If the system is unable to correct for the interference, the user is informed.

The machining limits are finally filed together with the first record containing the definition of the cutter used to generate the machining limit data. The surface together with the machining limit is displayed for checking. On completion of the machining limit definition, this information is stored in the boundary file and will form the constraints on the cutter movements when the surface is machined.

(j) Area clearance on surface
Before any attempt is made to generate the cutting sequence for machining a surface, the system first checks that the surface is properly defined, along with the cutter. If either the surface or cutter is not defined then the user is informed.

If both are correctly defined, then the system searches for the surface boundaries for machining. The user is then asked to select internal or external boundaries. The system then examines the cutter that has already been selected to check its suitability for machining the chosen surface. The user is then asked to select the cutter spacing and dominance (direction) of machining.

From a knowledge of the surface type, cutter geometry, cutter spacing and dominance of machining, the intersections with external and internal boundaries can be determined. This data is then checked for errors and erroneous data is removed.

The data generated up to this stage is 2D in nature. The cutter offsets are then calculated along the cutter path to generate the true 3D tool path. All this information is stored in an intermediate file.

If the component to be machined is initially produced by casting, then the data generated so far is enough to remove the excess metal, i.e. a finishing operation. However, if the component is to be machined from a solid billet then further data may be required such as roughing cuts.

(k) Roughing cuts

The intermediate file status is checked to ensure that it contains three-dimensional data. The user is then asked to define the height of the billet. If the specified height is less than the height of the component an error message is displayed. If the specified height is valid then the user is asked to define the depth of cut. The intermediate file is examined and cutter path extracted. The cutter location is then determined for each cutter pass. The z-height of each pass is reduced by the depth of cut. If the z-height reached is the minimum z-value of the surface, the operation is stopped and the final pass is added separately to the intermediate file. The cutter location data may be displayed and if the user is satisfied it is transferred to the manufacturing data file.

This procedure is satisfactory for generating the cutter location data for plane surfaces and 2½D components. For true 3D surfaces, the problem becomes more complex owing to the changing nature of the surface. A different method is used to determine the required cutter offsets.

A graphical illustration of the various stages involved in producing manufacturing data for 2½D components is shown in Fig. 12.9.

(l) Machining of 3D shapes

The procedure for machining complex surfaces, such as a mould for casting a wax pattern of centrifugal water pump vane, is as follows. First a surface is selected, the external boundary defined and the machining limits calculated as described previously. The cutter spacing and machining dominance are then defined. The surface is then divided into a number of cross-sections and each cross-section is divided into an equal number of points. The points on each cross-section are then jointed by fitting a cubic spline through them. If the component is particularly complex having for example a number of definable surfaces which can be joined together, then each surface can be separately specified and the relationship between each surface defined.

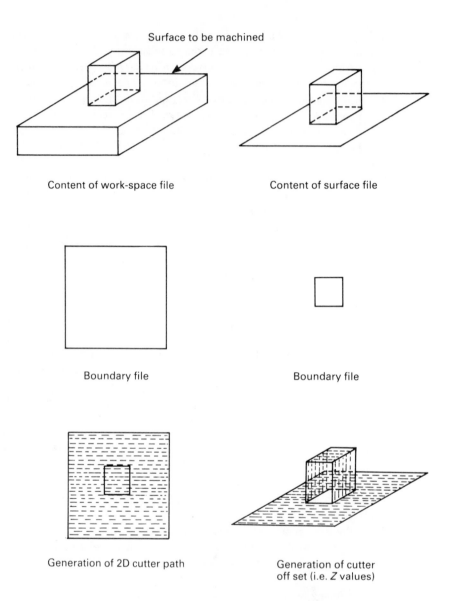

Fig. 12.9 – Generation of cutter paths for machining surfaces.

(m) Cutter offset

The surface file is checked and the splines, in the direction of machining dominance, are extracted. In order to determine the cutter offsets, assuming a ball-nose cutter, three splines are considered together. At each point on the middle spline four planes are determined using the outer two splines, with each plane being defined by three points. The direction cosines of the perpendiculars to these planes are calculated and the cutter offset determined from the selected point.

This procedure is repeated across the surface until it is covered by the correct cutter location paths. When all the cutter positions have been calculated, the system then tests for cutter interference with the part surface. That is it checks to see if there is any smaller cutting. If any undercutting occurs the user is informed and the procedure repeated using a cutter with a smaller radius.

(n) Control information

Some control functions are added to the process during the production of the machining data. These functions include coolant ON/OFF, spindle ON/OFF and dwell. The user may select any of these functions during the generation of the cutter location file. The appropriate miscellaneous code will be generated when the machine command tape or information is generated.

(o) Detection of erroneous data

The following checks are made to detect erroneous information:

(1) Prior to generating the cutter path the component is checked for overlapping data. If any is detected, the data are modified to avoid undercutting.
(2) The second inspection is to test that the cutter can reach all parts of the surface without undercutting. If any undercutting results from this source, the user is advised to use a smaller cutter.
(3) The third inspection checks for interference between the cutter assembly or tool holder and the billet or casting. If any fouling does occur then the user is asked to lengthen the cutter.

(p) Point-to-point machining

Point-to-point machining is available whithin the system for drilling operations. As previously described operations, checks are built into the system to avoid the production of erroneous manufacturing data. For example if the drill diameter chosen is larger than the hole it has to drill or if the drill length is less than the depth of the hole then the system informs the user.

12.3.3 Summary of CAD/CAM system operation

Operation of the CAD/CAM system is performed through extensive use of a menu which is important to the user-friendliness of the system. The menu is tablet-orientated for some workstations, particularly those containing storage tube displays, and display-orientated for workstations containing refreshed displays supported by some intelligence. The menu functions can be grouped into general areas such as Data Specification, Symbols, Display and Manipulation, Library, Manufacture and Miscellaneous. Details of these general areas are shown in Figs. 12.10 to 12.13. The method of proceeding through the menu has already been described in detail in the previous sections. A brief example of the preparation of data for the manufacture of a mould for casting a fibreglass pattern, of a vane for a centrifugal water pump, is now described in pictorial form in Figs. 12.14 to 12.21. Each of these figures shows three orthogonal views and an isometric. In Figs. 12.18 to 12.21 an enlarged isometric view is also included for clarity.

ISO METRIC ANGLE	ISO/ PERPEG TIVE		ENABLE VIEWS		ORTHOGONOL-VIEWS X-Y-Z Z-X		MIRROR	DISABLE VIEWS			LINE EDITOR		ALPHA- NUMERIC EDITOR	MACRO EDITOR
		ISO. CUBE					CLEAR WORK- SPACE	SAVE WORK- SPACE						
SET ISO. WINDOW	RESET ISO. WINDOW		PLOT WORK- SPACE				ERASE SCREEN	RESET WINDOW			ZERO FILE			DISPLAY FILE
				DISPLAY 1	DISPLAY 2	DISPLAY ORTHO- GONAL								
ROTATION X-AXIS Y-AXIS Z-AXIS			EXTRUDE X-AXIS Y-AXIS Z-AXIS					CONE GENERATION X-AXIS Y-AXIS Z-AXIS		SPECIAL SPLINE ORTHO- GONAL SURFACE BLEND SPLINE				
			EXTRUCE PLANE X Y Z					SOLID OF REVOLUTION X-AXIS Y-AXIS Z-AXIS			CURVE DISPLAY	FIT STORE		SEC./ CONTOUR
				JOY- STICK	DEBUG									

Fig. 12.10 – A display and manipulation of data menu.

GRID ENABLE DISABLE	Z GRID	RECOVER GRID	
ZERO GRID	Y GRID	FILE GRID	
GRID ORIGIN	X GRID	DISPLAY GRID	
3-D ELLIPSE			
TABLE MODE	3-D ARC	DISPLAY DELAY	OUTPUT SCALE
3-D FILLET	EYE POSITION	INPUT SCALE	
ALPHA-NUMERIC MODE	3-D SEMI-CIRCLE	SET Z DATUM	
POLYGON MODE	SET SKEW	ALPHA-NUMERIC SIZE	
DIMENSION MODE	3-D CIRCLE	INPUT ORIGIN	SET TOLERANCE
2-D RECT-ANGLE			
PLANE MODE	2-D CIRCLE		
LINE MODE	DISPLAY CURRENT MACRO	SELECT MACRO FILE	→
PLACE MACRO	DISPLAY MACRO FILE		
DEFINE MACRO	DELETE MACRO FILE	←	

Fig. 12.11 — A symbol, macro manipulation and data specification menu.

			DRAW ANGLE					OFFSET TO STORE	DISPLAY OFFSET
		POINT TO POINT		INTER. TO MAN	INTER. TO 3-D			GENE-RATE OFFSET	OFFSET TO INTER.
			ROUGH CUT	DISPLAY INTER.	INTER. STATUS	STORE TO INTER.			
		TOOL CHANGE		ELGA MILL	ADD Z TO INTER.	DELETES STORE			
ENTER/EXIT	MANUFACTURE MODE		AREA CLEAR		ZERO INTER.	DISPLAY STORE	INPUT SURFACE	DISPLAY SURFACE	
							INPUT BOUND-ARY	DISPLAY BOUND-ARY	
			RAPID FEED RATE	CUTTER		PLOT MAN.			
		DEFINE M/C		START POINT	ZERO MAN		C.L. TAPE		
			FEED RATE	MATE-RIAL		DISPLAY MAN.			
EXIT					CENTRE DRILL	INPUT SURFACE	DISPLAY SURFACE		
			SIMULT-ANUOUS OP.		BAR FEED		SELECT TOOL	LIST TOOLS	
				TURRET POSI-TION		CONTOR			
			DWELL		TOOL PATH	CUT SURFACE			
		COMP. DATUM		CUTTING STATUS					
			SPIND-LE & COOLANT						DEFINE COMP.

Fig. 12.12 – A manufacture mode menu.

BUTNUM & DRIVE MODE

<	+/−	E	C	EX
7	8	9		RESET
4	5	6		X
1	2	3		Y
0	.	=		Z

SURFACE SELECTOR

100	DELAY 10	1
ACCEPT	ADVANCE	RAPID SEARCH
DISPLAY SELECT-ION	LINE SEARCH	CHAIN
EXIT	RESET ALL	RESET SWEEP

45	15	5	1	EN	−1	−5	−15	−45

45	15	5	1	−1	−5	−15	−45

Fig. 12.13 — A joystick, surface selection, calculator and drive mode menu.

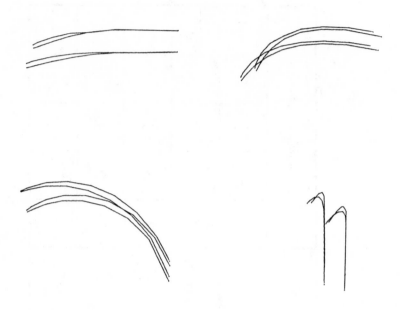

Fig. 12.14 – Input blade geometry, nine points per curve.

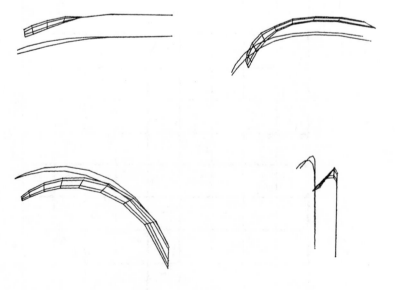

Fig. 12.15 – Generate the surface.

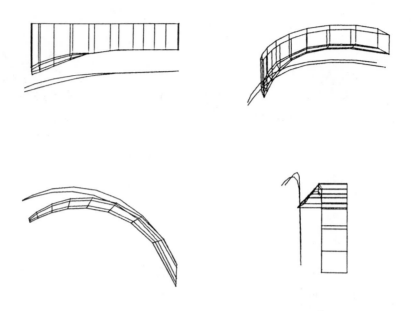

Fig. 12.16 — Add stand to blade. The stand is to permit location of the blade
pattern of the hub of the pump.

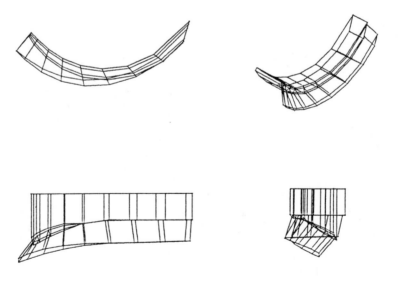

Fig. 12.17 — Further views of blade and stand.

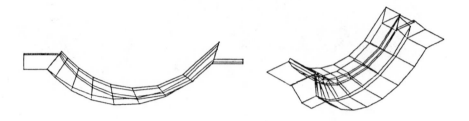

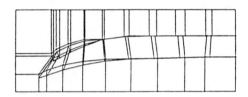

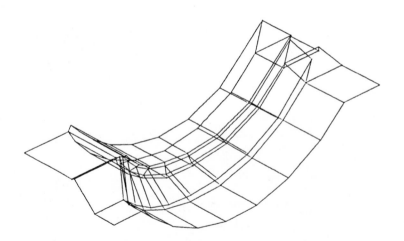

Fig. 12.18 − Specify die block for blade and stand.

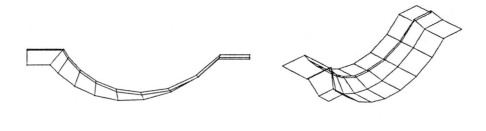

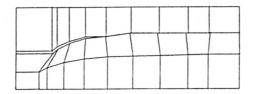

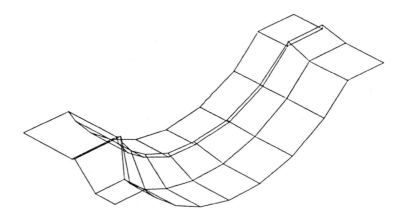

Fig. 12.19 — Select surfaces on die block for machining.

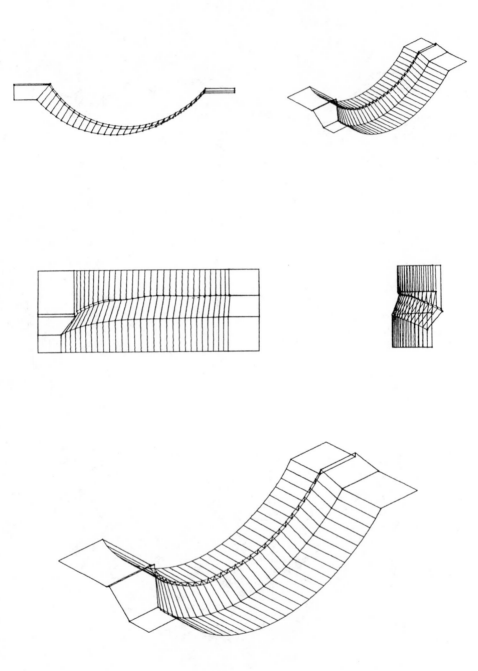

Fig. 12.20 — Blend splines of 30 points to give smooth definition of curved surfaces.

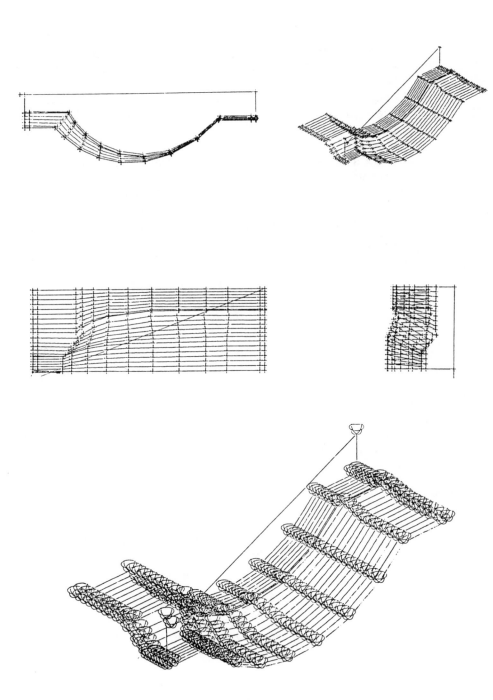

Fig. 12.21 – Generate cutter paths and test for under cutting. (Note finishing cuts only are shown.)

REFERENCES

[1] CAM-X User Manual, Ferranti Infographics Ltd., Edinburgh, UK.

[2] Coles, W. A. APT at the British Aircraft Corporation. CAM-I International Seminar, 1975.

[3] Craig, D. P., Besant, C. B., Hsiung, C. Y., and Williams, G. M. J. The use of interactive CAD techniques and NC techniques in mould design and manufacture. 18th International Machine Tool Design and Research Conference. Imperial College 1977.

[4] Olama, I. M. CADCAM methods for the design and manufacture of complex components. Ph.D. thesis University of London, 1985.

[5] Yi, C., Besant, C. B. and Jebb, A. Three dimensional digitising techniques for computer aided animation. Proceedings of an International Conference on Computers in Engineering and Building Design. CAD 76, Imperial College, 1976.

[6] Nutbourne, A. W. Curve Fitting by a Sequence of Cubic Polynomials *J. of CAD*, **1**, 4, 1969.

13

Industrial robots

The industrial robot is now becoming commonplace and it is a machine that can be programmed to perform a range of tasks automatically. It offers flexibility, since it is programmable, which is not the case with the older more conventional automated production line systems. The robot is a device that can easily be taught to perform simple tasks such as pick-and-place or spot welding and it does not require expensive retooling every time the task is changed. The robot is therefore a concept of a computer coupled with a flexible manipulator and this combination has helped to open the door on new manufacturing methods. In many applications robots are programmed by a teach-by-show technique using either a pendant or a textural language. Increasingly, though, there are moves to program robots using CAD/CAM methods with a level of automatic programming. Methods for programming robots are discussed in detail in the next chapter but it is necessary first to discuss robots themselves before going on to the subject of programming them.

13.1 BASIC ROBOT ELEMENTS

There are three basic components of an industrial robot and these are:

(a) Controller: the robot controller functions as the coordinating system of the robot and is based on a computer or a system of computers. In sophisticated industrial robots the control computer is capable of a level of artificial intelligence (AI) and not only runs the robot through its programmed moves but also integrates it with ancillary machinery or devices. The controller can also monitor processes and can make decisions based on system demands while at the same time reporting to a supervisory controller.

(b) Manipulator: the manipulator consists of the base and arm of the robot, including the power supply, being either electric, hydraulic or pneumatic. The manipulator is the machine that provides movements in any number of degrees of freedom. The movement of the manipulator can be described in relation to its coordinate system which may be cylindrical, spherical, anthropomorphic or cartesian. Depending on the controller, movement can be servo or non-servo controlled and can be a point-to-point motion or a continuous path motion.

(c) Tooling: the hand or gripper, often called the 'end effector' can be a mechanical, vacuum or magnetic device for handling parts. It can incorporate levels of compliancy to accommodate any slight misalignment. This can be in the form of passive compliance, whereby any correction is provided locally within the tool or active compliance where sensors provide additional positional information for the robot controller.

13.2 TYPES OF INDUSTRIAL ROBOT

There are many ways of classifying industrial robots, and they can be categorised by the work they perform, such as spray painting, welding, assembly etc. Three principal types can therefore be described which relate to various tasks.

(a) Simple robots: these are 'pick and place' devices and are generally used in materials handling. They are usually low in cost, easy to maintain and are fast and can dramatically increase productivity in medium- and long-run production applications. The automobile industry is a good example of the use of such robots and, more recently, an area such as packaging is becoming an application for robots.

Normally these robots are restricted to three or four non-servo degrees of freedom. Mechanical stops are used on each axis to set the limit of travel and operation is by pneumatics giving a repeatability to within ± 0.025 mm. Simple robots are very dependent on support equipment in terms of feeders and part presenters because of the limitations on available movements. Clearly a relatively simple control system, i.e. intelligence, is required for these robots since the main control is via on/off switching with little or no feedback.

(b) Medium technology robots: these are robots that contain a greater level of intelligence than simple robots and in particular contain a greater memory capacity with extensive teaching facilities. Such robots have four to six degrees of freedom and are usually servo-controlled in most of these axes of movement.

Medium technology robots are used for single machine loading and loading or similar tasks and are not capable of continuous path operations such as that required for arc welding. There are many tasks that could be automated with this type of robot.

(c) Complex robots: these robots possess flexible and programmable manipulators together with controllers that exemplify the high levels of artificial intelligence used in industrial automation. Such controllers can be interfaced with sensory and inspection devices that enable the robot to be taught even the most complex tasks such as assembly operations. The sophisticated industrial robot has the capability of being integrated into computer-controlled work cells and manufacturing systems.

The sophisticated robot has a large on-board memory capable of storing multiple programs and the ability to change programs automatically depending on the requirements of the system in which it is working. These robots have powerful textural languages that permit even the most complex tasks to be programmed and even CAD techniques can be used to program them. Typical robots in this class are Unimation PUMA and the R. D. Projects Type 8 and Type 35 (Figs. 13.1(a) and (b)).

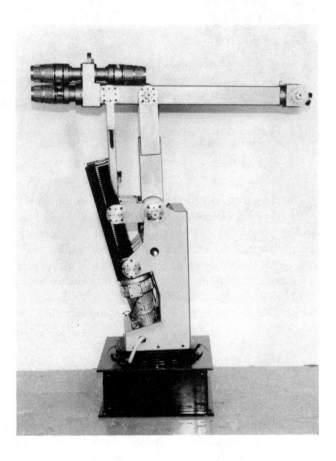

Fig. 13.1(a) − The R D Projects Type 8 Robot. This robot has six degrees of freedom and a capacity of 8 kg. (Courtesy of R D Projects Ltd.)

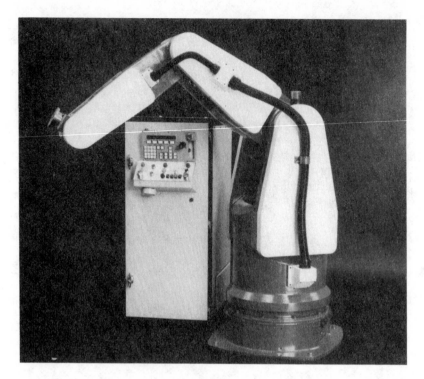

Fig. 13.1(b) – The R D Projects Type 35 Robot. This robot has five degrees of
freedom and a capacity of 35 kg (Courtesy of R D Projects Ltd.)

The physical configuration of robots can vary from the anthropomorphic to
the cartesian and various configurations are shown in Fig. 13.2. The anthropomorphic type is the most common since it is structurally stiff and can cover
large working envelopes. In some assembly applications a robot of the type
called SCARA is often used. This is normally a four-degree-of-freedom robot
and looks like an anthropomorphic robot turned on its side with the shoulder
and elbow joints operating the arms in a horizontal plane. The SCARA robot
is used for the simpler assembly tasks such as assembling a printed circuit board.

13.3 THE ROBOT CONTROLLER

The versatility of a robot arises from its multi-axis mechanical configuration and
the robot controller. The ability to re-program the robot controller so that the
robot can perform a wide range of duties gives the robot its flexibility. The use
of microprocessors in the controller is the key to this flexibility. The early
robots were point-to-point motion devices and programming them for anything
other than pick-and-place applications was time-consuming and tedious. Most of
the programming was done via a pendant using the teach-by-show technique.

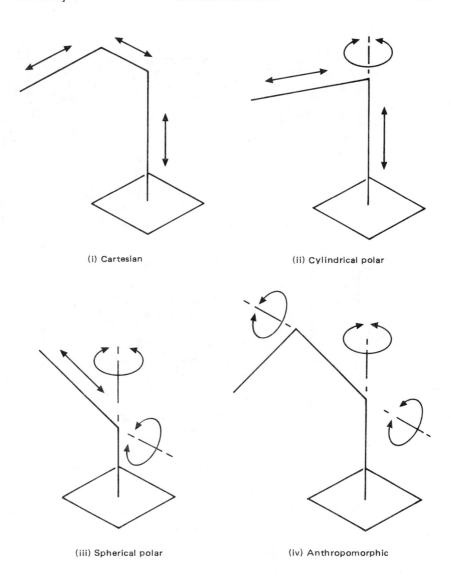

(i) Cartesian   (ii) Cylindrical polar

(iii) Spherical polar   (iv) Anthropomorphic

Fig. 13.2 – Standard robot primary axes configurations.

The amount of on-line programming and the lack of facilities to include environmental sensing, restricted the use of these robots to the less sophisticated industrial tasks. Knowledge of the robot environment, through the use of sensors, is a prerequisite for increased robot intelligence.

The features required of a modern robot controller are illustrated in Fig. 13.3. The controller contains various interfaces with both command devices and sensing units. These interfaces tend to be structured on various levels. The controller performs a number of important functions starting with the definition

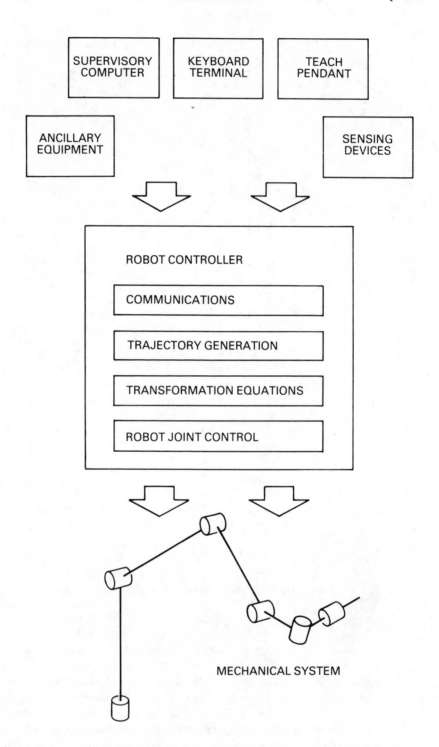

Fig. 13.3 – Features required of a robot controller.

of the trajectory of the robot gripper and interpolating movements with respect to time. It is then necessary to transform the trajectory of the gripper, which is in cartesian coordinates, into the base frame coordinate system or work space of the robot and to convert from the trajectory in the work space to the joint demands (angles) of the robot. Finally, it is then necessary to servo control each joint of the robot. Many of these tasks need to be performed in real time in order for the robot to perform its task at the requisite speed. If this is to be achieved on an economic basis then a single powerful computer would be too expensive. The most common way to achieve the necessary speed is to allocate a complete microcomputer to each task, i.e. distribute the intelligence.

There are two basic methods of achieving a multiple microprocessor system for control applications. One is to use a multiprocessor system to improve cost performance by the use of interconnecting processors in a tightly coupled manner, such that processors share resources and function under a single operating system. The second is to distribute the intelligence using a number of processors each handling a number of different but related tasks. The former option carries a higher software overhead owing to the requirement of a complex operating system. This, coupled with the limitations for hardware modularity and expandability, suggests that the distributed intelligence concept would prove more suitable for the robot control problem.

The Distributed Intelligence Microcomputer System (DIMS) [1] concept divides the control problem into small definable subsystems, each controlled by an individual processor. This system differs from the multiprocessor system in that only one microprocessor is used for each task as opposed to using a number. The DIMS concept distributes the intelligence throughout the robot controller by means of microcomputers each acting semiautonomously and communicating with other elements in the system. Each microcomputer has a dedicated task, ideally each task being isolated from that of any other microcomputer in the DIMS, ensuring hardware and software isolation.

The advantages of using the DIMS approach in robot control are that the cost of the system is low for the computing power that can be achieved. Furthermore, because isolated tasks are each controlled by a single microcomputer high reliability may be achieved. In addition the communication system between the various microcomputers, and hence the tasks, is simple, handling only limited data flow and synchronisation. Since each element is independent, the system can be designed to ensure that the failure of a microcomputer is unlikely to corrupt the whole control system.

The DIMS approach allows expandability of the controller in that processing power can be added in increments without causing major disruption to the rest of the controller. For example another degree of freedom or axis can be added by simply adding another computer. The standardisation that results from this concept reduces maintenance problems and provides for the interchangeability of hardware.

There are now a number of DIMS-based controllers on robots and the R D Projects robots utilise such a system that was developed in association with Imperial College [2].

13.4 ASSEMBLY ROBOTS

Automatic assembly by robots requires extensive data collection from sensors and processing of the data from the sensors in order to identify the world around the robot together with feedback of its interaction with the world. Strategies for robot applications in assembly involving adaptive control require a considerable amount of real-time processing. Where the sensor processing is considered along with functions such as interpolation, coordinate transformation and dynamic compensation, the overall processing task increasingly becomes a major problem. The solution to this problem, as already discussed, lies in the division of these tasks into independent problems which can be solved in parallel but in a coordinated and synchronised manner.

The control system strategy should to a large extent be independent of robot configuration and the system must be interfaced to a high level language for ease in robot programming.

A number of existing control systems already incorporate a level of parallel processing and are based on extensive use of a standard bus such as VME [3] or IEEE-488. The structure of the robot controller must be such that it can execute the high number of tasks which have different properties and priorities. The architecture of the system must permit high program execution speed, be fault tolerant, allow for parallel program interpretation and be flexible on the number of processor modules that can be utilised. Particular emphasis should be attributed to task arbitration to avoid processing bottlenecks. The basic requirements of the control system for assembly robots can be summarised as follows:

> Control functions are to be distributed on microcomputers.
> Modular hardware configuration.
> A bus system to allow for different levels of processing.
> Expandability without the need for hardware or software re-configuration.
> Parallel interpretation of problem-oriented programs.
> Use of existing special purpose hardware modules, i.e. floating-point processors, fast fourier transformers, interpolators.

13.4.1 Structure of assembly robot systems

The general structure of a sensor-based assembly robot is shown in Fig. 13.4. The system is divided into the following components:

— the mechanical manipulator;
— the axis control system;
— the tool controller;
— the sensor system;
— the robot controller;
— the programming system.

In most systems the robot controller performs functions such as multi-tasking and coordinate transformation and contains the interpreter.

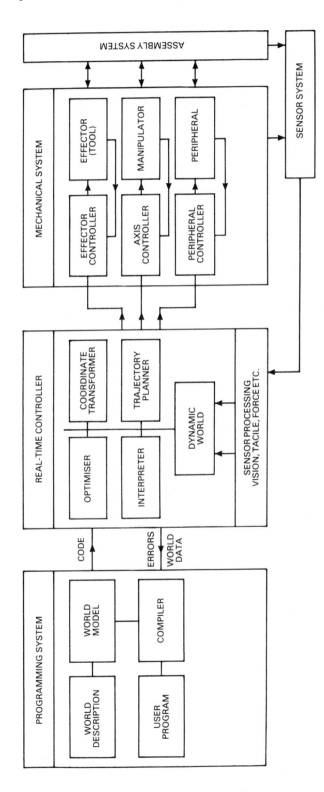

Fig. 13.4 — Schematic structure of an assembly robot.

13.4.2 Mechanical structure of robots

For general-purpose assembly work, a robot requires a mechanical system consisting of a kinematic chain with prismatic and rotary joint linkages. Six degrees of freedom are required in order to position the end effector at any point within the working space and allow it to assume any arbitrary orientation. In order to obtain high positional accuracy of the effectors, each joint has to be operated with a high resolution of 12 bits or more. For assembly robots positional repeatability less than 0.1 mm is required. Sensor feedback can do much to enhance high positional repeatability.

In the case where path control is important in order that the robot can follow a defined trajectory, the servo-control system must have access to dynamic compensation in order to provide the most suitable accelerations or retardations to the various links.

13.4.3 Axis control

A modern general-purpose robot should have path control such that the tool follows a defined trajectory. The trajectory is planned within the robot controller and ideally should operate in conjunction with sensory information in real time to overcome any disturbance caused by the following:

— gravity force;
— centrifugal force;
— Coriolis force;
— inertia;
— friction;
— reaction forces.

Ristić [4] has proposed a technique based on a self-tuning algorithm for overcoming disturbances. The self-tuner will operate in real time and will require minimal sensor information.

An Axis Control Computer performs the servo-loop closure calculations and there may be an axis control computer for each axis. Coordinate transformation, trajectory planning and self-tuning are all performed at a higher level than the servo-loop closure with a computer for each function.

Usually joint control is done in joint space coordinates (robot coordinates). The joint space trajectory is generated in cartesian coordinates and transformed into robot-dependent joint coordinates.

13.4.4 Control of the end effector (tool)

Control of the end effector refers to the control of the interaction between the robot and the assembly task. Thus the robot can be considered as a positioning device for the end effector. Most assembly tasks are compliant operations with active and passive adaptation.

The end effector is characterised by:

— the tool;
— the object to be handled;

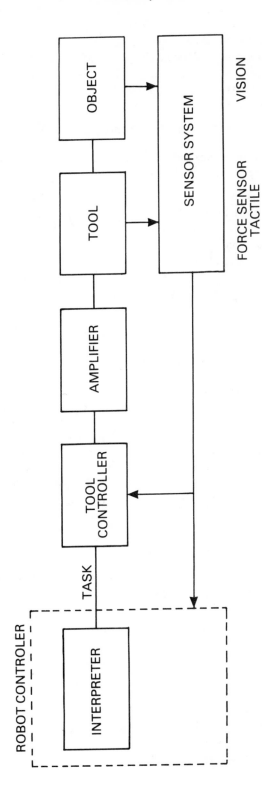

Fig. 13.5 — Structure of an end effector or tool controller.

- position and orientation of the object;
- reactive forces and moments;
- the parameters for grasping;
- friction between finger and object;
- compliant forces.

The general structure of the end-effector controller is shown in Fig. 13.5.

13.4.5 Sensor systems and processing of data

The sensor environment of an assembly robot can be classified as follows:

Internal sensors
- (i) joint position (encoders);
- (ii) velocity (tachogenerators);
- (iii) acceleration;
- (iv) force wrist sensor;
- (v) gripper force.

External sensors
- (i) vision;
- (ii) tactile array;
- (iii) safety sensor;
- (iv) geometric sensor (2D and 3D).

Internal sensors are part of the closed axis control loop and have to sample data at high speed. External sensors can work at various speeds depending on their type and they send their information to various levels in the hierarchy of the robot control system. The sensor system is divided into one of data acquisition and one of data processing. The data acquisition system samples the data and runs the digital converting routines. The data processing system uses the sensor information to extract vital information, e.g. pattern recognition from a vision sensor. The data flow may be bidirectional so that the sensor system can be initialised and controlled by the robot controller.

13.5 THE ROBOT CONTROLLER

The robot controller must perform the task of ensuring proper execution of manipulation sequences defined by user programs. The individual tasks may be summarised as follows:

- interpretation of execution of a user-defined assembly program in real time;
- multiple task scheduling;
- trajectory planning under sensor influence;
- coordinate transformation;
- effector control;
- control of peripherals;
- sensor control and data processing;
- bus management;
- communication with peripherals.

It is clear that a multiprocessor type control system is required for this application.

The type or size of processor required in a multiprocessor system may vary according to the task. Some tasks such as coordinate transformation or dynamic compensation may require a 16-bit or even a 32-bit microprocessor while others such as the control of peripherals may easily be performed with an 8-bit microprocessor.

13.5.1 Trajectory planner

The trajectory planner is that part of the robot control system which determines the motion of the manipulator under the influences of sensors. The parameters of the trajectory, such as start condition, velocity, acceleration, end points etc., are defined by the user program or modified by the sensor system in the trajectory planner. The following motions have to be considered in assembly applications:

— from point to point;
— on straight lines;
— on defined curves with specified velocities;
— tracking;
— to avoid collisions;
— under sensor control.

Certain interpolation algorithms are required to achieve these motions and these algorithms are dependent on the desired end effector motion. Typical interpolation algorithms are:

— linear interpolation for straight lines;
— linear interpolation with transition between straight line segments;
— interpolation using polynomials;
— interpolation between time variant points via time functions.

The trajectory generation is robot-independent and is performed in cartesian coordinates at a data rate of 100 Hz.

13.5.2 Coordinate transformation

If the end effector is to be moved along a calculated cartesian trajectory then it is necessary to determine for each joint of the robot the appropriate joint angle as a function of time. This is the transformation of world coordinates into robot coordinates. The reverse of this process is applicable when the joint angles are known such as when the teach pendant is being used. The coordinate transformation requires real time processing of a dedicated Motorola 68000-based computer, or similar machine, with the coding in assembly language in order to achieve the required processing speeds.

If sensor-controlled trajectories are considered the geometric conditions of the working space of the robot have to be identified by, for example, a vision system. The data of interest are:

- position and orientation of workpieces;
- distances;
- geometry of workpieces;
- path of workpieces.

These geometric data must be transformed from sensor coordinates into cartesian world coordinates.

If real-time control is to be achieved with a large degree of sensor feedback, then the requirements for high-speed processing become of paramount importance involving fields such as parallel processing.

13.5.3 Dynamic world processing

A reference to robot programming is normally a world model. This model can be generated in a number of ways and defines the relation between objects. The frame concept, as in languages such as AL, [5] is applied to describe the relation between an object and the end effector. In the case of a well-structured assembly operation the frames need only be defined once. However, in a time variant case, such as with conveyor tracking or robots with multiple arms, the frames have to be continuously refreshed in order to present actual states. Again this task involves a great deal of rapid processing.

13.6 THE STRUCTURE OF THE ROBOT CONTROLLER

The tasks previously described can all be attached to a hierarchy of control levels depending on the data flow and degree of processing. The task levels can be described as follows.

Level 1 Servo-loop closure and effector control

At this level the drive signals for actuators are generated to move the joints. Special control algorithms are applied to control the individual servo motors in an optimal manner. The input signals of this level are the desired joint trajectories in joint coordinates. There may be a processor for controlling each axis.

Level 2 Coordinate transformation

The transformation of the cartesian trajectory into robot specific joint coordinates is done at this level by a processor which is running robot configuration dependent transformation routines.

Level 3 Trajectory interpolation

At this level robot interpolation routines generate a continuous trajectory in world coordinates which can be executed by the robot. The inputs to these routines are parameters, such as points, velocities etc., which define the trajectory and motion along it. The routines are then used to describe the path in terms of interpolation functions such as cubic, polynomial or straight lines. At this level, the entire kinematics of the robot are calculated in cartesian coordinates.

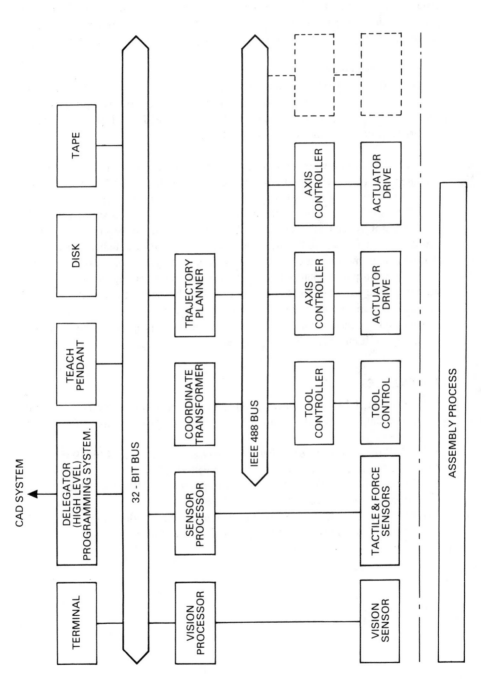

Fig. 13.6 — Hardware structure of robot controller.

Level 4 Trajectory control

Data from the user program, the world model and the sensor are combined to provide trajectory parameters. This level consists of strategies which optimise the parameters under sensor control to ensure precise program execution.

Level 5 Central controller

The central controller is the master or delegator. Its function is to distribute and synchronise the various tasks to be performed by the complete control system. With assembly applications the level of processing is high and a degree of parallel processing will be required at this level.

The actual hardware configuration of modern robots using the DIMS principle can vary according to the microprocessors used together with the bus structure. A structure for an advanced robot is shown in Fig. 13.6.

The simple and medium technology robots can be programmed by well-known methods such as teach-by-show using a pendant. However, with complex robots that are to be used for the difficult robot applications, such as in assembly, teach-by-show methods are inadequate. It is necessary, in many cases, to resort to CAD techniques to plan out the task for the robot making use of simulation methods. The program for the robot can then be determined from the CAD system and transferred to the robot. The programming of robots is an important subject and a considerable amount of research is in progress in this area. The topic is fully discussed in the next chapter.

REFERENCES

[1] Anderson, L. H. Distributed Intelligence Microcomputer Systems (DIMS). Proceedings 11th Int. Symp. on Ind. Robots. Tokyo, Japan, Oct. 1981, p. 497.
[2] R D Projects *Robot Controller: User Manual.* R D Projects Ltd, 1985.
[3] VME Bus Specification Manual, MVME BS/D1, Motorola.
[4] Ristić, M. The Dynamics of Anthropomorphic Robots. Ph.D. thesis. University of London, 1985.
[5] Finkel, R. A. AL, a programming system for automation. Memo AIM 213, Artificial Intelligence Laboratory, Stanford, November 1974.

14

CAD and the programming
of robots

A robot is by definition a flexible machine. It can be programmed to perform a variety of tasks without major hardware modifications. It derives its flexibility from a number of sources, one being its mechanical structure containing actuators which allow it to move within a large working envelope. The second is that it can adapt to its environment providing it is using appropriate sensors. The use of sensors will reduce the burden on the designer of the robot environment.

If the advantages of the flexibility of robots are to be attained, then a great deal of information processing and calculation may be necessary, some of it being performed in real-time. The organisation of this processing is based on the concept of robot programming which is embodied in a language so that the user may specifiy how action and sensing functions of the robot are to be applied in order to execute a given task.

A robot programming system is complex in nature mainly because it is attempting to satisfy the requirements of providing an easy-to-use system for the least expert users on the one hand while enabling complex tasks to be described on the other hand.

Different concepts are now evolving for programming robots in order to try and satisfy the criteria of an easy-to-use system that can describe complex tasks. In the short term, effort has been concentrated on the provision of robot languages which are situated within the controller of the robot. These are used via a teach pendant or a simple alphanumeric terminal. More recently efforts have been concentrated on using CAD techniques to aid robot programming, and in the longer term a level of automatic robot programming will be possible with the aid of AI technology. The use of CAD techniques will enable tasks to be described that are more complex than those that can be described with a teach pendant. Also the programming can take place off-line and away from the robot, thereby saving downtime on the robot while it is being programmed.

14.1 PROBLEMS ASSOCIATED WITH ROBOT PROGRAMMING

In programming robots it is necessary to determine the actions that are to be executed by the robot together with the sensing that may be required. These actions and the sensing must be embodied in an algorithm or series of algorithms which may be linked by conditional and loop statements.

There are two types of problem that must be considered in robot programming. They are as follows:

(i) Specification of positions
A robot is mainly a manipulator of objects. It can position a part or tool by moving it from one place to another. Therefore the programming is concerned with the defining of a sequence of positions or trajectories through which the robot must pass during the execution of its task. Care must be taken to ensure that the robot does not infringe any space occupied by any object otherwise there will be a collision. In many cases careful sequencing of events in the correct time is necessary to avoid collisions or for the robot gripper to be in the right place.

(ii) Dealing with uncertainty
The positioning of objects cannot be exact, nor can the actions of the robot. Both are limited by dimensional tolerances. Uncertainty is not a problem provided that it is lower than the precision requirements of the task. This normally means that the environment has to be carefully engineered, which can result in high-cost jigs.

Uncertainty can be alleviated to a large extent by using sensors. For example a vision system can give information on position and a force sensor on the interaction between surfaces which are in contact.

Both of the problems outlined above bear some relationship to each other. For example trajectory planning must take into account uncertainty, particularly if parts are being manipulated or located in confined spaces. A robot program must be able to deal with both of these problems and the solution must be acceptable in terms of time and reliability.

14.2 ROBOT-PROGRAMMING SYSTEMS

There are two main types of programming systems on robots: the 'teach-by-showing' method, and textual programming.

The 'teach-by-showing' method is the earlier and is based on extensive use of an interactive pendant where the programmer operates the robot for performing a sample execution of a task. The motion of the joints are recorded so that they can be played back later. This method is popular because of its immediacy and also the user needs no specialised knowledge. However, it does have its drawbacks such as its limitation to simple tasks such as pick-and-place. Furthermore it is fastidious to program a long sequence of actions and difficult to edit programs. Sensory data is also difficult to use and coordination with other systems is almost impossible.

CAD has been used with some success to simulate the 'teach-by-showing' method (Heginbotham *et al.* [1]), Dooner *et al.* [2]). Teaching by showing on a graphical display does have advantages over the pendant method since it improves personnel safety, for easy experimentation both with using other robots as well as with a variety of methods with the same robot. Different parts and different world arrangements can also easily be tried. However, this method also suffers from similar limitations to that described for the pendant 'teach-by-showing' method.

Some of the limitations of the 'teach-by-showing' method have been overcome by the development of symbolic languages known as manipulator-level languages known as manipulator level languages such as AL (Finkel [3]) and AML (Taylor *et al.* [4]). Most of these languages are an extension of languages such as BASIC or PASCAL. They include both the main features of such languages and additions appropriate to robotics to cover, for example, motion specification, world modelling and sensing. They give a large access to the basic primitives of the robot, so that they avoid the limitations of the 'teach-by-showing' method. However, they have the disadvantage that they require a user to be knowledgable in programming methods in order to take full advantage of the language.

For some time both of these techniques used for programming robots have been rivals. However, it has now been realised that they are complementary and some new systems combine the two methods to take advantage of the attributes of both. Typical of these combined systems are SRL-90 (Ahmad [5]) and IMAG (Bansard [6]).

(i) Functions of a textual programming language

There are now many textual languages on robots including: AL (Finkel [3]), AML (Taylor *et al.* [4]) and VAL Unimation [7]). Such languages provide algorithmic facilities such as symbolic variables, conditional and iterative statements, input/output on files and terminals and subroutines. They also contain robot-specific constructs which can broadly be divided into four classes, namely, world modelling, specification of robot motions and operations, sensory and interaction with other devices and flow of control.

The most widely accepted method for modelling the robot environment is to define objects relative to cartesian frames. These frames must be previously defined and known by the language interpreter. The frames include a fixed frame to serve as the reference frame for the user and a moving frame attached to the extremity of the robot manipulator. The user can declare the frames required in the program and can define their positions by applying transforms. This may be achieved by translations and rotations to already known frames. Rotations and translations can be defined using vectors.

Some languages, like AL and AML, permit the user to attach one frame to another. Two attached frames remain in the same relative position in the model of the world in the interpreter. If a motion statement is used to modify the position of one frame then the position of the attached frame is automatically changed in the world model. This is a particularly useful feature for defining positions of an object and grasping positions for the object. Thus, if the two frame positions are attached then the grasping position will always be in the same relative position to the object no matter what motion the robot gripper is given. The robot-programmer can also define motions directly in terms of object frames.

The trajectory for motion may be defined in terms of joint coordinates or more usually in terms of cartesian coordinates. The second possibility, which takes advantage of the world model, is usually more convenient. If the frame attachment facility is available in the language, then each motion statement applies for a frame F which the user considers is the most appropriate to describe the motion. F must be attached to the frame which is in turn attached to the extremity of the manipulator. The final position is defined by a frame F_n, where F_n might be a frame attached to an object. In order to reach F_n, the origin of F may be translated along a straight path and if F and F_n do not have the same orientation, the axes of F may be rotated around a fixed direction. The interpreter performs all the computations, involving trajectory planning and generation, cartesian to joint coordinates transformation and the generation of servo set points (Paul [8]). The user can control the speed of the motion and can describe non-linear trajectories by defining intermediate positions for F before it reaches F_n.

A motion may be subject to restrictions which are imposed by a boolean condition involving sensory data such as a force feedback. This condition is periodically evaluated during the motion and the robot is stopped when the condition or set criterion is met. Languages like AL also include options to control both the position and force exerted by the robot during motion, for example:

$$\text{move OBJECT to GOAL with force } (f_y) = \text{FY}$$

Synchronisation of several robots is offered in only a few languages. For example AL permits sections of programs to be executed in parallel. These sections may involve different robots operating asynchronously. Languages such as AML permit the user to get control back to its program once a motion has commenced and this allows another motion on another robot to be implemented in parallel.

(ii) Facilities for programming
The robot-programming language requires certain facilities to assist the programmer in implementing and debugging programs. The basic facilities that are available include interactive interpretation, programming-by-showing utilities, graphical simulation and interfacing with CAD databases.

(a) Interactive interpretation

Programming languages can be either interpreted or compiled in a lower-level language. Compilers offer higher performance at execution time but in robot applications they lack the immediacy in terms of implementing a new program. On the other hand the relative inefficiency of interpreters does result in severe restrictions such as in the execution of functions in parallel or in sensory interaction.

A sensible solution is to provide both a compiler and an interactive interpreter. The interpreter is used for interactive programming and if a graphic simulator is available then the execution of a piece of code may either be tried on the robot or on the simulated robot.

(b) Programming-by-showing

The programming-by-showing method involves the switching of programming control from the terminal to the teach pendant. The robot is then moved to a number of desired positions and as each desired position is recorded on the pendant the last position is compared with the new position and the appropriate move command is automatically generated and displayed on the terminal. Thus the pendant operator can show the robot a complete range of tasks and at the same time generate a program in the textural language of the robot. These teach-by-showing utilities can be easily adapted to various types of tasks, for instance; pick-and-place, palettising, point-to-point welding and assembly.

Such utilities can be used by operators with no programming experience as well as being used by experienced programmers. It is important that such utilities cannot only generate positions in a file, but textural code that can be edited and inserted into another program.

(c) Graphical simulation

Graphical simulation is a tool for building robot programs using the teach-by-showing utilities. It is also a powerful tool for debugging.

The use of graphical simulation for debugging robot programs is widely established — Heginbotham *et al* [1], Laugier and Pertin-Troccas *et al.* [9], to name a few. Manipulator trajectories are computed using the language interpreter and are sent to the simulation program. This program runs on a world model which includes a geometric representation of objects and the robot. The simulator uses the information sent by the language interpreter to update the world model which is graphically displayed.

Graphical visualisation does require that the user has access to commands to define how the scene is to be viewed, i.e. view-point, zoom, etc.

The biggest problem associated with robot simulation is in sensor interaction. Some simulators include commands for emulating the data from sensors. However, research in this field is still in its infancy. Using a teach-by-show method implemented in a textural language with a graphical simulator can present difficulties. These difficulties arise from teaching a simulated robot on a two-dimensional screen which is not as easy as moving the actual robot.

(d) Interface with a CAD database

Modelling the environment of the robot may be more convenient if explicit geometrical data defining objects is available. Then the positions of objects and the robot can be described by symbolic relationships. The position of the frames attached to the objects in the CAD database are automatically computed from these relationships. The explicit geometric models are available only during the development phase of a program. At execution time the robot system knows only the frames attached to the robot.

The kind of facility is provided by languages like RAPT (Popplestone *et al.* [10]) and LM-GEO (Mazer [11]). LM-GEO which is a preprocessor to a more classical manipulator-level language LM (Miribel [12]). The conventional languages such as AL, AML and LM require that the programmer must reason quantitatively in the three-dimensional world and the robot arm is often used to define positions. New languages are seeking to describe tasks independently of the end-effector motions and operations. The problems that require a solution are:

— *geometric,* including computing theoretical positions of objects at each stage of a task such as in assembly, automatic grasping positions and collision-free paths;
— *physical,* such as those relating to the correct use of force sensor data, for example in performing an insertion, and touch sensor data for controlling micro-displacement of objects during motions;
— *strategic,* including scheduling the execution of subtasks and allocating robots to subtasks.

14.3 AUTOMATIC ROBOT-PROGRAMMING

The goal of automatic robot programming is to transform automatically the description of a manufacturing task from data provided by a CAD system into a robot program, the execution of which will make the robot accomplish the task. For example, in an assembly task a CAD system will hold in its database a definition of the parts together with a description of the assembly relations to be achieved amongst them. As many organisations are introducing CAD technology into their work programmes it is becoming increasingly important that information that is already in computer format is used effectively and not regenerated. It is in this context the automatic robot-programming is important. Furthermore some assembly tasks are complex and are difficult to describe to robots using teach-by-show methods or very slow using a textural language.

The main thrust in the development of automatic robot programming is in the mechanical assembly of parts. It is a task that involves both action planning and environmental planning.

Action planning involves process planning and phase detailing. Process planning consists of ordering the assembly tasks to be achieved and assigning a robot or more than one robot to perform the tasks. Phase detailing consists of planning the robot actions to achieve a given task.

World planning includes the definition of a work area, the selection of part feeders, fixtures and the robot end effectors.

Most research in the automatic robot programming field has concentrated on phase detailing. The result has been a number of 'task-level' programming systems such as LAMA (Lozano-Perez [13]). A task-level program consists of a sequence of target assembly relations between parts together with models of the parts. It is different from a manipulator program in that the robot actions necessary to achieve the tasks are not specified.

Early research on the translation of task level programs into manipulator level programs was extremely difficult because of such problems as specifying safe position and dealing with uncertainty. These difficulties lead to a break-down of the problem into a series of subproblems associated with phase detailing such as group planning, transfer motion planning and part-mating planning.

Grasp planning consists in computing the grasping position of the gripper in relation to the part together with the approach and withdrawal trajectories (Lozano-Perez [14] and Laugier and Pertin [15]).

Transfer motion planning consists in computing a collision-free trajectory for moving a part from its initial position to its final position (Brooks [16]).

Part-mating planning consists in synthesising a sensor-based program for controlling motions of the robot when achieving a specified relationship between two parts. (Dufay and Latcombe [17] and Lozano-Perez *et al* [18]).

Recent proposals for uniting these subproblem solutions into an overall automatic robot programming system have been made by Lozano-Perez and Brooks [19].

14.3.1 Computing safe positions

The most fundamental problem is given the object to be grasped and a set of obstacles, determine the continuous collision-free path to the target.

Much of the research devoted to a solution of this problem is based on an explicit representation of the world model together with a set of 'test' algorithms for determining a collision-free path amongst the objects in the world model. One of the more recent methods based on explicit representation of free space is by Brooks [16]. In a two-dimensional representation of the world, use is made of pairs of edges of the obstacles to compute the splines of generalised cones, called freeways, representing the free space. The possible motions of a moving object are along the constructed freeways. At each point along the freeway only a certain number of orientations of the object may be permitted. Motion segments consist of translations, with possible reorientation, along segments of splines. A complete path is constructed by moving from one spline to another at points of spline intersections. At points of intersection the object must satisfy an orientation criterion for each freeway. The method has been extended by Brooks [16] to find collision-free paths for robots with five and six revolute joints. Freeways are determined for the gripper and the upper arm. The method works well in relatively uncluttered surroundings.

The work of path finding can be applied to phase detailing such as grasp planning, transfer-motion planning and part-mating planning. However, when looked at in depth, there are some differences in these latter activities and the phase detailing.

Grasp planning includes the additional difficulty of choosing the grasp position of the jaws on the part. The approach and withdrawal are short trajectories which are computed. The gripper and the part also move close to obstacles during this phase.

Transfer-motion planning requires the computation of a long path sufficiently far from all obstacles because transfer-motions are usually executed at high speed and as such may be inaccurate. In general these motions are performed in an uncluttered environment.

Part-mating planning involves short trajectories executed in highly constrained environments local to the end effector. The importance in dealing with geometric uncertainty becomes paramount.

14.3.2 Dealing with uncertainty

The problem of dealing with uncertainty particularly manifests itself with part-mating in assembly operations. If there was no uncertainty then mating of two parts would consist of a sequence of goal-oriented motions so that the parts could achieve the correct relationship. However, because of uncertainty with the motions or uncertainties associated with the parts to be mated, it becomes necessary to resort to sensory data, such as force or position data, for determining whether the correct relationships are being achieved. Unwanted relationships must naturally lead to corrective motions.

Several techniques have been evolved for dealing with uncertainty. One approach is to compute possible errors associated with the mating of two parts from tolerances in the CAD database of the world model. The task of entering a pin into a hole, for example, can then be assessed and a decision made on what code is required for a local search of the hole. This approach to programming suffers from a number of drawbacks. Firstly, many errors are difficult to estimate and error propoagation techniques based on the 'worst-case uncertainty' hypothesis may lead to unrealistic over-estimates. Secondly, the techniques developed so far for propagating errors are only applicable to linear programs and their extension to iterative ones would require a solution to fix point computation problems.

These drawbacks have resulted in a second approach to part-mating based on inductive learning from experiements (Dufay and Latombe [17]). The method consists of assembling input partial local strategies described as rules into a complete manipulator-level program by processing the execution traces of several attempts to carry out the task. The execution traces are generated during the 'training phase'. Then the input rules are used to build plans in the form of sequences of situations and motions.

Let us consider a task such as inserting a pin into a hole. A first plan, called a ground plan, may contain a single motion starting from the point where the pin and hole are aligned, with some uncertainty, and the pin translated along its axis until its end touches the bottom of the hole. After each motion of the plan is executed by the robot, the system checks for position and force sensory data to determine whether the planned actions have been achieved. If this is positive then the next motion is planned. If this is not the case then rules are invoked for proposing a corrective plan starting from the current situation and having the expected situation as its target. For example, if the pin touches the chamfer instead of entering the hole, then the corrective plan will consist of first moving the pin along the force vector perpendicular to the axis of the pin followed by moving the pin along its own axis. The corrective plan is then patched into the ground plan and execution is resumed by the robot.

The assembly of the part-mating program happens during the second phase of the approach called the 'induction phase'. It proceeds through iterative transformations of the execution traces of motions and situations. The result is a flow chart, including branch points and cycles, of the part-mating program.

14.4 CONCLUSIONS

Robots have evolved into very flexible machines particularly when used with sensors such as vision, force and tactile. In order to make use of the potential flexibility that robots could offer for dealing with complex tasks, considerable efforts are being made in the development of robot languages that are of a general-purpose manipulator type and that interact with the environment.

In the latter part of the 1970s research has concentrated on the provision of a level of automatic programming systems based on an interaction with CAD technology. The aim of such systems is to minimise the expertise required in programming robots. The trend has been to break down the complex problem of automatic programming into subproblems such as grasp planning, transfer-motion planning and part-mating planning.

REFERENCES

[1] Heginbotham, W. B., Dooner, M. and Case, K. Rapid assessment of robot performance by interactive computer graphics, 9th International Symposium on Industrial Robots, Washington D.C., 1979.

[2] Dooner, M., Taylor, N. K. and Bonney, M. C. Planning robot installations by CAD, Computer-Aided Design Conference, Brighton, March 1982.

[3] Finkel, R. A., AL, a programming system for automation. Memo AIM 213, Artificial Intelligence Laboratory, Stanford, November 1974.

[4] Taylor, R. H., Summers, P. D. and Meyer, J. M., AML, a manufacturing language. *The International Journal of Robotics Research*, 1, 3, 1982.

[5] Ahmad, S., Robot level programming languages and the SRIL-90 language. IEEE International Conference on Robots and Automation, Atlanta, 1984.

[6] Bansard, J. P., Le système LM-EX. IMAG, Grenoble, June 1983.

[7] Unimation 1980. User's guide to VAL: a robot programming and control system. Unimation Inc., Danbury, Connecticut, Version 12, June 1980.

[8] Paul, R. P., *Robot manipulators; mathematics, programming and control.* The MIT Press, 1981.

[9] Laugier, C. and Pertin-Troccaz, J. Graphic simulation as a tool for debugging robot control programs. International Symposium on Design and Synthesis, Tokyo, July 1984.

[10] Popplestone, R. J., Ambler, A. P. and Bellos, I., RAPT: a language for describing assemblies. *The Industrial Robot,* September, 1978.

[11] Mazer, E., Geometric programming of assembly robots (LM-GEO). International Meeting on Advanced Software in Robotics, Liège, May 1983.

[12] Miribel, J. F. and Mazer, E., *The LM reference manual,* Vol. 7. ITMI, Meylan, France, October 1983.

[13] Lozano-Perez, T., The design of a mechanical assembly system. AI TR 397, Artificial Intelligence Laboratory, MIT, 1976.

[14] Lozano-Perez, T., Automatic planning of manipulator transfer movements. *IEEE Transactions on Systems, Man and Cybernetics,* **SMC-11**, 10, 1981.

[15] Laugier, C. and Pertin, J., Automatic grasping: a case study in accessibility analysis. International Meeting on Advanced Software in Robotics, Liège, May 1983.

[16] Brooks, R. A. Planning collision-free motions for pick and place operations. 1st International Symposium on Robotics Research, Bretton Woods, August 1983.

[17] Dufay, B. and Latombe, J. C., An Approach to automatic robot programming based on inductive learning. 1st International Symposium on Robotics Research, Bretton Woods, August 1983.

[18] Lozano-Perez, T., Mason, M. T. and Taylor, R. H., Automatic synthesis of fine-motion strategies for robots. 1st International Symposium on Robotics Research, August 1983.

[19] Lozano-Perez, T. and Brooks, R. A. In: *An approach to automatic robot programming in solid modelling by computers: from theory to applications* (Edited by J. W. Boyse and M. S. Pickett). Plenum Press, New York, 1984.

15

Automated guided vehicles

An automated guilded vehicle (AGV) is a robot type of vehicle that is used to carry objects from one place to another and can be programmed to travel in a predetermined path. All functions performed during its task, such as stopping and starting at workstations, require no human intervention. AGVs have been applied in manufacturing industry for many years and their use will rapidly increase since modern types of AGV offer considerable advantages over conveyor systems both in terms of flexible use and cost. Initial applications of the AGV, then known as 'robot-tugs', were in piece/part distribution. Here the AGV was a tractor pulling a number of trucks. The last decade has seen the development of load-bearing vehicles rather than tractors.

There are three main applications of AGVs. The largest application, which is still growing, is in automated high-density warehouses where AGVs supply the link between the high-reach stacker cranes and the goods receiving and dispatch areas. The second use is in automotive plants for the movement of engines, transmissions and body shells. The third application is as feeders to and within workshops based on advanced manufacturing technology such as flexible manufacturing systems (FMS). This application, although widely publicised, is not as yet well exploited. However, this application will grow and the demands on the AGV technology will grow with it since requirements for accuracy in positioning and for handling features mounted on the AGV are more stringent than those generally required in warehousing or automotive assembly. There is a growing class of specialised applications for AGVs such as in the handling of radioactive materials and waste. These applications will also grow as the AGVs become more flexible in their use.

We have seen with robots that the position of the end effector is mathematically determinable, relative to a fixed frame of reference origined at the base since the lengths of the links that connect the end effector to the base are known. A mobile robot or AGV, however, is in a moving frame of reference, thus its position must be determined relative to a fixed frame of reference somewhere in the surroundings. This fundamental difference between a fixed robot and AGV or mobile robot implies that in order to control the vehicle reliably either of the following three conditions must be met.

(i)   The motion of the vehicle must be heavily constrained, by fixing the paths of the vehicle, off-board the vehicle.

(ii)   A fixed frame of reference must be provided for the vehicle, so that it can refer continuously to this frame using on-board sensors, and therefore know its precise position in the surroundings. Hence the vehicle can detect and correct for deviations in its path, in comparison to the commanded path stored on-board.

(iii)   A fixed map of the surroundings must be put on board the vehicle, in which the position of the vehicle is known. The vehicle would then periodically produce a 'current map' of the surroundings, during motion, using on-board sensors. The 'current map' would then be compared with the fixed map to determine the current vehicle position. Using this technique the vehicle could dynamically recognise its surroundings and avoid obstacles, hence determining its own path to the commanded destination.

There are currently AGVs with fixed paths for guidance, but new free-ranging AGVs are in the final stages of development. Both types require a supervisory computer to control and monitor AGVs when used in any significant numbers. The various types of AGV are now discussed along with the all-important supervisory computer which makes up the complete AGV system.

15.1 VEHICLES GUIDED BY OFF-BOARD FIXED PATHS

15.1.1 Wire-guided AGVs

At present this is the most widely used guidance technique for industrial AGVs. This method uses a network of buried cables, as illustrated in Fig. 15.1. The cables are arranged in the form of complex closed loops, and each loop carries a different frequency a.c. signal. The loops are arranged in such a manner that at junctions in the network, different 'arms' of the junction are characterised by different cable frequencies. Small magnetic plates, fixed to the floor at all the exits of each junction, enable the AGVs to detect junctions during motion. Magnetic plates may also be present at the two ends of sharp bends in the network. This allows AGVs to approach these bends at top speed, reduce speed, negotiate the bend, and then proceed at top speed once again. By using different frequencies in the network, usually in the range 7 to 15 kHz, a 'passive' network results. If only one frequency was used, some form of active switching would be needed at each junction to direct the AGV traffic along the required paths, making the system complex and prone to failure. Hence, in a passive network an

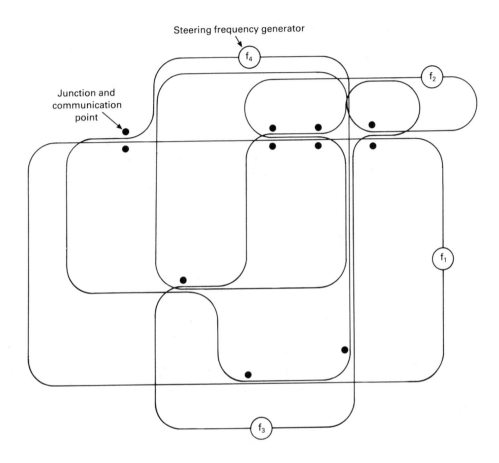

Fig. 15.1 – A typical wire-guided AGV network.

AGV is programmed to follow a sequence of cable frequencies, that correspond to the sequence of network paths, which connect the starting location to the destination.

In such networks as well as 'steering cables', 'communication cables' are also present. The communication cables link each communication point, see Fig. 15.1, to the network supervisory computer. An AGV identifies itself at a communication point and reports its status to the supervisory computer. Thus the supervisory computer can track the progress of each AGV in the network, and can 'block' the network sections already occupied by AGVs, to avoid collisions. If the network is large, containing many AGVs, a hierarchical control structure is usually adopted, as shown in Fig. 15.2.

The actual 'inductive guidance' technique used by wire-guided AGVs, is illustrated in Fig. 15.3. It essentially consists of two 'steering coils' fixed underneath the AGV, a few centimetres above the ground, which pick up the a.c. steering coil signal via electromagnetic induction. These two signals are then

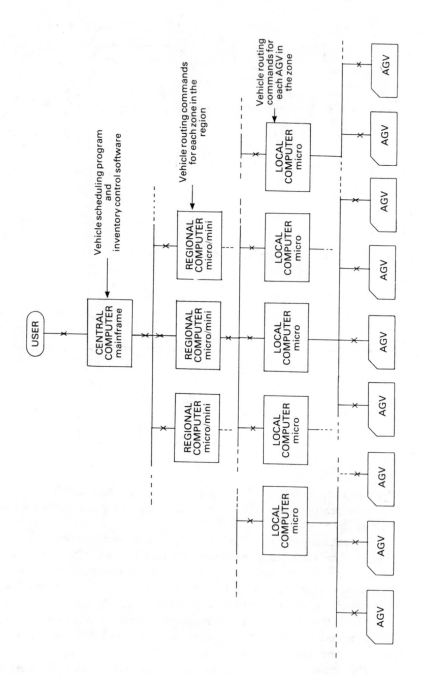

Fig. 15.2 — A typical control hierarchy of a wire-guided AGV system.

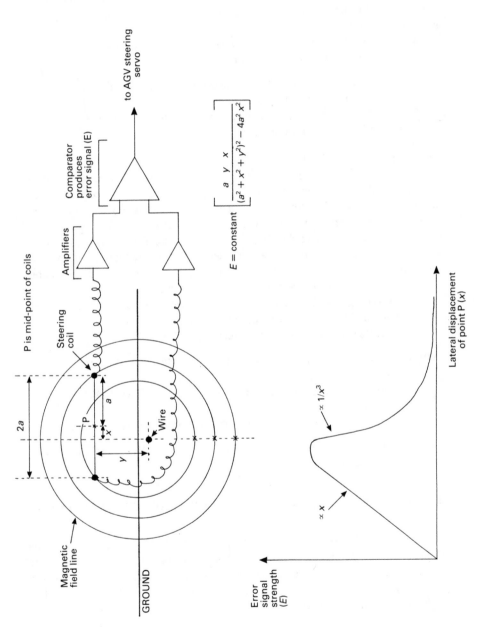

$$E = \text{constant} \left[\frac{a \; y \; x}{(a^2 + x^2 + y^2)^2 - 4a^2 x^2} \right]$$

Fig. 15.3 — Wire-guided AGV steering principles.

amplified, filtered and mutually compared, to produce an error signal. This error signal is then used via a servo system to steer the AGV. The error signal can be shown to be [3] linear with respect to lateral deviation (x) of the steering coils, as shown in Fig. 15.3.

Wire-guided AGV systems are popular in industry because they are fairly reliable, but they suffer from two major drawbacks.

(i)   AGV paths are not easily alterable because the steering cables are buried in channels about 3 mm wide and about 1 cm deep.
(ii) The cost of laying cables into the ground is quite high, about 40 dollars per metre, and is also labour-intensive.

15.1.2 Painted-line guided AGVs

These are generally found in light engineering or office environments. As systems they are similar to wire-guided AGVs, except that their guidance technique is different. These AGVs follow lines on the floor, which have been painted using a visible or invisible fluorescent dye. As an AGV traverses a line, it shines UV light onto it, thus causing the dye to fluoresce. Two photosensors, fixed in a similar manner as the coils in a wire-guided AGV, then detect the intensity of the fluorescence. These two signals are then amplified and mutually compared, to obtain signal suitable for servoing the AGV back onto the line.

The advantage of this guidance technique over wire guidance is that paths can be fixed quickly and are easily alterable using a dye solvent.

The disadvantages of this guidance technique are:

(i)   networks must be kept fairly simple, since junctions are not as easy to manage as in the wire-guided case;
(ii)   through wear and tear the dye gets erased and so the lines must be repainted from time to time;
(iii) lines can be obscured by objects thus disenabling the vehicle guidance.

Owing to these disadvantages, these AGVs are not suitable in medium to heavy engineering industries.

15.2 VEHICLES GUIDED BY ON-BOARD, SOFTWARE-PROGRAMMABLE PATHS

These vehicles represent the 'state of the art' in AGV research. The Imperial College AGV and vehicles built by various people [4–7], fall into this general category. The paths of these vehicles are software-programmable and therefore are easily alterable. Various novel techniques exist for the guidance of these 'free-ranging' vehicles, as will now be described.

15.2.1 Various guidance techniques for 'free-ranging' vehicles

(i)   Dead-reckoning. This consists of periodically measuring the precise rotation of each vehicle drive wheel, using for example optical shaft encoders. The vehicle can then calculate its expected position in the surroundings, if it knows its starting point of motion.

(ii) Position referencing beacons. These beacons are fixed at appropriate locations in the surroundings. The precise positions of these beacons are known to the vehicle. As the vehicle moves it utilises some on-board device to measure its exact distance and direction from any one beacon. Hence the vehicle can calculate its own precise position in the surroundings.

(iii) Inertial navigation. This consists of setting up the axis of a gyroscope, parallel to the direction of motion of the vehicle. If the vehicle deviates from its path, an acceleration will occur perpendicular to the direction of motion and this is detected by the gyroscope. This acceleration is then integrated twice to yield the position deviation from the path, which can then be corrected for by a servo mechanism. Note, however, that since acceleration is detected, path deviation at constant velocity cannot be corrected.

(iv) Optical or ultrasonic imaging of the surroundings. This consists of generating an 'absolute map' of the surroundings in which the position of the vehicle is known, and storing this on-board the vehicle. The vehicle then periodically generates a 'current map' of the surroundings, as it moves, using an on-board video camera or ultrasonic transducers. Various objects in the 'absolute map' are then recognised in the 'current map' and by cross-correlation, estimates for the position of the vehicle are obtained. These estimates are then statistically averaged to obtain an overall estimate for the position of the vehicle.

   Note, that an ultrasonic map of the surroundings is obtained by emitting a narrow ultrasonic pulse, in all directions from the vehicle, and simultaneously activating an on-board timer. The time taken for echoes of the pulse to be detected at the vehicle, from all directions, is then measured. Since the speed of sound in air is known to the vehicle, the distances and directions to objects in the surroundings can be determined.

(v) Optical stereoscopic vision. This consists of viewing the same point on an object in the surroundings, using two on-board video cameras. Precise measurements are then made on the stereoscopic image of the object. The angular disposition of each camera is then measured and, since the inter-camera distance is known to the vehicle, the distance and direction to the object can be estimated. If the object is recognisable in the 'absolute map', as described in (iv), the 'absolute position' of the vehicle can be estimated. Repeating this procedure with several objects, an overall estimate for the vehicle position can be obtained.

Out of the guidance techniques just described, dead-reckoning and inertial navigation are the easiest to implement. However, inertial navigational equipment such as gyroscopes are expensive and bulky. The axis of a gyroscope also tends to drift with time, thus giving rise to errors. The biggest disadvantage of inertial navigation is that acceleration, rather than velocity, is sensed, so that vehicle deviations at constant velocity cannot be detected.

   Dead-reckoning, as described in (i), suffers from the problems of drive wheel slippage. If this occurred at a drive wheel, the encoder on that wheel would register a wheel rotation, even though that wheel is not driving the vehicle relative to the ground.

All guidance techniques utilising imaging of the surroundings suffer from complex, time-consuming scene-analysis calculations. Optical imaging additionally suffers from the requirement of critical ambient lighting. Ultrasonic imaging suffers from very 'fuzzy' images due to stray multiple reflections of the transmitted pulse. Also, the speed of sound in air is not constant, but it varies with temperature. Since temperature gradients can be present in the surroundings, these would add to the 'fuzziness' of the images.

In general, imaging does provide scope for more intelligent vehicles, which can dynamically change their paths because of the presence of obstacles but at present this technique seems to be hampered by the lack of fast image processors.

15.3 THE FREE-RANGING AGV

There is no doubt that the free-ranging AGV will soon be a reality since it offers much greater flexibility than the wire-guided systems and also allows the AGVs to be much more densely utilised. For example, assembly operations on products such as computers are likely to take the form of islands of assembly robots with free-ranging AGVs bringing components to be assembled and removing finished assemblies or subassemblies. High density utilisation of AGVs will be of paramount importance in such applications. The free-ranging AGVs also permit rapid rescheduling of tasks.

In order to understand the AGV concept a little more, a free-ranging vehicle under development at Imperial College will be described and compared with other concepts.

15.3.1 Free-range AGV concept

The following concepts were considered when designing the Imperial College AGV [8].

(i)   The steering mechanism should be kept simple, resulting in simple vehicle control. 'Differential steering' was chosen as shown in Fig. 15.4 since this allows for a vehicle which could operate in both directions with equal performance. It is recognised that the 'tricycle arrangement' as shown in Fig. 15.4 is more suitable for a vehicle in certain applications but it is more difficult to reverse than the differentially steered vehicle.

(ii) On-board vehicle path storage should be kept in a simple format, enabling a large number of paths to be utilised, for a given on-board memory. A factory layout can be set out for an AGV network as a series of nodes where the nodes represent stations where loading or unloading takes place or even intermediate points to assist with path definition. The vehicle should contain an on-board computer that determines its path between nodes in terms of straight lines and rotation about the centre of the vehicle. A vehicle will get path instructions from a supervisory computer in terms of a series of nodes and a time of arrival and departure from each node. A vehicle path is shown in Fig. 15.5.

(iii)   A dead-reckoning method should be continuously used to compute the vehicle position and this should be periodically verified by using an independent position measurement system. Optical shaft encoders should be used to monitor continuously the position of the vehicle and these were to be driven by spring-loaded wheels mounted outboard of the two driving wheels, as shown in Fig. 15.6. This is to reduce the effect of slippage of the driving wheels which could cause errors in the odometry system. The alternative measuring system could be provided by a variety of techniques and several methods were proposed using either an on-board scanning laser onto active beacons (Fig. 15.7) or a fixed laser reflecting off the vehicle.

(iv)   In a multiple vehicle system, the position of each vehicle must be monitored regularly by a supervisory computer for management purposes. Thus a communication system must exist between the vehicles and supervisory computer. Most systems utilise either an FM or an infra-red link.

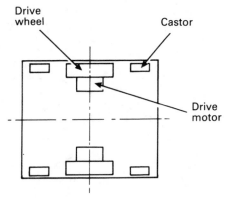

Differential
steering wheel arrangement

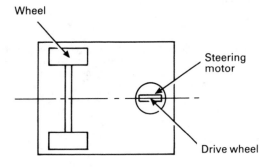

Tricycle-type wheel
arrangement

Fig. 15.4 – AGV wheel and drive layouts.

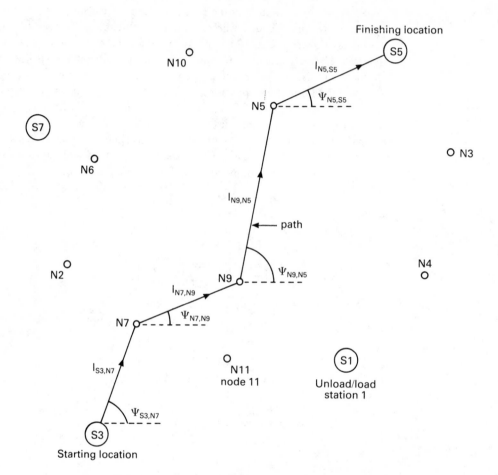

Fig. 15.5 – Path storage form of ICAGV. The sequence of paths $\ell_{i,j}$ and rotations $\psi_{i,j}$ represent the shortest route from S3 to S5. $\psi_{i,j}$ and $\ell_{i,j}$ are elements of the AGV on-board path library.

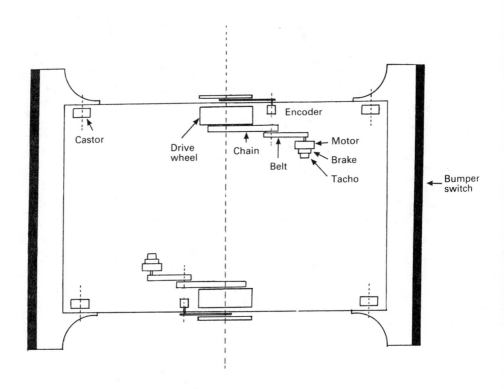

Fig. 15.6 – AGV mechanical structure.

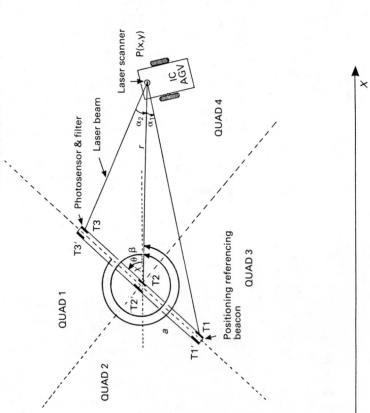

T2 is at $(x2, y2)$, x is known, and α_1, α_2 are measured, then $P(x,y)$ is given by:—

$$x = x2 + r\cos(\theta)$$
$$y = y2 + r\sin(\theta)$$

where :—

$$r = \frac{a \sin(\alpha_2 + \beta)}{\sin \alpha_2}$$

$$\beta = \tan^{-1}\left[\frac{2 \tan\alpha_1 \tan\alpha_2}{\tan\alpha_2 - \tan\alpha_1}\right]$$

In QUAD 1
$$\theta = x + |\beta|, \quad \beta = |\beta|$$

In QUAD 2
$$\theta = x + |\beta|, \quad \beta = \pi - |\beta|$$

In QUAD 3
$$\beta = \pi + |\beta|,$$
$$\theta = x + |\beta|$$

In QUAD 4
$$\beta = 2\pi - |\beta|, \quad \theta = x + |\beta|$$

Fig. 15.7 — Position determination by laser triangulation.

Fig. 15.8 — Imperial College free-ranging automated guided vehicle (ICAGV).

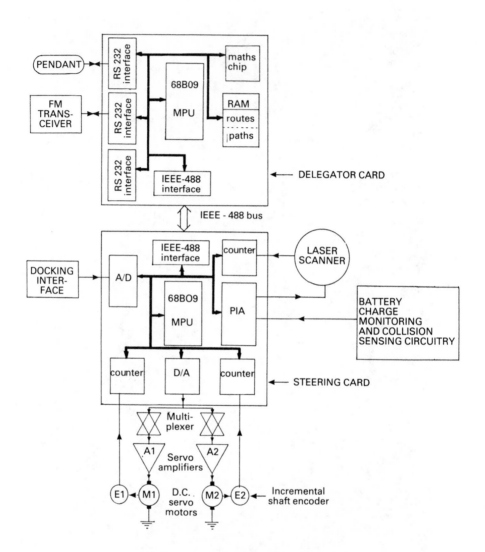

Fig. 15.9 – Control structure of ICAGV.

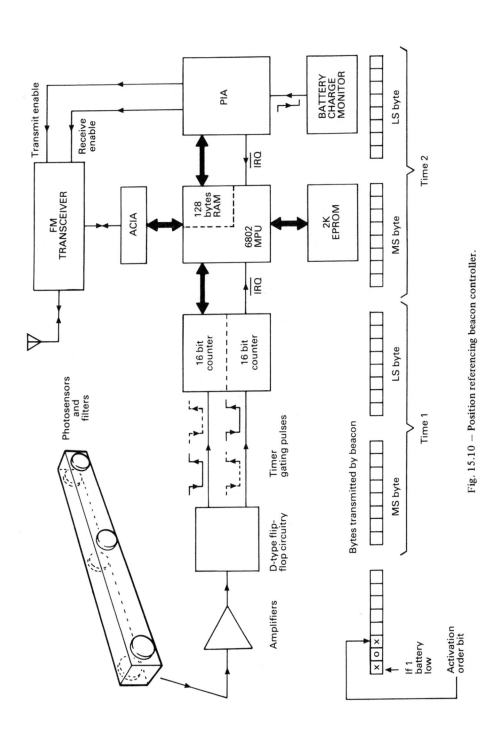

Fig. 15.10 — Position referencing beacon controller.

(v) Accurate docking is important in many AGV applications and sensors should be provided to facilitate this action.

(vi) Safety from vehicle collisons must be assured. There is a need to ensure that obstacles are sensed well in front of the vehicle (2 m) as well as providing a safety bumper containing an emergency switch.

15.3.2 The vehicle

A free-ranging AGV was built at Imperial College to meet these criteria. The AGV was powered by two 800 W Mavilor d.c. servomotors with integral disk brakes and tachos. The motors were powered by Infranor servo amplifiers from four 12 V lead-acid batteries. Odometry was provided by two 1000 ppr Muirhead optical encoders. The AGV is shown in Fig. 15.8 and is capable of carrying a load of 1000 kg.

The on-board control computer is a multiprocessor unit similar in concept to that used on the R D Projects robot discussed in Chapter 13. A schematic of the control structure is shown in Fig. 15.9. The controller is made up of two computers communicating via an IEEE-488 bus. One computer, the 'steering card', provides closed loop control of the motors while the second computer, the 'delegator', provides for communication with the supervisory computer or a pendant.

Collision-sensing is provided by two ultrasonic sensors and in infra-red sensor fore and aft.

15.3.3 Position determination via fixed beacons in the surroundings

The position-determination system using a scanning laser onto fixed beacons is shown in Figs. 15.7 and 15.10.

This method relies on the fact that, if there are three equidistant, co-linear points T0, T1, T2 then the position of another co-planar point $P(x,y)$ can be calculated by measuring the angles α_1, α_2, and using the equations shown in Fig. 15.7. This method was implemented by using three photosensors as T0, T1, T2, which was coupled to an FM transmitter as shown in Fig. 15.10. A lower power laser on board the the vehicle scans the surroundings periodically. Each time LASER light falls on either of the three sensors, a radio message is transmitted, in the form of a bit stream, that identifies the activated beacon and the activated sensor. By receiving and decoding this message, the vehicle is able to measure α_1, α_2. Then using the maths chip on its delegator card, the vehicle is able to compute its position rapidly, since the positions of all the beacons are known to the vehicle.

15.4 THE SUPERVISORY SYSTEM

Conventional wire-guided AGV systems have path networks which consist of interconnected loops around which the vehicles always travel in the same direction. The problem of scheduling the AGVs in such a system is not particularly complex, since all that is necessary is to ensure that a vehicle does not attempt to join a loop at a junction at the same time as another vehicle is passing the junction. A free-range AGV system, however, can have a large complicated network of interconnecting, bidirectional paths and hence requires a much more intelligent supervisory system.

The supervisiory controller system has to perform a number of tasks. It has to deal with the generation of transportation demands. It must determine vehicle routes, check for conflicts and perform rerouting if necessary. It must communicate with all the vehicles and finally it must simulate the vehicle system together with the provision of an animated display of the simulation.

15.4.1 Supervisory fleet control

Controlling a fleet of AGVs is not an easy task and it is important to discuss this subject in more detail. Most of the discussion will be aimed at free-ranging AGVs, although some aspects of the subject matter will be applicable to the wire-guided AGV since the latter is a simplified special case of the free-ranging AGV system.

In a free-ranging AGV system it would be possible for each vehicle independently to determine how to get from one point to another by the most efficient route. However, this would inevitably lead to two or more vehicles deciding to use the same path or occupy the same path junction simultaneously. They should, of course, detect each other before they actually collide, but would then have to find alternative routes to their respective destinations. From such considerations it is clear that finding routes for the vehicles is best done with a complete knowledge of paths and nodes that are available for use. Hence the need arises for an overall scheduling algorithm.

The scheduling algorithm must be able to accept a transportation demand, such as take a pallet from station A to station B, and a priority. Also it has to decide on the vehicle to be used, the route to be used and communicate all the appropriate information to the vehicle.

The vehicle dispatching rules for a free-ranging AGV system need not differ significantly from those for wire-guided systems. Simulations performed on wire guided systems serving flexible manufacturing systems (FMS) environments give an indication of the best vehicle selection algorithms, although they do indicate that if the transportation system is kept busy at all times, then the vehicle selection problem reduces to one of waiting for any vehicle to be free. The Imperial College AGV scheduler developed by Broadbent [9] is able to schedule an arbitrary number of future journeys for each vehicle, so that vehicle selection rules will critically affect system performance.

When a vehicle has been selected, the scheduler must then find a route for it. The shortest route is found, taking account of any maximum speed restrictions which may exist on certain paths. This is achieved using the Dejkstra optimal minimum spanning tree algorithm [10]. Once the route has been calculated, a timetable for that route is created from a knowledge of the acceleration rate, speeds and cornering capabilities of the vehicle and its retardation. The timetable is used to produce a matrix containing the occupation times of all the nodes involved in that journey and, by inference, the occupation times of the paths between the nodes. This is then compared with a global future node occupation matrix to determine whether the new route conflicts with any journeys already under way. If a conflict is found then the journey priorities are examined. If the new journey is not of a higher priority than the existing journey with which it conflicts, then an analysis of the type of conflict is performed. Three types of conflict are distinguished:

(a) a collision along a path by vehicles travelling in opposite directions;
(b) a collision along a path by vehicles travelling along a path in the same direction but at different speeds;
(c) a collision at a path junction by vehicles approaching it along different paths.

When a conflict has been classified, a new route and timetable are calculated, with appropriate action taken to avoid the conflict. Type (a) conflicts are overcome by finding a new route not involving the path where the conflict occurs. It is here that the real benefit of having a highly redundant path network becomes clear. Type (b) and type (c) conflicts are overcome by slowing the new vehicle to let the previously scheduled journey proceed.

When transportation demands are re-scheduled it is necessary to re-check the new route and timetable since new conflicts may have been introduced later in the journey. The process of rescheduling and re-checking continues until either a free route is found or it is discovered that the network is too busy to allow the new vehicle to satisfy its demand. In this latter case, or if the new journey is found to be of high priority when the first conflict is found, the scheduler will investigate the possibilities of re-scheduling the already scheduled vehicle which causes the conflict.

A route for a vehicle therefore consists of a series of nodes, together with an arrival time for each node. Once calculated the route is transmitted to the relevant vehicle which then has to execute its timetable to be sure of completing its journey with no conflicts.

There will be occasions, such as when an obstacle is encountered or a breakdown occurs, when vehicles will be unable to complete their journeys as planned. The supervisory computer will be informed as soon as such an incident occurs. The supervisory computer then re-schedules the vehicle.

REFERENCES

[1] Morris, E. W., Developments in guided vehicle systems, possibilities and limitations and the economics of their operation. Proceedings of the 1st International Conference on Automated Guided Vehicle Systems, June 2–4, 1981, pp. 67–77.

[2] Sommer, J. Digitron's automated guided vehicle systems are controlled by standard software: A field-proven approach. Proceedings of the 1st International Conference on Automated Guided Vehicle Systems, June 2–4, 1981, pp. 95–101.

[3] Larcombe, M. H. E. Tracking stability of wire guided vehicles. Proceedings of the 1st International Conference on Automated Guided Vehicle Systems, June 2–4, 1981, pp. 137–144.

[4] Fujiwara, K. Development of guideless robot vehicle. Proceedings of the 11th International Symposium on Industrial Robots, October 7–9, 1981, pp. 203–210.

[5] Iijuma, J., Elementary functions of a self-contained robot 'Yamabico 3.1'. Proceedings of the 11th International Symposium on Industrial Robots, October 7–9, 1981, pp. 211–218.

[6] Nakamura, T., Edge distribution understanding for locating a mobile robot. Proceedings of the 11th International Symposium on Industrial Robots, October 7–9, 1981, pp. 195–202.

[7] Fujii, S., Computer control of a locomotive robot with visual feedback. Proceedings of the 11th International Symposium on Industrial Robots, October 7–9, 1981, pp. 219–226.

[8] Premi, S. K., Automated guided vehicles. Ph.D. thesis, University of London, 1985.

[9] Broadbent, A. J., Besant, C. B., Premi, S. K., and Walker, S. P. Free ranging AGV systems: Promise, problems and pathways. Advance Manufacturing Summit, Birmingham, 1985.

[10] Tanenbaum, A. S., *Computer networks.* Prentice-Hall, 1981, p. 38.

16

Flexible manufacturing systems

We have already discussed such topics as NC and CNC, robots and automated guided vehicles. These individual technologies can all be brought together to form a system or systems. Such systems are known as 'flexible manufacturing systems' (FMS) and are aimed at the provision of solutions to manufacturing problems. The objectives of FMS are as follows:

(i) To provide flexible manufacturing facilities for a family of workpieces.
(ii) To adapt the flexibility and productivity of CNC to the manufacture of medium volume quantities.
(iii) To provide manufacturers the benefits from maximising a combination of operations at a single location.
(iv) To increase the utilisation of facilities through the inherent flexibility offered by FMS.
(v) To provide better management control from the integration of computers, NC and automated materials handling.

FMS really means that it is possible to have a manufacturing system that can produce any part from a selected family of parts on a random basis without incurring system downtime for changeover. Clearly such a system requires advanced planning to ensure the availability of preset tooling and that part programs are fully tested. In practice, flexible manufacturing systems eliminate work-in-progress inventory by removing batch parts processing.

Flexible manufacturing systems offer considerable benefits to manufacturing industry as a result of the integration of systems and in particular from the fact that changes can rapidly be made through computer software rather than changes through hardware. Some of these benefits are as follows.

(a) *Minimisation of direct labour.* FMS utilises very little direct labour and this is often only in a supervisory role. The machines and materials handling system associated with FMS are all unmanned. Both machines and materials handling systems are under the direct control of a supervisory computer.

(b) *Minimisation of lead time.* Parts or workpieces that are produced via batch processing techniques suffer from downtime when no work is being performed on them. Parts can spend a large percentage of their time on the shopfloor in containers waiting to be processed. FMS minimises work-in-progress since it provides for a rapid throughput of work on the part.

(c) *Minimisation of inventory.* We have just seen that FMS leads to a rapid throughput of work on the shopfloor owing to more efficient handling and manufacture of parts. The time therefore that parts or workpieces spend on the shopfloor is reduced dramatically by the FMS concept and consequently inventory is minimised.

(d) *Minimisation of special tooling.* The machines used with an FMS are universal in nature and the specialised tooling is conceptually in software in most cases.

(e) *Minimisation of indirect labour.* With most systems, materials handling and inspection are automated and therefore the requirement for indirect labour is small.

(f) *Maximisation of flexibility.* FMS has one really big advantage over other systems in its flexibility of use. It can cope with changes in terms of product, engineering content, or technology. FMS is a modular concept and can be reconfigured, expanded or contracted as the necessity arises.

The justification for using the FMS concept must be made on cosideration of the range or family of parts to be manufactured, the quantities and the effect of machining parts in a single location.

16.1 FMS APPLICATIONS

FMS is particularly suitable for manufacturing complex parts requiring considerable machining in steels, cast iron or aluminium in the 10 to 1,000 kg range. Parts would be milled, bored, turned, faced, drilled, tapped etc. in FMS applications and many machined features would have to be considered with close attention to accuracy and geometric relationships.

Clearly a considerable investment is required in setting up FMS and the correct mix of parts coupled with an adequate volume, is necessary to justify such expenditure. The job-shop approach is usually considered for the low-volume manufacture of parts. Such parts would be produced in batches and would be moved around various work centres as required. This approach is flexible for small batches and many parts can be machined concurrently with the correct planning. There is little special tooling required for a given part and production mix can be changed rapidly.

However, the job-shop approach is costly since the output per machine is low owing to setting-up time, material handling and dependence on direct labour. In-process inventory also tends to be high which further adds to costs.

If the volume is very high ($>10^4$ parts per year) then special tooling with transfer lines can be justified. Multi-spindle automatic machine tools could be used and the cost per part would be low. Such systems require very high investments and the automobile industry is a typical example of such lines. These transfer lines are highly dedicated to particular parts and changes cannot easily be accommodated without incurring major costs and disruptions. Accurate market forecasting and careful planning is required before embarking on this type of production.

For the mid-volume (100 – 10,000 parts per year) manufacture of the same kind of parts, the choice has been between the job-shop approach or a modified transfer line. The result has been unsatisfactoy in terms of cost-effectiveness since both systems of production were being used in an inefficient manner because the quantities were unsuited to both systems. The FMS concept incorporates the flexibility of the job-shop and high productivity of the transfer line for production in the mid-volume range.

16.2 EQUIPMENT USED IN FMS

We have already stressed how FMS is made up of standard-type machines and technology but the essential feature is that the machines are brought together and controlled using computers to operate as a system. Thus it is the system integration coupled with automated materials handling which is the hallmark of FMS.

While most machines within FMS are standard CNC machining centres there are occasions where specialised machines are introduced to fulfil a specific function. Most of the machines in FMS utilise randomly selectable heads to perform functions such as drilling, tapping, boring, facing, grooving etc. Automatic inspection stations are often included in FMS.

Materials handling is an important feature in FMS and AGVs are now commonplace as feeders to these systems. They permit palletised parts to be moved from station to station in a flexible and simple manner. This type of layout leaves the floor area clear, permits easy access and can be readily expanded. The AGVs can themselves be used as part of the workstation with assembly operations being performed on the AGV platform.

Robots are playing an increasing role in FMS and add flexibility to the materials handling. It further reduces the requirement for special loading fixtures which are expensive and reduce flexibility. Future FMS technology is looking to a closer integration of flexible machines. For example, assembly operations could be performed on islands containing robots which are fed by free-range AGVs in a densely packed configuration. Some of the assembly operations may even take place on the moving AGVs as they pass close to a robot.

Within an FMS is a controlling computer which coordinates the functions of the various flexible machines in the system. It acts as the link between the CNC systems within the FMS, the robot, the gauging station and the AGV supervisory system. It performs the functions of planning and scheduling, simulation and health monitoring. It also acts as a storage facility for CNC part programs with editing facilities.

16.3 MATERIALS HANDLING

The material handling system moves parts from a load/unload area, such as a store, to and between the machine tools or robots that make up an FMS and returns finished parts back to the store area. The loading and unloading at the store end is performed by stacker cranes that work in conjunction with the AGVs. Loading and unloading at the FMS stations is either performed by a robot or shuttle system. The FMS materials handling system is flexible and can handle a wide range of parts. This is quite different from transfer lines which tend to be inflexible to part changes.

An FMS materials handling system must have the following attributes:

(a) Provide automatic, random movement of palletised parts through a system of workstations involving interaction between robots or shuttle systems on machine tools.
(b) Operate in a general machine shop environment.
(c) Permit easy accessibility to systems for maintenance.
(d) Permit integration with existing systems with the minimum of physical disruption.
(e) Be safe with full protection of personnel.
(f) Be expandable on a modular basis and be compatible with computer control systems.
(g) Provide the required performance at minimum cost including low first cost and maintenance, ease of control, ease of scheduling and adaptability to various workstation layouts.
(h) Serve as a workstation in its own right, e.g. cleaning, assembly or inspection.

16.4 ECONOMICS OF FMS TECHNOLOGY

The economics behind FMS technology is extremely important since the investment in FMS is relatively high. The approach to the economic justification of FMS is analogous to that of the stand-alone CNC machining centre. The philosophy is similar in that the maximisation of the combinations at a single location is a high priority.

In considering the economic justification of FMS it should be remembered that FMS is a considerable technological leap from conventional machine shop or transfer line technologies. We have already discussed the advantages that the FMS concept can bring to industry provided that the correct application is found in terms of a family of parts and volume.

The various alternatives that must be considered for manufacturing parts in the mid-volume range are as follows:

(a) Standard machinery with special tooling.
(b) Special dedicated machinery.
(c) Transfer line.
(d) CNC stand alone.
(e) Flexible manufacturing system.

When considering the manufacturing requirements the following priorities should be adopted.

(i) Minimisation of direct labour.
(ii) Minimisation of work in progress.
(iii) Maximising inventory turn.
(iv) Maximising flexibility.
(v) Minimising factory expense.

These are all high priority considerations that must be carefully determined in terms of relative merits. There are secondary considerations which are as follows.

(i) Minimisation of tooling and fixture investment.
(ii) Reaction to market changes.
(iii) Minimisation of set-up time.
(iv) Minimising indirect labour.
(v) Minimising lead times.
(vi) Minimising batch size.
(vii) Maximisation of ability to match capacity with requirements.

Finally there are tertiary groups for consideration but these are of a much lower priority than the two previous groups. The tertiary group is as follows.

(1) Minimisation of setting up.
(2) Minimisation of indirect labour.
(3) Maximisation of equipment utilisation.
(4) Reaction to engineering changes.
(5) Minimisation of capital investment.
(6) Minimisation of marketing lead time for forecasting.
(7) Minimisation of production control.
(8) Minimisation of industrial engineering requirements.

Justification of flexible manufacturing systems, as with numerical control systems, can be approached on a direct labour comparison basis. Whatever approach is adopted, manufacturing objectives must be matched with available alternatives with long-term planning. Considerations for investment in FMS technology may arise when a significant requirement for replacement of existing equipment is due. Generally, replacing existing equipment with similar equipment or even NC systems can prove more expensive than going for FMS technology.

17

Process planning

In CAD/CAM technology the organisation of databases is extremely important. It may be that a company produces a large number of manufactured components so that when a designer is creating a new design of product, every effort is made to use existing components to increase standardisation and reduce inventory costs. Components or parts are therefore grouped and identified in order that the designer can easily retrieve the requisite information from the database. Parts are also grouped according to their similarities to be compatible for similar manufacturing processes. Thus a large number of parts or components may be grouped into a relatively small number of families. This grouping of parts is a manufacturing philosophy known as 'group technology'. The grouping of components into families of similar design and manufacturing characteristics results in manufacturing efficiencies from reduced set-up times, lower in-process inventories, better scheduling, improved tool control and the use of standardised planning procedures. Furthermore, machines are often grouped in a certain way to take advantage of group technology to facilitate work flow and parts handling.

In any computerised design and manufacturing system, the ability to code parts in a way that aids retrieval and planning is of fundamental importance. Parts classification and coding is concerned with identifying the similarities among components and relating these similarities to a coding system. Part similarities are of two types, one related to design attributes, such as geometric shape and size, and the second to manufacturing attributes that involves the processing steps required to make the part. In general there is often a correlation between design and manufacturing attributes. However, this is not always the case and most of the more recent coding systems are devised to allow for differences between the design of a part and its manufacture.

17.1 PART FAMILIES, CLASSIFICATION AND CODING

We have already seen that part families are built up from parts that have similarities of geometry, size and manufacturing processes. The coding system, however, must be such that differences within a family can be identified. For example two parts may have a similar geometry and size but their materials and tolerances on size could be completely different leading to different manufacturing processes.

The part family is extremely important to the concept of 'computer-aided process planning' (CAPP) not only for design retrieval systems but also for the layout of machines for manufacture and the routing of workpieces. Where batch production is involved, parts have to be moved around from one machine to another and this results in significant amounts of materials handling and long manufacturing lead times. The result is that machines are often grouped together in cells where each cell will specialise in the manufacture of a particular part family.

The grouping together of machines in cells relies on a proper parts classification and coding system which is not easy to achieve. The classification procedure can be achieved by randomly sampling parts made within a company. This is not satisfactory since there is a risk that the sample may not be completely representative of the entire population. A better method is to adopt one of the many classification and coding systems that are internationally recognised.

These coding systems all have a number of fundamental common features which make up their general structure. These features can be grouped as follows.

Design attributes:	Basic external and internal shapes, length/diameter ratio, material, major and minor dimensions, surface finish.
Manufacturing attributes:	Major and minor processes, major dimensions, length/diameter ratio, surface finish, machine tool, operation sequence, batch size, fixtures, cutting tools.

Systems in computer-aided planning are essentially based on the manufacturing attributes but these also contain some of the design attributes.

The coding structure is usually hierarchical in nature with a chain like structure at each level of the hierarchy. This is similar to the database structures for CAD/CAM systems discussed in Chapter 11.

17.2 COMMON CLASSIFICATION AND CODING SYSTEMS

There are many such systems that are available commercially and only a few will be briefly discussed here. It is important to assess the objectives before a system is chosen for a particular application. The cost of introducing a system is high in terms of training and installation. It must also fit in with the existing computer systems used within the company and, finally, management must be educated and supportive of its introduction along with union cooperation.

(a) The Opitz system

This parts classification system is one of the most famous and was developed at the University of Aachen by Opitz [1].

The Opitz code is made up from the following digital sequence:

12345 6789 ABCD

The basic code is made up from nine digits which can be extended by a further four. The first nine digits refer to both design and manufacturing information related to a part. The first five digits, 12345, are called the 'form code' and describe the primary design attributes while the next four, 6789, refer to the manufacturing attributes. The extra four digits, ABCD, are the 'secondary code', and refer to the production operation and its sequence. Consider the 'form code' which is made up of five digits. Each digit can be numbered 0 to 9. Digit 1, classifies parts according to rotational or non-rotational. If they are rotational then 0, 1 and 2 refer to various length/diameter ratio. Digit 2 refers to the external shape of the part such as 7 for a cone or 8 for an operating thread. Digit 3 refers to the internal shape such as holes, grooves and threads. Digit 4 identifies plane surface machining, such as an external spline, 4, or an internal slot, 6. Digit 5 specifies auxiliary holes and gear teeth.

(b) The MICLASS system

MICLASS stands for Metal Institute Classification System and was developed by TNO in the Netherlands [2]. MICLASS was developed to standardise certain engineering processes such as in design, production and management of production.

The MICLASS classification number can range from 12 to 30 digits. The first 12 digits are universal and can be applied to any part. These digits cover the following attributes:

Digit	Description
1	Main shape
2 & 3	Shape elements
4	Position of shape elements
5 & 6	Main dimensions
7	Dimension ratio
8	Auxiliary dimensions
9 & 10	Tolerance codes
11 & 12	Material codes

A further 18 digits may be used to classify additional information covering such items as batch size, operational sequence and cost data. The MICLASS system is computer-orientated and allows for parts to be classified interactively. It is a universal system in that the program can be used with a variety of languages, English, French, German or Dutch and it operates in imperial as well as metric units.

17.3 PLANNING FUNCTIONS

The planning of manufacturing processes involves the sequencing of individual manufacturing operations which may consist of machining or assembly operations. Besides the process planning are the functions for determining the manufacturing or cutting conditions for machine operations. In the past this has been a clerically labour-intensive activity which is now becoming very computerised. The early techniques involved a planning engineer who, given an engineering drawing of a component would decide on the type of machine tool, if machining was required, or process if assembly or some other operation was required. A planning sheet would then be completed detailing each machining stage or process, including the specification of special tooling as well as standard tools. The process or production engineer would organise the scheduling of machines and flow of materials and components in order to optimise production facilities. Such techniques were not only labour-intensive but nearly always did not give the optimum solution since there could be considerable differences amongst process and planning engineers as to what constituted an optimum solution.

Automated process planning systems based on the use of computer programs has sought to integrate logic, judgement and experience into a viable problem-solving method. This method is based on the characteristics of a given part from which the program can generate the manufacturing sequence. Computer-aided process planning (CAPP) permits the generation of manufacturing procedures and routes which are both consistent and optimal. There are two principal CAPP systems, one being the 'retrieval-type system' and the other the 'generative system'.

17.4 RETRIEVAL-TYPE PLANNING SYSTEM

Retrieval-type computer-aided process planning systems use parts classification and coding together with group technology as a starting point. Once the family of parts have been properly classified and coded, a standard process plan for manufacture is created for each family of parts. The standard process plan is stored on computer ready for retrieval for the process planning of a new part. In some cases the standard plan may have to be modified to cope with a new part. This happens when the manufacturing requirements of the new part differ slightly from the general requirements specified for the family of parts. Sometimes the retrieval CAPP system is known as a 'variant system', owing to the modifications that are made in the plan.

The detail procedures used in the retrieval system are illustrated in Fig. 17.1. The sequence is commenced by entering the part code number to the computer. The CAPP program then searches the part family file to determine if a family exists for the part. If an identified match is found then the standard machine sorting and operation sequence are retrieved from the computer for display. The standard plan can then be edited in order to make it compatible with the new component. The formatter then prepares the appropriate paper document and creates a new number for the new plan.

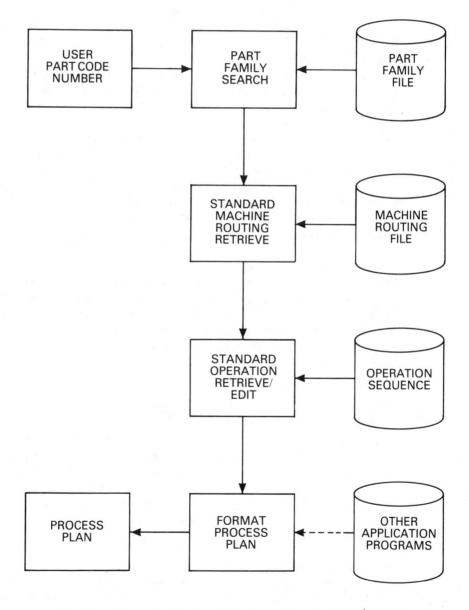

Fig. 17.1 – Information flow in a retrieval-type, computer-aided process plan.

17.5 GENERATIVE PROCESS PLANNING SYSTEMS

Generative process planning makes use of the computer to create an individual plan without the requirement for human intervention. Technological and logistical decisions are made by the computer, using a set of algorithms, in order to achieve a viable manufacturing plan. The generative CAPP system starts from basic data on component geometry, material and other information relating to manufacture, and synthesises the design of the optimum process. In practice systems based on generative CAPP fall short of this ideal and are generally limited to a small range of manufacturing processes. Furthermore, human intervention is often utilised during decision making or for checking purposes. This is done interactively as the program is being run.

Typical industrial generative CAPP systems are GENPLAN developed at Lockheed [3] and CAR developed in Japan by Akiba and Hitomi [4]. The CAR system will now be described in some detail to illustrate the principles of generative process planning. CAR seeks to determine the best process route or sequence of operations automatically when given the basic geometric data and properties of the part together with the initial shape of the raw material. Unit patterns consisting of basic elements of geometrical shapes to be removed by machining operations are identified by the user and are displayed in an interactive mode on the display of a CAD system. CAR then produces the optimal sequence for removing each unit pattern. The CAR flow system is made up of four programs which are shown in Fig. 17.2. The system is as follows.

(a) Processing geometrical shape

The original shape of the raw material for a part is first specified interactively using CAD techniques. Unit patterns that are suitable for producing the final shape of the part against the initial shape are retrieved along with additional information such as dimensions and accuracy. After receiving all the necessary information the program fits the unit patterns on the specified positions of the blank and eliminates those portions. The technological feasibility of operations are then checked.

(b) Sequencing of operations or process route

Once the unit patterns have been specified and checked, the program then automatically determines the optimum sequence of operations. This is done in connection with the machine tools to be used by arranging the unit patterns with a primary routing program which decides on the principal machinery of constitutional shape elements. A secondary routing program is used for supplementary machining of technological surface elements such as precision machining, drilling or screw cutting. The sequence of operations and the proportions to be machined are then displayed for checking purposes.

(c) Deduction of production times

Production times are calculated for each operation and are based on standard time data. All the necessary data required for these calculations are on file within the computer.

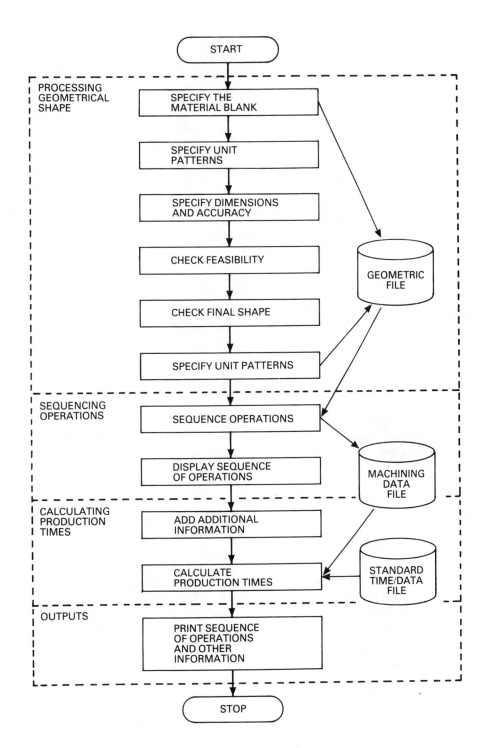

Fig. 17.2 – Flow chart for the CAR system.

(d) Output documentation

The results obtained in the procedures discussed are printed out as a recommended sequence of operations together with any other information required for the manufacture of the part.

17.6 MACHINABILITY INFORMATION

The goal of increased automation typified by the concepts of 'computer-integrated manufacture' (CIM) and 'flexible manufacturing systems' (FMS) is unmanned operation. The economic justification of the development of unmanned systems is very great [5]. However, the full potential of such systems will be lost unless their overall performance is monitored and optimized to the conditions prevailing. It is no good keeping a process going if it is producing scrap. Occassional scrap becomes increasingly significant as batch sizes become smaller because of reductions in product life cycles and the drive to reduce in-process inventory. Alternatively, cutting conditions needed to ensure reliable operation might significantly reduce productivity.

The optimization of a particular process, usually on economic grounds, requires 'experience' in order to select the best operating conditions. Traditionally selection of the operating conditions was left to the experience or judgement of the process planner, foreman or machine operator. Although feedback was possible, it was neither systematic nor reliable. Personal judgement was undesirable because it had no scientific foundation. However, by collecting the experience of more than one person and using systematic analysis of large quantities of data, handbooks of machinability information were compiled. These cutting recommendations were often based on laboratory experiments. Although these represented an improvement over personal judgements, they also suffered some drawbacks. The values were usually conservative, did not coincide with a particular product line of machine tool and were not of a form compatible for use in a computerized database. To overcome these difficulties, efforts have been directed to the development of computerized machinability data systems, linked to computer-aided process planning (CAPP) systems. However, these systems all rely on accurate and reliable data, otherwise they are being 'built on sand' [6]. With unmanned operation of machine tools there is a need for information feedback on more than just machine utilization or status. The performance of the cutting tool, and hence the cutting process needs to be monitored. In fact success with unmanned machining will depend on automatic sensing to monitor performance and make compensations, combined with automatic updating of machining 'experience' data.

17.7 PROCESS OPTIMIZATION

Optimization processes have been developed to improve the operational performance of NC machine tool systems [7]. The two distinct methods of optimization are adaptive control (AC) and machinability data prediction. Adaptive control optimizes an NC process by sensing and logically evaluating variables that are not controlled by position or velocity feedback loops. Essentially, an AC system monitors process variables such as cutting forces, tool temperatures or motor torque and alters the NC commands so that optimal metal removal under safe conditions is maintained. The selection of suitable speeds and feeds is essential for the application of NC to machining.

The selection of feedrate and speed are the responsibility of the process planner. Machinability data is usually chosen so that one or more of the following criteria is satisfied.

(1) Tool life   — the cutting tool lasts for a specified period of time.
(2) Surface finish   — a specified surface smoothness is achieved and maintained.
(3) Accuracy   — tool deflection and vibration are below a specified maximum.
(4) Power consumption   — power consumption is maintained below a certain level.
(5) Economic criteria   — a maximum production rate or minimum cost per piece is achieved.

The criteria used will depend on the process and the state of the process. For example, for a roughing cut, power consumption and economic criteria will be

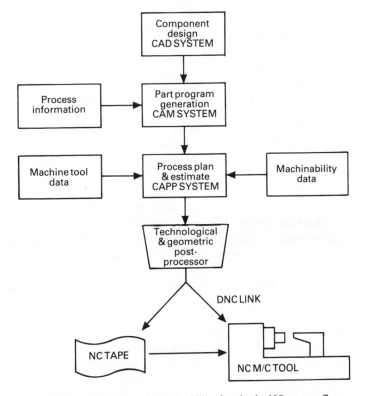

Fig. 17.3 — Acquisition of machinability data in the NC process flow.

the most important; whereas, for a finishing cut, accuracy and surface finish will be more important.

A typical process schematic showing the relationship between the machinability data and the NC process flow is illustrated in Fig. 17.3. The geometric description of the component is assessed from the CAD system. Additional information is required about the manufacturing process before the CAM operator can define the cutter paths and produce an elemental breakdown of the operation steps. Using the CAPP system the planner can select optimum machine operating parameters using the machinability database as 'experience'.

17.8 MACHINABILITY DATA SYSTEM

Process planning should include the selection of cutting conditions that are to be used in various machining operations. The cutting conditions consist of speed, feed and depth of cut. Depth of cut is usually predetermined by the workpiece geometry and operation sequence. Therefore, the problem reduces to one of selecting the proper speed and feed combination. Machinability data systems are intended to solve this problem.

The objective of a machinability data system is to select cutting speed and feedrate given that the following characteristics of the operation have been defined.

(a) Type of machining operation.
(b) Machine tool.
(c) Cutting tool.
(d) Workpiece.
(e) Operating parameters other than feed and speed.

For each of these operating characteristics there are a multitude of different parameters. A partial list is presented in Table 17.1.

There are several methods available to solve the feed/speed selection problem:

(1) to rely on the skill and experience of the process planner, foreman or machine operator;
(2) to use information compiled into machining handbooks;
(3) to use computerised machinability data systems.

Of these, computerized machinability data systems offer the greatest benefits. Efforts to develop machinability data systems date back to the 1960s and are continuing today. Some of these systems have been developed by individual firms to meet their own specific requirements. The importance of these systems has grown with the increase in the use of NC machines and the economic need to operate these machines as efficiently as possible. The importance of computerised machinability data systems will continue to grow with the development of integrated manufacturing data systems.

Table 17.1
Characteristics of a machining operation

---

(1) Type of machining operation
    (a) Process type — turning, drilling, milling, grinding etc.
    (b) Roughing operation verses finishing operation

(2) Machine tool parameters
    (a) Size and rigidity
    (b) Horsepower
    (c) Spindle speed and feedrate levels
    (d) Conventional or NC
    (e) Accuracy and precision capabilities
    (f) Operating time data

(3) Cutting tool parameters
    (a) Tool material type (HSS, cemented carbide, ceramic)
    (b) Tool material composition
    (c) Physical and mechanical properties (hardness, wear resist.)
    (d) Type of tool (single point, drill, milling cutter etc.)
    (e) Geometry
    (f) Tool cost data

(4) Workpart characteristics
    (a) Material — basic type and specific grade
    (b) Hardness and strength of work material
    (c) Geometric size and shape
    (d) Tolerances
    (e) Surface finish
    (f) Initial surface condition of workpiece

(5) Operating parameters other than feed and speed
    (a) Depth of cut
    (b) Cutting fluid, if any
    (c) Workpiece rigidity
    (d) Fixture and jigs used

17.9 COMPUTERISED MACHINABILITY DATA SYSTEMS

Computerised machinability data systems can be classified into two general types:

(1) database systems;
(2) mathematical model systems.

17.9.1 Database systems

These systems require the collection and storage of large quantities of data from the laboratory experiments and shop experience. To collect this data, cutting experiments need to be performed over a range of feasible conditions [8]. For each set of conditions, the total cost and component costs are calculated. To use this data the user would have to enter certain descriptive information to identify the type of machining operation, work material, tooling and so on. The printout would consist of recommended cutting conditions.

17.9.2 Mathematical modelling systems

Instead of simply retrieving cost information on operations that have already been performed, the mathematical model systems attempt to predict the optimum cutting conditions for an operation. The prediction is generally limited to optimum speed, given a certain feedrate.

A common mathematical model to predict optimum cutting speed relies on the familiar Taylor tool-life equation [9].

$$VT^n = C$$

where V = surface speed (m/s), T = tool life (min), and C, n = constants.

By combining this equation with the conventional cost breakdown per component, an equation for the speed to give minimum cost machining can be derived. In a similar way and equation for the cutting speed that yields maximum production rate can also be derived. These or similar equations are used in the predictive-type machinability data systems to determine recommendations that approximate optimal cutting conditions.

The potential weakness of these systems lies in the validity of the Taylor tool-life equation. Taylor's equation is an empirical relationship derived from experimental data that contains random errors. There are also dangers in extrapolating the Taylor equation beyond the range which data has been collected. However, other models can be used that better approximate to particular data.

17.10 MACHINABILITY DATA COLLECTION SYSTEM

The next logical step with regard to computerised machinability databases is to try to automate the collection of the machining data. This has several advantages. To begin with, the existing machining data collected from laboratory condition experiments could be verified and correlated with the shop experience. In this way the effects of the conditions peculiar to the shopfloor could be determined and included in factors used to convert old experimental data. By automating the collection, processing and storage of machining data the cost of acquiring this information could be reduced. Better use could be made of existing data by processing it and presenting it in a compact and usable form. A more systematic approach should produce more accurate analysis of the data. The use of such a system could mean that the database could be made 'dynamic', that is continuously changing to reflect the conditions on the shopfloor.

The ability to monitor overall performance means that it should be possible to notice changes in performance and take corrective action. Another possibility of machine tool performance monitoring would be the ability to record the suitability of particular machine tools for certain jobs. For example, one machine tool may be able to produce parts to a higher accuracy and surface finish than a seemingly identical machine. By knowing the capacity of individual machines it would be possible to match job specifications to particular machines considering more than just size and power requirements. The outline specification of a system to monitor, analyse and store relevant machining data has been proposed by Steen [10] and will be discussed next.

17.10.1 Adaptive optimization
The proposed system would be based on the storage of mathematical models of tool wear. See Fig. 17.4. A simple system might work something like this.

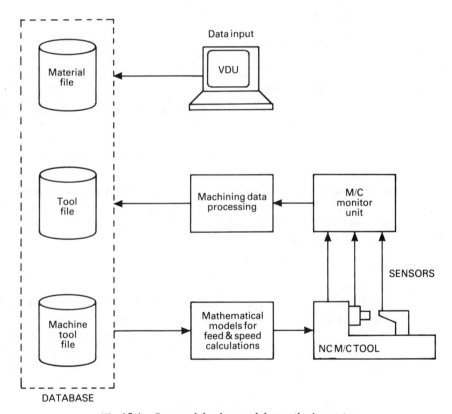

Fig. 17.4 – Proposed database and data gathering system.

The tool wear curve can be broken down into three main stages as shown in Fig. 17.5. The simplest form of mathematical model for this would be a straight line approximation with slope B_1 and an intercept on the flank wear axis of B_c. The linear model of flank wear $w(t)$ as a function of time, t is given by:

$$w(t) = B_0 + B_1 \cdot t + E.$$

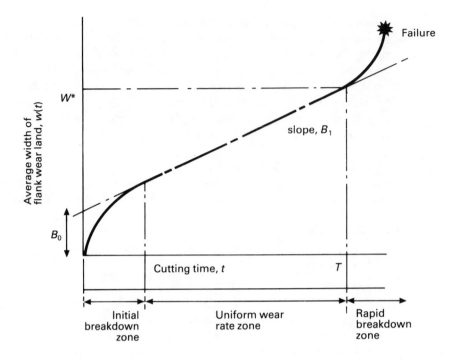

Fig. 17.5 – Tool-life curve for constant cutting conditions.

where E is a random variable normally distributed about mean zero and with variance σ^2. This approximation is not unreasonable as the region of linear wear normally accounts for the largest period in the life of a tool. A more general model of the form,

$$w(t) = B_0 + B_1 \cdot t + B_2 \cdot t^2 + E$$

can be used for better approximation.

Given cost and time data and estimates of the parameters n and C from prior tool-life tests, an initial minimum cost cutting speed (assuming this is the criterion used), V_{min}, can be determined and production begun at this speed. See Fig. 17.6, together with definitions of V_{min}, n and C which are given below:

average component cost,

$$C_c = C_0 t_m + \frac{t_m}{T}(C_0 t_c + C_t) + C_0 t_n$$

where

$$
\begin{aligned}
C_0 &= \text{operating cost,} & \text{£/min} \\
C_t &= \text{tool cost,} & \text{£/edge} \\
t_m &= \text{machining time,} & \text{min} \\
t_c &= \text{tool-change time,} & \text{min} \\
t_n &= \text{handling time,} & \text{min} \\
T &= \text{tool life,} & \text{min/edge}
\end{aligned}
$$

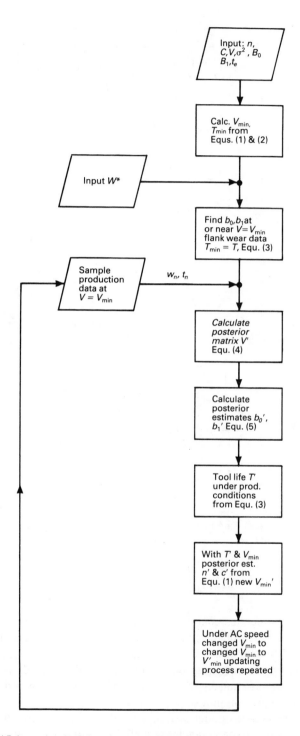

Fig. 17.6 – An algorithm for machining process optimization under adaptive control.

$$V_{\min} \frac{C}{\left[\left(\frac{1}{n}-1\right)\left(t_c+\frac{C_t}{C_0}\right)\right]^n} = \frac{C}{\left[\left(\frac{1}{n}-1\right)t_c\right]^n} \tag{1}$$

$$t_e = t_c + \frac{C_t}{C_0} \text{ , min/edge}$$

$$T_m = \left(\frac{1}{n}-1\right)t_e \tag{2}$$

$$T = \frac{w^* - B_0}{B_1} \tag{3}$$

$$V^1 = V + Z^T Z = \begin{bmatrix} V_{11}+K & V_{12}+\sum_{i=1}^{K} t_i \\ V_{11}+\sum_{i=1}^{K} t_i & V_{12}+\sum_{i=1}^{K} t_i^2 \end{bmatrix} \tag{4}$$

$$b_1 = \begin{bmatrix} b_0{}^1 \\ b_1{}^1 \end{bmatrix} [v^1]^{-1} \cdot [V \cdot b + Z^T w] \tag{5}$$

where

$$Z = \begin{bmatrix} 1 & t_1 \\ 1 & t_2 \\ \vdots & \vdots \\ 1 & t_n \end{bmatrix} \qquad w = \begin{bmatrix} w_1 \\ w_2 \\ \vdots \\ w_n \end{bmatrix}$$

$t = $ time, min

$w = $ wear land, mm

As production continues, sample data about tool wear will become available, and application of the Bayesian learning model will give updated tool-life parameter estimates n' and C'. From these posterior values $V_{\min}'$ and $T_{\min}'$ for the minimum cost cutting speed and tool-life are determined. As further production data becomes available the cutting speed can be continuously updated to match the prevailing conditions.

The basic method described has wider application to other processes such as milling and grinding providing a matheamatical model of the cutting process can be defined.

REFERENCES

[1] Opitz, H. *A classification system to describe workpieces,* Pergamon Press, Oxford, England.

[2] Organisation for Industrial Research. Miclass, Migroup, Miplan, Migraphics, Waltham, Mass., USA.

[3] Hegland, D. E. Out in front with CADCAM at Lockheed — Georgia. *Production Engineering,* November, 1981.

[4] Akiba, H. and Hitomi, K. Computer-aided routing system for manufacturing industries. Report, Ozaka/Kansai Institute of Information Systems, 1975.

[5] Schaffer, G. Sensors: the eyes and ears of CIM. Special Report No. 756, *American Machinist,* July, 1983.

[6] Ermer, D. S. The effect of experimental error on the determination of optimum metal cutting conditions. *Trans. ASME,* May, 1967.

[7] Pressman and Williams. *Numerical control and computer-aided manufacture,* Chapter 9, Process Optimisation.

[8] Evershiem, W. *Computer-aided determination and optimisation of cutting data and cutting time.* CIRP, 1981.

[9] Ermer, D. S. A Bayesian model of machinery economics for optimisation by adaptive control. *Trans. ASME,* August, 1970.

[10] Steen, W. Machinability data from production experience. Internal Report, Imperial College, May, 1984.

18

Factory management

18.1 INTRODUCTION

In general, the ultimate aim of any manufacturing company, or, for that matter, any company, is to make profits through the sales of goods or services. To achieve the aim of making maximum profits, a company must provide goods or services that can satisfy the requirements of the customer while inventory investment and plant operating cost are kept to the minimum. Certainly, it is not an easy task to strike a balance in order to achieve all these objectives together, as they are conflicting and interdependent. Furthermore, the efforts necessary to achieve each objective are organised separately by different departments in the company. The sales department generally requires that sufficient inventory is available to meet the immediate needs of customers and the production process must be flexible enough to be easily adapted to satisfy the changing requirements of the market. On the other hand, the finance department tries to reduce as much as possible the capital needed for inventory. In addition, the manufacturing department prefers consistent schedules for production with little or no overtime required and infrequent changes of products. As can easily be imagined a change in one objective will affect the others and it is undoubtedly a difficult problem to try and coordinate the various departments to achieve these conflicting and interdependent objectives for a profitable operation.

With the arrival of the computer, many people began to realise the tremendous potential it offered to coordinate the activities of these departments to form an integrated production management function. In some cases, this potential is taken even a stage further with the production management function being incorporated into a fully integrated CAD/CAM system that covers all a company's activities involved in making the products.

18.2 COMPUTER-ASSISTED FACTORY MANAGEMENT

The management of a manufacturing operation can generally be divided into the following typical functions.

(a) *Material requirements planning.* Material requirements planning, MRP for short, is the method used to derive the master schedule from the forecast, sales orders, or both. The master schedule is the foundation of all the operations. MRP handles ordering and scheduling of inventories such as raw materials, sub-assemblies and component parts. MRP identifies the individual components and sub-assemblies that make up each end product, and indicates the required quantities and when to order them so that they are available when needed for assembling the final product. This topic will be described in greater detail in section 18.3.

(b) *Inventory and production control.* Inventory and production control functions ensures that sufficient products of each type are available to satisfy customer damands while capital investment in inventory is kept at a minimum. It keeps track of movements of raw materials, components, sub-assemblies and assembled end products to avoid potential shortages and delays. The levels of the various product sales, production and inventory are constantly monitored so that an optimum can be achieved to prevent the danger of inventory shortages and the expensive cost of maintaining excessive inventory.

(c) *Capacity planning.* Capacity planning is concerned with the manufacturing of products from raw materials and components using the labour and equipment available in a company. It determines and plans the adequate amount of workforce and machines needed to meet the master production schedule. With capacity planning, preventive actions can be taken to eliminate potential bottlenecks in advance. Also, areas where human and equipment resources are underutilised can be corrected to make them more efficient. The function of capacity planning is to ensure that the production plan can be achieved by appropriately adjusting plant capacity.

(d) *Shopfloor control.* No matter how much detail has gone into producing as accurate a master production schedule as possible, the plant will not, unfortunately, operate exactly as planned, owing to unpredictable events. Machines may break down and thus cause an accumulation of work behind some critical machine. Many other unexpected production problems may occur and interrupt the master schedule. With hundreds or possibly thousands of current work orders in a shop, it is an enormous problem to try to make sure that work orders will ultimately meet the original schedules. The function of shopfloor control is to document each work order so that they can all be identified and monitored as they progress through the factory. The current status of work orders is reported as information feedback to show any potential production delays and thus appropriate actions can be taken in time to overcome the problems.

(e) *Engineering*. The designs of various products are frequently changed to meet different market requirements, and modified to improve their functions. Whenever alterations are made to the design of a product, they must be recorded and controlled with effective dating. In addition, every bill of materials affected by the product design changes will be correspondingly updated, together with cost estimates for the changes. This function generates up-to-date bills of materials that can be used for both engineering and manufacturing.

(f) *Cost control*. The cost of manufacturing a particular product must be monitored to ensure that it is not unusually high. Cost control analyses the value of inventory, work-in-progress, labour and so on to compute the actual costs that occurred during production. Cost breakdowns of all parts or sub-assemblies may be produced if necessary. The actual costs are compared with the standard costs to check for any excessive variances so that any problem areas can be found and explained. Then, corrective actions can be taken to eliminate, or at least reduce, these variances.

(g) *Purchasing*. To manufacture products, the company will have to order raw materials and/or purchase components from subcontractors. The function of purchasing is to work out the detailed requirements of raw materials and components for every job, and place the purchase orders at the right time and in the right quantity. It will also keep track of the deliveries of purchase orders and check their quality. Purchasing is often responsible for validating supplier invoices and reviews cost variances against standard costs.

(h) *Sales and marketing*. Typically, the sales and marketing department of a company receives an order from a customer and initiates the procedure to make the products. It will promote and sell the products for the company. Other important functions are to review current and overdue shipping costs and status reports of sales orders to prevent customer backlogs. It arranges the despatching of sales orders by assigning orders to shipments and validating quantity and date of shipments.

Many companies have already used computers to help perform each of the factory management functions mentioned in previous sections. Some have even integrated these functions, using computers, into a single system to provide a closed-loop order planning, monitoring, and distributed communications for factory management. A large number of production management systems can deal with inventory control, material requirements planning, shopfloor control and cost control.

Computer-assisted factory management systems are usually organised into modules, each of which performs a specific factory management function, as shown in Fig. 18.1. The modular design of these systems enables the company easily to modify, update, and expand them to accommodate growth and changing needs. They offer an on-line and interactive means of monitoring every stage of the manufacturing process by linking almost every element of company operations and activities, together with a combination of closed-loop information management, data collection and feedback functions.

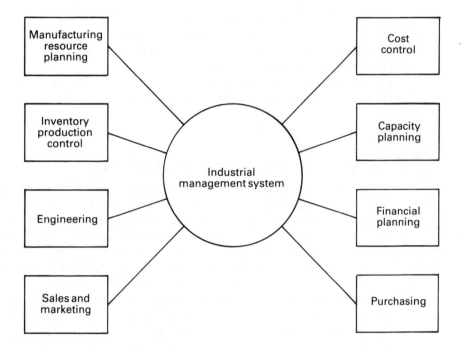

Fig. 18.1 – Organisation of a typical factory management system.

In an industrial environment, production requirements are always changing, so computer-assisted factory management systems are self-adjusting to continuously replan and reallocate existing inventories and other resources. Thus, inventories needed to manufacture the products are minimised as much as possible, given the master production schedule, lot-sizing policies and manufacturing lead times. For example, when new information is added to the system, it can automatically update every schedule so that inventory balances, parts lists and work order status reports reflect the latest changes. When there is an increase in the cost of some materials, the system will compute exactly how the rise will affect the budget. When there is a delay in the delivery of some components, the system will work out how much production schedules will slip, how overtime and machine schedules can be organised to make up for this delay, and how much the costs of manufacturing will be affected by it.

One useful advantage of these systems is that management is able to forecast and see the possible effects of proposed changes in strategy on operations. The company can at least have the chance to examine the implications of their critical decisions prior to being firmly committed to them, and, therefore, better manufacturing planning decisions can be made.

With the kind of advantages described before, it is not difficult to imagine the following significant benefits that can be derived from the application of computer-assisted factory management systems.

(a) Effective long-term planning.
(b) Minimum levels of inventory and less inventory obsolescence.
(c) Better selection of raw material and component suppliers.
(d) Optimum balance of production.
(e) Increased productivity and profits.
(f) On-time deliveries of products and better customer service.

18.3 MATERIAL REQUIREMENTS PLANNING (MRP)

Computers are now often employed to perform the various factory management tasks such as those described in the previous section. More commonly, they are used to carry out inventory control and production scheduling in systems that have come to be known as material requirements planning (MRP) systems. Material requirements planning is the technique used to produce the master production schedule for end products from the demand forecast, sales orders, or both. Then, the master schedule, which forms the basis for all the operations, is converted by MRP into a detailed schedule for raw materials, components and sub-assemblies used in the end products. This detailed schedule outlines the required quantities and when to order each raw material, components and sub-assemblies so that they are delivered in time to meet the master schedule for the assembled final products.

Although MRP is relatively simple and straightforward in concept, it is rather complicated to apply because of the sheer magnitude of the data involved. The master production schedule has to be drawn up in terms of month-by-month or week-by-week delivery requirements for assembled final products. Each of the products may consist of hundreds of individual component parts which are produced out of raw materials. Some of the component parts may be made of the same raw material. The component parts are put together into simple sub-assemblies. In turn, these sub-assemblies are assembled into more complex assemblies. The process of assembling continues until the final product is reached. Each production and assembly step involved takes a certain length of time. All of these factors must be taken into account when performing the MRP computations. As can be imagined, each separate computation is fairly easy, but the amount of data to be processed is so enormous that it is in practice very difficult to apply the technique of MRP without a digital computer.

Figure 18.2 shows schematically the structure of a typical requirements planning system which usually contains four basic subsystems:

(a) the master production schedule;
(b) the bill of materials (BOM) file;
(c) the inventory status file;
(d) the material requirements planning software package.

As explained previously, the master production schedule indicates what final products are to be produced, how many of each of these products is to be produced, and when the products are to be ready for customer delivery. The master production schedule must be formed based on an accurate forecast of demand for the firm's products, together with a realistic estimate of its production capacity.

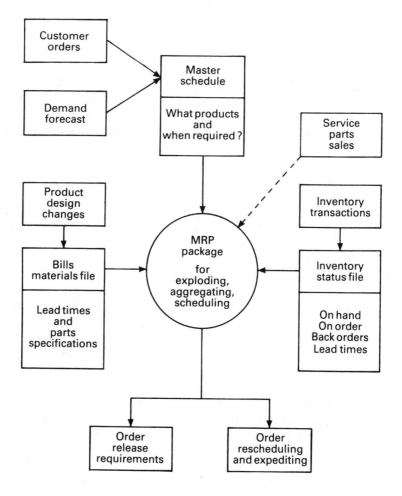

Fig. 18.2 – Structure of an MRP system.

Product demand that influences the master production schedule can be classified into three categories. The first category comprises actual customer orders for certain products which must be delivered before a specific date, as promised to the customer by the sales department. The second category is demand for various product lines as forecast from past demand using statistical methods and from estimates provided by sales staff and other sources. This forecast often makes up a large portion of the master schedule. The third category is demand for individual parts which are stocked by the firm's service department so that they can be used as replacement parts for the repair of existing sold products. As the requirements of repair parts do not represent demand for final products, this third category is usually not included in the master schedule, as indicated by the dotted arrow in Fig. 18.2 connecting directly to the MRP package.

The bill of materials specifies the composition of a finished product, and outlines the structure of a product in terms of its component parts and sub-assemblies. This product structure is important for the computation of raw materials and component parts required to manufacture each end product listed in the master schedule. A bill of materials file is a file that contains all of these product structures. An example is given in Fig. 18.3 to illustrate the simple structure of a product. The product structure is typically in the form of some kind of pyramid. The items at each successively higher level are called the parents of the items in the level directly below. Product P1 is the parent of sub-assemblies S1 and S2 which in turn are the parents of components C1, C2

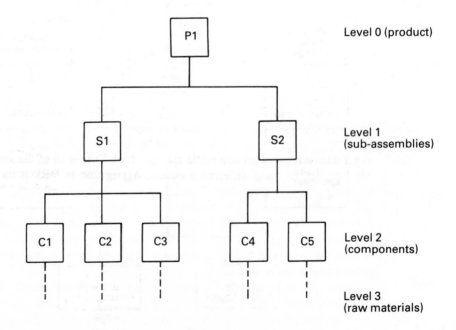

Fig. 18.3 – An example of product structure.

and C3, and components C4 and C5 respectively. The number in brackets associated with each item indicates how many of that item are used to make its parent. At the bottom of the structure are usually the raw materials, which are now shown in Fig. 18.3, used to produce the various components.

The inventory status file provides accurate current data on how much inventory is on hand and/or order. It basically contains a record of actual inventory level, lead time, ordered inventory and back order information.

The master schedule, the bill of materials file and the inventory status file are all inputs to the material requirements planning package. The MRP package processes the information in these input files using various computer programs to calculate the net requirements for each period of the planning horizon. Essentially, the MRP package performs the necessary explosion, aggregation and eventual scheduling of production for components.

The way that MRP works is that it begins from the master production schedule which specifies a period-by-period list of requirements for assembled final products. From this list, each final product is 'exploded' by the MRP package into its raw materials and component parts according to its product structure specified in the bill of materials file. The requirement for an end product is commonly referred to as independent demand because the demand for the end product is not related to demand for other products. The requirement for raw materials, component parts and sub-assemblies resulting from the explosion of end products is often referred to as dependent demand because the demand for these items is directly related to the demand for some end product. This demand dependency is due to the fact that the item is a constituent part of an end product. The demand for an end product must often be forecast, while the forecast for raw materials, component parts and sub-assemblies would not be necessary. Once the master schedule for end products has been established, the demand for raw materials and components can be calculated directly. For instance, the demand for motorcycles in a given period can only be forecast, but once that quantity is determined the total number of tyres to be made can be easily computed since each motorcycle requires two tyres.

After the explosion of the end products, the process of aggregation follows. Very often, many basic raw materials and components are used to produce more than one product. For example, the same type of metal rod stock may be used to produce screws on an automatic machine tool. Then each of the screw types may be used on many different products. Aggregation in MRP is the process of adding up the items used in common in various end products. The total quantities for each common-used item are aggregated into a single net requirement for the item. The advantage of aggregation is that it helps the accounting of requirements for raw materials and components. More importantly, it enables the company to effect economies in ordering the raw materials and manufacturing the components.

Once the requirements for each raw material and component have been calculated, MRP must determine when to commence work to make the products, taking into account the ordering lead times and manufacturing lead times for the required items. An ordering lead time for an item is the time required from the initiation of the purchase order to the actual receipt of the item from the supplier. For a raw material, the ordering lead time is relatively short, say several weeks, because it is usually stocked by the supplier. In the case of a component which must be fabricated by the supplier, the ordering and lead time could possibly be as long as a few months. The manufacturing lead time is the time required to manufacture the component part using the various machines specified on the route sheet. This manufacturing lead time is made up of the time taken to process the item by the machines and the non-productive time which must be allowed for waiting at each machine and for transferring the item between machines. The MRP package determines the starting date for assembling the sub-assemblies by offsetting the due delivery dates for these items by their respective manufacturing lead times. Then, the due dates for components must be similarly offset by their manufacturing lead times. Finally, the due dates for raw materials used to produce components must also be offset by their respective ordering lead times. To perform this lead-time-offset calculation, MRP usually has to obtain the data from the inventory status file and from the route sheet.

A typical format of output from MRP for each item is shown in Fig. 18.4. It basically consists of the following five items:

(a) Gross requirements – the total expected demand for a raw material or final product component during each time period.

(b) Quantity on-hand – the amount of an item already available at the beginning of each time period.

Component:	Week					
Lead time:	1	2	3	4	5	6
Gross Requirements						
Quantity on hand						
Net requirements						
Planned order receipts						
Planned order release	☐					

Action bucket

Fig. 18.4 – Format of MRP output.

(c) Net requirement – the difference between gross requirements and the expected quantity on-hand for each time period.

(d) Planned order receipts – the quantity of an item expected to be received by the beginning of the period in which it is indicated. If the company follows a lot-for-lot ordering procedure, this quantity should be equal to the net requirements. If an economical-lot-size ordering procedure is adopted, this quantity may exceed the net requirements. Any quantity in excess is carried forward to projected quantity on-hand in the following time period.

(e) Planned order release – the quantity of an item to be ordered or produced in each time period in order to ensure availability for the next-higher level of production or assembly. This quantity indicates gross requirements for items at the next-lower level in the production or assembly chain. The box for planned order release for the current period is called the action bucket because it is this box that initiates the required amount of ordering activity.

In general, the MRP package outputs a variety of information that can help a company in the planning and management of factory operations. These outputs include order release notice for placing orders that have been planned by the MRP package, reports of planned order releases in future periods, rescheduling notices for changes in due dates of open orders, cancellation notices for cancellations of open orders as a result of changes in the master schedule, and reports on current inventory status. Apart from these outputs, the MRP package can optionally generate other useful outputs which include (a) various performance reports to show costs, item usage and actual versus expected lead times, etc., (b) exception reports to indicate deviations from schedule and orders that are overdue, scrap and so on, and (c) inventory forecasts to show projected inventory levels (both aggregate and item inventory) in future periods.

As the benefits of material requirements planning are many, it is not difficult to explain the growing application of MRP in manufacturing industries. An increasing number of manufacturing companies either already use a computerised MRP system or are considering implementing one. Some of the major benefits of MRP are described below.

(a) *Low inventory levels.* As MRP sorts out priorities, it provides a better production schedule which can reduce average lead time and inventories. Very often, the actual machining time of a product is only a small fraction of the time that the product spends on the shopfloor. Most of the time it waits for parts and sub-assemblies to catch up. MRP produces proper priorities which ensure all parts and sub-assemblies are available on time. Consequently, this minimises waiting times and lead times, as well as inventories.

(b) *Avoiding possible delays.* As the MRP package can actually simulate alternative production schedules, late deliveries can be predicted and promised delivery dates can be confirmed. It is thus possible to detect delivery delays even before actual production commences.

(c) *Expediting and de-expediting orders*. Consider the situation where a customer decides to defer an order. As a result, the production schedule must be delayed. MRP then suspends the processing of component parts, and releases the machines for other outstanding jobs. This can prevent excessive in-process inventory of raw materials.

(d) *Long-term planning tool*. Apart from being an effective tool for material control and for scheduling purposes, MRP can also be used as a long-term planning tool. MRP can simulate various annual production plans, and check budgeting, manpower, and purchasing implications using the data stored in the bill of materials file, the inventory status file and the annual production requirements.

18.4 MANUFACTURING RESOURCE PLANNING (MRP II)

Manufacturing resource planning, often called MRP II for short, is a progressive evolution of the simpler material requirements planning, or just MRP for short. Material requirements planning has changed significantly over the years. Initially, It was merely an improved ordering method which made use of computers to perform the requirements planning calculations when computers first became available. Before the arrival of computers, the calculations used to be performed by hand and required tremendous amounts of time and effort to complete. Using computerised MRP, great improvements could be achieved in the ordering of raw materials and components as a result of the speed and accuracy with which the requirements planning calculations could be performed.

    This computerised MRP worked very well, but it could not cope with an unrealistic master production schedule. This was a production schedule that did not take into account the limitations imposed by plant capacity and other constraints. MRP would simply generate schedules and requirements that could not be achieved by the factory. Because of these problems, priority planning was incorporated into the computations of MRP in order to overcome them. Priority planning enables MRP to determine not only what materials should be ordered but also when those materials would be required. Material requirements could then be scheduled period by period, typically weeks or even days. With the priority planning incorporated, MRP can easily deal with urgent jobs by increasing their priorities, and also make it possible to de-expedite jobs by reducing their priorities.

    MRP with priority planning is still not complete without some feedback information relative to the execution of the priority plan. This close-loop MRP represents a major improvement by integrating the various functions in production planning and control such as inventory management, capacity planning, shopfloor control etc. into a single system. Closed-loop MRP also means that feedback data from suppliers of raw materials and components, the production shop and so on, allow potential problems to be detected and then corrected during the implementation of the production plan.

Although closed-loop MRP represents a significant improvement in the planning of material requirements, it can be further enhanced by linking together the closed-loop MRP system with the financial system of the company. The combination of these two systems is often known as manufacturing resource planning, or just MRP II.

MRP II includes two basic characteristics in additon to closed-loop MRP.

(a) *An operational and financial system.* The operational and financial system makes MRP II a system for everyone in the company. It covers all aspects of the company business which includes sales, production, engineering, inventories and accounting. MRP II uses financial data as the common medium for the operations of each independent department and between separate departments. Through this common medium, departments in the company are able to work closer together and the company management can obtain the necessary information to manage it successfully. For example, inventories on hand such as raw materials and work-in-process can be converted and evaluated into their equivalent cost. Similarly, other operating data can be translated into money terms so that a better picture of the company's financial performance can be obtained.

(b) *A simulator.* MRP II can be used as a simulator to answer 'what if' questions. MRP II can simulate alternative production plans and management decisions which are currently being considered. This enables the company to examine carefully the probable outcomes before committing itself to them.

REFERENCES

[1] *CAE systems and software annual,* Penton IPC Publications, 1982.
[2] Groover, M.P. and Zimmers Jr., E.W. *Computer-aided design and computer-aided manufacturing,* Prentice-Hall, 1984.
[3] Menipaz, E. *Essentials of production and operations management,* Prentice-Hall, 1984.
[4] Buffa, E. S. *Modern production/operations management,* 7th edn, John Wiley, 1983.
[5] Orlicky, J. *Materials requirements planning,* McGraw-Hill, 1975.
[6] Groover, M. P., *Automation, production systems, and computer-aided manufacturing,* Prentice-Hall, 1980.

19

Implementation of CAD/CAM system

19.1 INTRODUCTION

Although CAD/CAM began in the early 1960s and has gradually become known to the people in the research and academic world since then, it is only the last ten years that industry at learge has started to learn about it and has thought seriously about implementing it in their design and manufacture activities. With this popularity of CAD/CAM in industry, there are now numerous computer companies supplying different CAD/CAM equipment of different capabilities at various price ranges Since the early days of CAD/CAM, new, more powerful and Sophisticated hardware has come on the market. As a result, it can be quite a daunting task to try to choose from among those available commercially the right kind of CAD/CAM facilities that meet exactly the requirements of a particular industrial company. Furthermore, because CAD/CAM is a relatively new field, there are not that many people in industry who know the subject well enough to be in a psotion to select suitable CAD/CAM facilities to satisfy the specific needs of a company.

The main purpose of this chapter is to address the aforementioned problem, commonly encountered by industrial companies that wish to introduce CAD/CAM into their product design and manufacture processes. This problem often associated with implementing a CAD/CAM system is especially true and acute in a small industrial company which usually has very little in-house knowledge and experience of CAD/CAM. The subsequent sections of this chapter attempt to alleviate this problem by describing the various aspects that need to be considered in order to ensure a successful implementation of CAD/CAM. It is important to emphasise at this point that there is no hard-and-fast rule for the implementation of CAD/CAM; it depends very much on the circumstances of the company concerned. However, it would be fair to stress that the following points are crucially relevant considerations if the implementation of a CAD/CAM system is to be a success:

(a) financial justifications for the purchase of a CAD/CAM system;
(b) specifications of the CAD/CAM system required;
(c) evaluation and selection of CAD/CAM systems;
(d) installation of the CAD/CAM system selected;
(e) support services for the CAD/CAM system;
(f) future expansions of the CAD/CAM system.

19.2 SETTING UP A COMMITTEE

CAD/CAM is by now a very well-known and popular means of improving the productivity of an industrial company, and, indeed, many companies have already implemented such a system for designing and manufacturing their products. However, this should not be the main reason for a company to take the plunge into CAD/CAM just to follow the technological trend in industry. In other words, do not let overenthusiasm for CAD/CAM overtake business sense. Although the prices of CAD/CAM systems are gradually coming down, they are still very expensive and represent a major capital investment for most companies. In addition, it entails a substantial reorganisation for all the departments involved, therefore, a decision on whether or not to implement CAD/CAM must be taken seriously and cautiously.

When a company starts to consider implementing CAD/CAM, one of the very first questions it should ask itself is that whether or not a CAD/CAM system is really necessary for it to maintain and improve its business now and in the future. This is not an easy question to answer, but it is worthwhile thinking carefully about it before going any further with the idea of implementing CAD/CAM, because this can subsequently avoid any needless waste of time, effort and money by the company. By trying to answer this question, it may turn out that the company decides against the idea after all — maybe because it is not essential for the company to have it, or, at least, not at the moment. Or it may be that a CAD/CAM system is needed, but the company cannot afford it, or cost-justify it. Whatever the outcome and reasons, this question forms the first and vitally important stage for a successful implementation of CAD/CAM.

As already mentioned, CAD/CAM is cost-intensive and affects the working of many departments in the company, so it is of paramount importance to get together everyone concerned even at this early stage. All the departments affected directly or indirectly should be informed right from the beginning about the intention of introducing CAD/CAM into the company. It would be an advisable idea to set up in the company some kind of committee devoted entirely to the task of implementing CAD/CAM. The committee should consist essentially of representatives from various departments who would participate in this task on a part-time basis apart from their normal duties. Typically, some members of the committee would come from the design department, drawing office, manufacturing department, management and workforce on the shopfloor. The number of representatives in the committee from each of these departments depends on their size and the structure of the company.

By forming this committee, it ensures a more organised approach to the implementation of CAD/CAM. Also, every department involved has the opportunity to make known their views on the matter so that they can be taken into account in the final decision. It is vitally important that nobody is left out in the cold in the process because it is going to create more problems than help in the long run. Obviously, the committee members must cooperate with each other and work very closely as a team.

19.3 JUSTIFICATIONS

In an attempt to decide whether or not CAD/CAM should be implemented, the CAD/CAM committee must first justify the investment required. The justification process may vary considerably from one company to another, but, in the case of CAD/CAM. a good and convincing justification should contain three basic aspects. Generally speaking, a CAD/CAM system would have to be considered and justified financially, technically and socially.

Financially, one would try to compare the costs incurred with the benefits accrued from CAD/CAM. There are various different means of performing this comparison. Many advocate the use of productivity ratios to estimate the relative benefits and costs of CAD/CAM. Productivity ratio has been defined as the ratio of time taken by the existing manual method to that taken using CAD/CAM for the same work. The first part of such a comparison study is to review the present procedures employed in the company, and the workload volumes are analysed and categorised according to the type of products and the various activities involved in designing and manufacturing these products. Then, the CAD/CAM committee will have to consider for each of these products the kind of design and manufacturing activities where the techniques of CAD/CAM can be usefully applied. For each activity chosen to be replaced by CAD/CAM, estimate the amount of time saving that could be achieved if CAD/CAM was implemented, or, alternatively, determine the corresponding productivity ratio which will give the same indication, as both time saving and productivity ratio are related.

Although it is difficult to calculate accurately timesavings or productivity ratios at this stage, their estimates should be as close and realistic as possible. There are several methods that are commonly used to estimate productivity ratios.

(a) CAD/CAM system users. Much can be gained from talking to existing CAD/CAM system users who very often can tell you their first-hand experience of using CAD/CAM. More importantly, they can also explain how they implemented their CAD/CAM system so that you have the benefits of their experience

and know what kinds of problems and pitfalls to watch out for in order to avoid them. Besides, if they are cooperative enough, they may provide some performance figures of their own system which can help you estimate the productivity ratios. Better still, if they are really helpful and you know them well, they might even let you run some tests on their system for the purpose of estimating productivity ratios. This may sometimes be possible if it does not interfere with their normal work, and it can be arranged for the time when the system is not overloaded such as outside office hours and weekends. Useful performance figures are the actual job times on the system together with the associated preparation work and estimates of the equivalent times using manual methods.

(b) CAD/CAM publications. There is not a large number of books, magazines and sales literatures on the subject of CAD/CAM. They usually contain information on how to implement CAD/CAM and estimates of productivity ratios. However, these estimates should be viewed with caution as they may be just claimed possibilities rather than results obtained from actual tests.

(c) Benchmark tests. Benchmark tests are a good means of gaining data for estimating productivity ratios, but they are generally expensive in terms of company resources. It is not unusual to involve a number of different people in benchmark tests that extend over many weeks. Also, an actual CAD/CAM system is required to run the tests whose results are highly dependent on the skill of the people who take part. For this reason, benchmark tests are not recommended as a suitable method for estimating productivity ratios although they can be usefully employed to assess technical capabilities when various potential CAD/CAM systems are compared.

(d) Consultants. As there are already many CAD/CAM consultants in the market, it is possible to call in one to perform the estimation of productivity ratios for you. Although this consultancy will cost the company money, it is unlikely to stretch the already strained resources of the company too far. In fact, some companies, especially small ones, do not really have the in-house knowledge and experience to carry out these tasks, so consultants from outside the company are the only way. However, it must be remembered that not all consultants have actually been involved with the installation or operation of a CAD/CAM system. Hence, the performance data and other information quoted are not always what they have obtained first-hand but what they have read about.

CAD/CAM systems are particularly useful for the design and manufacture of components that have similar features in geometric shape and manufacture processes. In situations where components can be arranged into families of parts and CAD/CAM has been applied, it results in substantial time savings than is possible with manual methods. To estimate the productivity ratio in these circumstances, say, for the activity of draughting, a sample of drawings should be studied to assess the proportion of each drawing which contains similar items that are repetitively designed. These items are potential candidates for families of parts which could be automatically produced once they have been created in a CAD/CAM system. Here, it may be necessary to rationalise the design and drawing standards in order to maximise this proportion. For example, if the proportion that can benefit from using this family of parts technique is not more than about 75 per cent and the remaining 25 per cent is added either manually or through a CAD/CAM system, the maximum overall productivity ratio will be around 4:1 based on the reasonable assumption that the 25 per cent added has a productivity ratio of 1:1. However, in the calculation of this productivity ratio, the standard for each family of parts is assumed to have already been established in the system. If a more realistic figure is desired, the time and cost of setting up the standard must be taken into account.

Typically, the productivity ratio required to justify CAD/CAM is about 3:1, or a time saving of about 67 per cent. As the cost of CAD/CAM systems is gradually becoming less expensive, a lower productivity ratio is sufficient to achieve an acceptable cost justification. In addition, as the number of workstations in a CAD/CAM system increases, the average cost per workstation decreases, and thus the productivity ratio needed to achieve a certain payback period diminishes. This shows in effect that, for a large CAD/CAM system with up to, say, six workstations, if enough workload is there to utilise it fully, the average productivity ratio required to accomplish a satisfactory financial return is much less than that required by a system which is based on one workstation.

So far, only the methods for assessing the advantages and benefits of CAD/CAM have been discussed. It is equally important to estimate the costs involved in investment in CAD/CAM. Careful analysis of the principal costs incurred shows that they can generally be categorised as non-recurring and recurrring.

Non-recurring costs are the costs incurred when the system is first set up. They include many items such as the initial feasibility investigation, the hardware and software, personnel training and installation site. Also, it should account for the time lost due to the transition from manual methods to CAD/CAM.

Recurring costs are those incurred when the system is operational. They cover items such as maintenance, system support personnel and insurance. There are also items more difficult to estimate such as consumables, e.g. print-out paper, paper tape and drawing paper, etc., and system downtime.

The above cost items are mentioned here only as a guide and it is not exhaustive by any standards as different applications will need a different structure of costs. These cost items are often used to produce a cash flow analysis which can help to decide on the economic viability of CAD/CAM, and later plan the implementation programme if the purchase of a CAD/CAM system goes ahead.

Apart from justifying CAD/CAM financially, the same has to be done technically and socially. However, this is not so easy because the reasons are less tangible and cannot be translated directly into productivity ratios, time savings or cost reductions which can somehow be expressed in terms of figures.

CAD/CAM allows a component to be designed and manufactured, based on only one single definition of it in the central database of the system. Consequently, the resultant product will be more accurate and of higher quality. CAD/CAM can also make possible the standardisation of parts and processes which permits changes in specification to be performed quickly and other areas affected by these changes are updated automatically. It is quite common these days to use a CAD/CAM system for the simulation of prototype testing and verification of some manufacturing processes. If CAD/CAM is used, many of these processes are made easier if not eliminated. At least, it becomes unnecessary to have to actually produce the physical hardware such as a prototype just for the purpose of testing. As a result, product updates can be carried out speedily and cheaply; thus the product development lead times are greatly reduced. This factor is a distinct advantage, especially in a competitive market where the life cycle of products is relatively short. Sometimes, CAD/CAM is the only solution in a certain situation particularly in the area of analysis. For instance, the finite element method is a powerful and popular technique used to assist in the design of mechanical components for the analysis of stress and strain under both static and dynamic conditions. However, without the aid of a computer, it is impractical, if not almost impossible, to apply finite element analysis to component design, therefore, CAD/CAM becomes an absolute necessity if reliable analysis of a design is required.

There are many technical advantages to be gained from the application of CAD/CAM such as those mentioned in the last paragraph. As can be imagined, these advantages are rather qualitative in nature, which makes it very difficult to quantify and express directly in financial terms. Another such advantage is the fact that the purchase of a CAD/CAM system helps improve the image of the company. Many see CAD/CAM as a symbol of progressiveness and of being forward-looking on the part of the company. Such a dynamic company is often considered in industry as a desirable business partner. Sometimes, a customer may also have a CAD/CAM system which is found to be compatible with yours. This can eliminate the problems of transferring information such as drawings between the two companies as a direct data link can be easily established, or at least the transfer of data can be efficiently achieved through magnetic tape or disk. This situation can be a critical factor in obtaining a contract with a customer. One effect of using a CAD/CAM system is that the morale of the workforce will be boosted, provided that the introduction of such a system is not seen by the employees as a route to major redundancy and that they are involved throughout the implementation process. The reason is that a significant investment like CAD/CAM is viewed by the workforce as a commitment to stay in business and a determination to improve its competitiveness in the market. Despite the fact that CAD/CAM can enhance the image of a company both internally and externally, it must not be purchased and used only for the purposes of promoting the company's image if it does not actually provide any real beneficial effects on the design and manufacturing activities of the company, because CAD/CAM is expensive and it can well end up in disaster which will give the company a bad name and only achieve the opposite effect.

The use of CAD/CAM makes it possible to follow the trend of demands by the workforce for a shorter working week without causing a reduction in the productivity of the company, because CAD/CAM is much more efficient than traditional methods. Many industrial companies have to cope with a fluctuating demand pattern, which means that for some periods of the year they may be relatively underloaded with high peaks for other parts of the year. CAD/CAM provides the kind of flexibility for a company to meet changing demands, and to avoid the need for overtime working to deal with peak demands. Excessive overtime is undesirable in that it will cause interruption to the employees' social life and will lead to inefficiency during normal working hours. In addition, the premium rates for overtime will incur extra labour costs for the company.

In summary, this section attempts to outline some of the points that need to be considered before deciding to implement CAD/CAM on a company. The financial, technical and social aspects of CAD/CAM have been briefly covered. Obviously, there are many other things that could be relevant, but they depend very much on the particular application and business environment in which the company operates. Only when CAD/CAM has been justified in all three aforementioned aspects should the implementation be proceeded with. No doubt the CAD/CAM committee must satisfy itself that it is really in the interest of the company to introduce CAD/CAM. Also, it is probably fair to say that, although the points discussed here in this section are important considerations, a certain amount of common sense is just as valuable.

19.4 SPECIFICATION AND SELECTION

Obviously, this stage of specification and selection would not happen at all if the CAD/CAM committee could not justify the idea of purchasing CAD/CAM. If, on the other hand, the committee has decided to introduce CAD/CAM into the company, the next stage is to specify the CAD/CAM requirements and select a system that can best meet those requirements. There are generally three approaches to acquiring a CAD/CAM system.

(a) Purchase a whole turnkey CAD/CAM system which is one that a vendor will deliver, install and test for the customer until it is ready for use. It is called 'turnkey' because all a customer has to do is to turn the key to switch on the power in order to use it.

(b) Purchase graphics devices and software to build a CAD/CAM system on an existing computer system in the company.

(c) Purchase just the computer and graphics hardware required, and develop the associated CAD/CAM software within the company.

The first choice seems to be the most popular approach for most companies to acquire a CAD/CAM system becuase they usually do not have the necessary knowledge and experience in-house. It is only logical in this situation that they buy an entire CAD/CAM system, and let the vendor deal with any problems rising during implementation and support both the hardware and software of the system afterwards. If the company has already had a computer system, it would be sensible to extend the existing system by adding graphics hardware and soft-ware for CAD/CAM. However, this means that the computing department of the company will have to be involved, so prior agreement and arrangement with them are particularly important in order to avoid any major interruptions to their normal schedule. Probably, additional staff will be needed in the computing department to handle the extra workload. The third approach entails developing a complete CAD/CAM system from scratch by the user company. This means that the company would require some specialists with expertise in CAD/CAM. These CAD/CAM experts must be good programmers and understand graphics so as to solve application problems. Unfortunately, people with this combination of skills are rare, and they are usually in great demand by the CAD/CAM system vendors themselves. Another problem with this third approach is that even if competent programming staff is available, it will still take time to develop a project as difficult and complex as a CAD/CAM system. Such sophisticated projects could well run into years to complete because the learning process involved is often long and thus expensive. On the other hand, the turnkey CAD/CAM vendors have already invested this time and effort and have finished moving up the learning curve, so any CAD/CAM capabilities that a user company might desire can be satisfied by the vendors from their existing available product lines. This ensures that waiting and delay are kept to a minimum, and the desired capabilities can be run almost as immediately as they are delivered. Besides, CAD/CAM systems are the business of the CAD/CAM vendors who must spend a considerable amount of their resources in developing their CAD/CAM technology continuously to stay in competition. The scope and extent of their work cannot really be matched by any user company whose main line of business is not CAD/CAM.

For the above reasons, the rest of this chapter will be presented with the assumption that the first approach is adopted by the user company, that is, an entire turnkey CAD/CAM system is purchased, although the principles and techniques involved can still be applied to some degree in other approaches.

The CAD/CAM committee formed during the justification stage should remain the same throughout this and subsequent stages if possible, because, as soon as members are removed and new members are added to the committee, fresh ideas but contrary to those considered before will be put forward. Consequently, the project will end up forever discussing the numerous options and features that are worth including, and no real progress can be made.

To specify the requirements for a suitable CAD/CAM system, the members of the CAD/CAM committee must be sufficiently familiar with the technology itself apart from the activities of the company. As members come from various departments, it is not unusual to find that not everyone in the committee is knowledgeable enough in CAD/CAM. In such a situation, the first would be to educate the members so that they could understand most of the technicalities involved. This education can take the form of attending courses and exhibitions. Many universities and colleges of further education now organise courses, seminars and demonstrations on the subject. If they are relevant, it is well worth attending them. There are also CAD/CAM trade exhibitions scheduled throughout the year. They are useful sources of information as well because a large number of CAD/CAM system vendors and graphics hardware suppliers are all there under one roof to show their latest products. It is an excellent opportunity to view all the available hardware for CAD/CAM in the market and possibly have a demonstration too. Maybe, there is a chance to talk to some experienced salesmen and their support staff who can give some preliminary ideas as to what kinds of CAD/CAM capabilities are most suitable for the company. If the company knows any related organisations that have already installed CAD/CAM systems, arrange a mutually convenient time to see their system if possible. Ask them about all the different aspects of the system, how they cost-justified it, and how the final selection was determined. Generally, find out how they tackled the task at that time and thus benefit from their first-hand experience. They can probably provide some advice on how to watch out for and avoid pitfalls, and make suggestions on how they would proceed if they were in your situation. It is important that the entire CAD/CAM committee should visit these sites together, and not just the odd people who happen to be available on that day. The reason is that afterwards the whole committee can discuss and learn as a team to the benefit of the project which belongs to every member who takes part in it.

This CAD/CAM education process plays a very important role right from the beginning as it enables everyone in the committee to have a better understanding of CAD/CAM. Thus, members can appreciate the powers and limits of CAD/CAM system capabilities and will have realistic expections for the final chosen system.

Prior to producing a specification of CAD/CAM requirements, the first major task is to thoroughly review the flow of activities for a product through the company. This task might have already been performed to some extent during the justification stage, but it might need to be repeated in more detail. The review process can start from the design office, through the drawing office, and to the shopfloor. It is necessary to investigate all activities of design and manufacture to see if CAD/CAM can advantageously replace or improve existing methods. The investigation can even be extended into the sales department. For example, the sales staff might need to tender for contracts. If a CAD/CAM system is used, high quality tender drawings can be produced, and tenders can be made more frequently with more options in the proposals.

Obviously, the potential CAD/CAM application areas are enormous and often depend on the product-related activities of the company. The following are some of the typical areas for engineering companies:

(a) Design
        draughting
        schematic drawings
        design analysis
        standard components
        parts listing
        coding and classification
        design testing, etc.

(b) Manufacturer
        NC programming

jig and fixture design
sheet metal developme nt
process planning
robot programming
factory layout
production control etc.

At the end of the review of company operations, a specification of CAD/CAM requirements is compiled from the information collected. This specification is important for evaluating a potential CAD/CAM system because it defines the selection criteria which should be specific and will not be changed subsequently unless it is very necessary. If the requirements are not specified properly at this stage, the danger is that everyone will become confused after looking at many different systems and there are no guidelines by which a potential system can be judged.

The specification should contain clear descriptions outlining the existing activities and information flow, together with the functions required from a CAD/CAM system. They should be expressed in such terms as total drawing throughput, type and quantity of standard components, total component output, interfaces to other CAD/CAM systems, interchange of data to and from other systems, etc. It would be helpful if the time scale and the sequence in which the functions are to be implemented could be specified as well. There is a common mistake in trying to specify the requirements in terms of hardware such as computer system capabilities, disk capacities, graphics screen sizes, number of workstations and so on. This kind of specification is to be avoided at this stage as the hardware needed to form a CAD/CAM system whose perform-ance meets closely the requirements will gradually become more well defined during the course of system evaluation.

Once this specification of requirements is completed, the selection pro-cedure can begin by sending for information from various CAD/CAM suppliers. It is important not to pre-judge suppliers too much at this stage, although by now there is a fairly good idea of the kind of facilities required. Just go ahead and write to different suppliers who may have a suitable system to offer. Who knows, there may be a surprisingly good bargain to be had. Names and addresses of CAD/CAM suppliers can be obtained from exhibitions and magazine articles and by word of mouth, etc. From these sources, build up a comprehensive list of potential suppliers to whom an introductory letter is sent together with a copy of the required specification and invite them to reply in writing with an outline of a proposed solution. Ask them to give a rough idea of the scale of the budget and the time scale for the decision, purchase and installation. Other extra information that is useful includes the size of the supplier in terms of staff, number of installations and industries in which they are used, typical entry cost for a system with a single workstation, ability for future expansion and the variety of application programs offered with the system. A large proportion of the suppliers approached will respond probably within about two weeks, some might decline the invitation and some might not even reply at all. Those responses that come back should normally be about two to three pages long briefly describing the proposed system to satisfy the stated requirements.

From these replies, form a long list of suppliers. It is very likely that this list contains far too many suppliers for detailed evaluation individually. By now a set of minimum facilities required has already been established, a sensible method to proceed is to compare this batch of system proposals with the minimum requirements and eliminate any suppliers who cannot really meet them. This approach may seem crude and ruthless, but it is a convenient way to cut down the list of suppliers to a realistic and more manageable length. The list should ideally come to no more than about ten suppliers because it is hard to imagine that a large number of suppliers can offer a suitable system with the right combination of capabilities at a given budget allocation. The resultant list should not include suppliers who provide systems that are at both extremes of the capability spectrum because the evaluation should only concentrate on one particular band of capabilities. If this is not the case, someting has gone wrong somewhere along the line. Maybe, the specification of requirements was not prepared correctly at the initial stages, or it was not properly used as a filter during the process of eliminating unsuitable suppliers.

Then invite all suppliers on the updated list to come to the company for a preliminary meeting which should not last more than two hours. In this meeting, a brief presentation should be given on the company, its products and requirements. The supplier should introduce their company, outline the proposed system and explain the reasons why this system can do the job. It is worth emphasising once again that all the committee members must attend these meetings which are documented for future reference at the end of the meeting after the sales people have left.

After having met with the supplier for the first time at the user company, the next step is to make a visit to each supplier on the list. Prepare to spend a working day at each company to really try to find out about them and their products. The day should be used to give the supplier a chance to present the organisation of the company, their product strategy and, in particular, the proposed system in detail. This formal presentation should last about a couple of hours, with plenty of question time afterwards, and should be followed by a general demonstration of their system. Inevitably, the demonstration will be well prepared and rehearsed beforehand to give the best possible impression of the features on the system such as operation methods, speed and facilities etc. Even though this is a canned demonstration, it is still not a good idea to miss it. At some points during the demonstration, you might like to stop and ask questions that are relevant to your own applications. Very often, the supplier will not be too keen to sidetrack from this standard demonstration, so nominate a member to take down questions and notes so that they can be asked and discussed later on. Normally, there should be time scheduled at the end of the day to summarise what has been seen and to discuss any specific requirements and outstanding problems.

When all the prospective suppliers have been visited, the collected information will have to be thoroughly analysed in order to reduce the length of the list further into a short list which typically contains between three and five names. Then, each supplier that appears on this short list will be examined in greater detail in subsequent stages.

To compile this short list, some methods must be devised to compare the systems proposed by various suppliers. As each user company has different CAD/CAM requirements, it is difficult to outline a general standard method for comparing CAD/CAM systems. However, some experts advocate the use of benefit/cost ratio which is some sort of quantitative measure to relate the value of a system to its cost. The benefit/cost ratio is not really an accurate measure of the relative 'goodness' of a system because of the inherent imperfection that it is the sum total of items determined by the objective judgements of the members of the CAD/CAM committee. Despite this unavoidable shortcoming, the benefit/cost ratio provides a systematic and quantitative way of handling the selection process.

To calculate the benefit/cost ratio for a system, the cost and the estimated benefit of that system must first be obtained. The total cost of a system either is given or can be deduced from the supplier's system proposal, so it is simple and there is no great problem. The difficulty comes when the benefit is to be estimated because it cannot easily be quantified. The common procedure to work out the total benefit of a system can be simplified and explained as follows. The specification of requirements compiled in previous stages must contain a long list of CAD/CAM features that the company needs. Some of these features might have been marked important, less important or desirable, depending on how much weight is assigned to each feature. Groovers and Zimmers [3] suggest a fairly comprehensive list of features for selecting a CAD/CAM system. Typically, these features can be categorised under the headings of cost, quality, after-sales support, programming, system management, system upgrades and developments, standards, applications and so on.

For each required feature listed, give it a score out of 10 points, or some suitable scale, for every system according to the committee's assessment of its performance with regard to that particular feature. The assessment will be essentially based on the information provided by the supplier and from demonstration. Then the score is multiplied by a weighting factor which depends on the importance placed on that feature by the committee. The product of this multiplication is taken as the real score of the system for that feature. The whole assessment procedure is carried out for every feature required and for every system considered. The total benefit of a system can be determined by adding all its real scores for different features. Although this is an organised and quantitative method of evaluating CAD/CAM systems, the main drawback is that the score does not represent accurately the benefit of a system because a certain amount of arbitrariness and objective human judgements are involved. Nevertheless, it provides a means of reflecting the relative strength of various systems, and thus enables a rough comparison between them.

After the determination of the cost and benefit of a system, its benefit/cost ratio can now be easily calculated by dividing its estimated benefit by its total cost. When this ratio has been worked out for all the systems, the top three to five systems are selected into the short list. In awkward cases, where a decision has to be made on systems with very close benefit/cost ratios, the scoring may have to be reviewed for those systems because of the inherent inaccuracies of the assessment method. This will ensure that only the most suitable systems are included in the short list for further evaluation.

Once the short list has been drawn up in the order of their benefit/cost ratios, the next step is to arrange visits to the users of these shortlisted systems. Find out the names and addresses of existing users from the suppliers, especially those with similar applications to your own. If a user group exists for a system, contact the group and explain your situation so that they can put you in touch with some existing users who are members of the group. The purpose of these visits to user sites is to provide an opportunity for you to see the system in live action and performing some useful tasks for a company. In addition, you can talk to the people there about their experiences of using the system. A lot of valuable information can be gained from their comments because they have actually been through the process of implementing the system, overcome the problems involved and used it for their work. They surely have a better feel of the tasks you have in hand, and probably in a better position to give advice. For example, they can point out the strengths and weaknesses of the system, the problems encountered during delivery and installation, quality of after-sales support and so on. All these are some of the crucial factors to help you make a decision on the choice of system. Another advantage is that they can give you guidance on designing a benchmark test of the system for the final stages of selection.

From these visits, a better understanding of the system is obtained, and maybe some kind of scoring scheme is devised to give each system some points, depending on the opinion of the committee on its suitability. The positions of the systems in the short test are then adjusted according to their combined scores of benefit/cost ratio and those from the visits. Normally, these visits are mainly used to confirm their positions in the short list which should remain more or less the same, even after taking into account the results of these visits, unless something was seriously wrong with the initial assessment.

When the short list has been confirmed or rearranged, whichever the case may be, after the visits, the next stage can then begin for the benchmark test. As benchmark tests are an expensive and time-consuming exercise for both the supplier and the prospective user, they should only be carried out when the prospective user is serious about purchasing a system. The purpose of the benchmark is to validate the claims for a system by its supplier. The supplier should realise that if a benchmark test on their system is requested, it is very close to a purchase order, depending on the success of the test. Usually, a benchmark test is performed on the system that comes first on the short list. If the benchmark test is satisfactory, the system is selected. If not, the second system on the short list is benchmarked, and even if this fails, the next one on the list is put through a benchmark test and so on until one system passes the test.

preparing a good benchmark test is not an easy matter, and here are some suggestions that are well worth noting. The main objective of a benchmark test is to determine how effectively a system will perform with the type of work required by the prospective user. A benchmark test is different from an ordinary demonstration which merely shows the features outlined in the system specification. It should basically be a comprehensive exercise typical of the tasks for which it is intended to be used after purchase. For example, if the company designs and manufactures products based on the use of families of parts, the benchmark test should include the creation of families of parts for only one of different item types. The technique used for creation should be left to the supplier, but it should take place live. An assembly is formed by calling together appropriate individual parts and a parts list compiled. Other information such as dimensions and annotation should be added and the drawing produced on the plotter to your specification. During the test, maybe ask to enlarge certain portions, cross-hatch sections, show isometric views, modify and delete items, etc.

The above example is a very simple test and serves as just an indication of the ease and speed of drawing creation. The exact content of a benchmark test will vary according to the nature of the applications for the system. It might take the form of, say, NC part-programming, or geometric modelling, etc.

One important precaution is to make sure that the system under test is a similar configuration to that being proposed. In other words, it has the same processor, same number of, or more, workstations, same number of, or fewer disk drives of same or lower capacity, same version of the system and application software, and same graphics peripherals. If they are different in any way from the proposed system, ask for the reasons and make allowances for these differences when examining the results of the benchmark test.

Most people seem to agree that a benchmark test should ideally last no more than two days; but they appear to have differing views as to whether or not full details of the benchmark exercise should be given to or discussed with the supplier in advance. Quite honestly, there is no simple answer to this problem, but the advice is to use your own judgement on what is involved in the benchmark. Information about a benchmark test should not be disclosed if, by telling the supplier beforehand, it will enable them to make preparation for the benchmark to the extent that it will affect the intermediate stages of the process needed to achieve the benchmark. If only the end results, irrespective of the methods used, are of interest from the benchmark, it does not matter too much if the supplier knows prior to the actual benchmark test.

An important aspect of a benchmark test is to gather time information about a system. The time taken by a benchmark is commonly used to compare with that taken by the existing manual techniques. This time comparison can reflect the relative advantage of the two methods. One thing about this benchmark result is that it depends to some degree on the levels of skill of the operator. If an experienced operator is employed for the benchmark, the result obviously tends to be better. On the other hand, if a novice operator is used as when the company first buys a system, the time taken will be relatively long. Thus, suitable adjustments will have to be made to the results when assessing them. Another factor worth considering is how susceptible the speed of the system is to the number of tasks running on it simultaneously. Some systems might work very fast when there is only one task executing on them, but the response becomes drastically reduced when they are loaded with several other tasks. One way to check this is to stimulate a loaded system by initiating a given task at several workstations at once and measuring the completion time compared with that when a single workstation is running the task.

19.5 INSTALLATION

Selecting a CAD/CAM system is a long and tedious process as described in the previous section, but installation is not a trivial task either. Careful thought and planning is needed to ensure the eventual success of the installation. This is a crucial stage of the implementation as serious problems can arise later on if the installation process is not properly carried out.

Before the system is actually delivered to the company, adequate preparation must be made in order to allow a smooth, efficient installation and operation. For a start, a suitable site for the system must be allocated within the company. The chosen location must be large enough to accommodate the equipment in the configuration of the system. Besides the size of the site, the access to it is also very important. Easy access can save an enormous amount of time and trouble in physically moving the system off the lorry and into the office. Other considerations with respect to access are the height and location of the external loading door, and check whether or not there are any restrictions such as narrow passages, bends, or obstructions along the route to the site.

Once a site with suitable size, location and access has been chosen, its layout must then be planned appropriately in order to arrange the different pieces of hardware of the system on the site in an efficient manner. At this point, it may be useful to seek the advice from the installation manager or supervisor of the system supplier who can provide guidelines on the most suitable layout for the particular configuration of the CAD/CAM system.

The exact dimensions and weights of the individual units usually required in order to plan the layout accurately. Apart from the basic hardware of the system, items such as storage cabinets for listings, magnetic tapes, disk packs and other accessories must be considered as well in the planning. If purchase of extra hardware is planned for the near future, special space allowances must also be included in the layout.

When planning the layout, it is vital to bear in mind that people do not want to be disturbed when working, but sometimes people on the same design project might like to work close together and discuss their problems. The layout must be planned in such a way as to be able to cater for both needs of personal privacy and group projects.

Computers, plotters and lineprinters are usually very noisy when operating, so it is advisable to site them somewhere well away from the design area, and, if possible, partitioned off. The partition often has an inspection window through which the system and peripherals can be observed to check if, for example, the lineprinter is printing, or the plotter has completed a drawing. An additional advantage of housing the plotter and lineprinter within the partition is that it reduces the traffic in the design area of people going to get their drawings and listings. This human traffic can be an irritating distraction to some users.

Figure 19.1 illustrates an example layout of installtion which can probably satisfy most of the above conditions. Workstation 1 has a much larger area which is suitable for working in a group. It is also situated in the corner of the room, so their discussions will not cause too much distraction to other users. Individual users can use workstations 2–4 which are sectioned off to provide some privacy. In this example, the partition for the system manager's office will also have an inspection window so that users coming through the door will be inspected to ensure that access to the CAD/CAM facilities is limited strictly to authorised personnel. This kind of security is essential and necessary to prevent

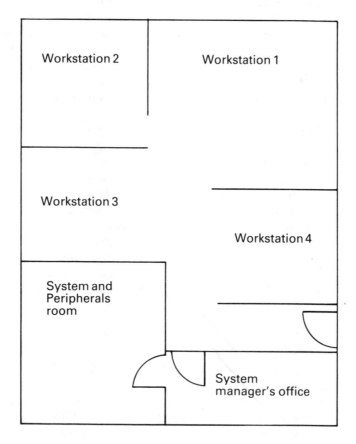

Fig. 19.1 – An example layout of installation.

any unauthorised use of the system. More importantly, if the work being designed is of a sensitive and confidential nature, it can help to prevent the information from being stolen or copied.

Lighting in the working area is always an important factor. Results have shown that it is quite common to develop eye-strain for people working for prolonged periods at workstations which are not suitably located and illuminated. It is sometimes difficult to see the screen clearly because of the problems of reflected glare from sunlight through some external windows or pale-coloured walls and floors. An ideal lighting scheme that can avoid these problems is one which has dimmer switches to adjust lighting directly above or behind the user workstations, dark-coloured furniture and surroundings, and probably drapes or blinds on all external windows to diffuse the light.

Although raised flooring is not a necessary requirement for the operation of a CAD/CAM system, it does help in cabling and cooling the installation. Also, as all cables are hidden under the false floor, there is no such potential hazard as trailing cables on the floor which can cause people to trip over and fall. Accidents of this nature can occasionally produce serious bodily injuries.

When planning an installation, it is worth remembering the importance of environmental control for the system. A system can run reliably only if it operates in the right environment. Any problems of heat, humidity, dust and electrical interference can often influence its performance. To overcome these problems, most systems are placed in air-conditioned rooms with air filtration units that purify air. Humidity regulation systems must also be incorporated to prevent condensation and build-up of dry air which will cause a static electricity problem.

In general, the power supply to the system usually comes from the mains and has to be stable and interference-free. Any variations or sudden fluctuations in the electrical supply can cause the system to fail. To avoid these problems due to electrical instabilities, the cabling system must be specially designed to resist interference of any kind and to reduce exposure to spurious noise. One solution is to isolate the cabling from any electrical device or installation by using raised flooring which provides a sealed protective invironment, minimises the length of any cabling run and facilitates maintenance.

For any installation, it will not be complete without some kind of safety and fire precautions. Adequate protection against accidents should be provided to comply with the normal fire and safety regulations.

19.6 CAD/CAM PERSONNEL

No doubt, a successful implementation of a CAD/CAM system depends very much on the technical capabilities of the system hardware and software; but it must not be forgotten that it is still the intellectual effort of people that design and manufacture a product. Therefore, a CAD/CAM system is virtually useless without qualified and skilled staff to operate and run it. Staffing considerations are an all-important factor contributing to the success of a CAD/CAM implementation; so CAD/CAM-related personnel must be carefully chosen and proper training programs must be provided for different levels of personnel. Additionally, the psychological effect on people of changing from traditional drawing boards to graphics displays must be considered, and the trade union will inevitably have to be consulted on matters concerning staffing.

In general, each installation of a CAD/CAM system will require a manager to be in overall charge of the running of the system and some operators or designers to use the system for productive work. This is the minimum requirement of staff to man a CAD/CAM operation, although extra personnel may be needed depending on the size of the particular installation and the organisation of the company concerned.

The function of the system manager is to look after the basic day-to-day operation and efficiency of the CAD/CAM facility. A person in this position should have a good general understanding and working knowledge of all the system facilities, although not necessarily to the same standard as an expert in any specific area, such as geometric modelling and NC part-programming. Typically, a system manager will have to be responsible for the security, accounting and maintenance of the system apart from managing it.

As far as the security of the system is concerned, the manager must ensure that only authorised personnel are allowed to use the CAD/CAM facility. Most systems allocate a unique code to each individual user as a password to gain access to the system. The log-on and password procedure can be used to prevent unauthorised access and usage of the system. The system manager should also be responsible for making sure that no data in the system should be lost or corrupted, either intentionally or unintentionally. For example, in the case of an integrated CAD/CAM system, information created by one department will often be used by another department. If the second department wish to modify the information for their own purposes, they should be allowed to do so only on a copy of that information and not the original. All system software and work must be backed up regularly on magnetic tapes to safeguard against any loss of data through disk or system failure. It is a good practice to develop some standard procedures for backing up the CAD/CAM system. These procedures can be semi-automatic or manual, and their frequency varies from installation to installation, but is typically daily, weekly or monthly, depending on the speed of valuable work produced and stored in the system.

The accounting of system usage is also an important function of the system manager. It is necessary for obvious reasons to keep a record of how long the system is actually used and for what it is used. This information is useful for estimating the cost of a particular phase of a job and for calculating the utilisation of the system to determine whether or not an extension or upgrading of the current system is desirable.

Another responsibility of the system manager is to make sure that every user can use a workstation whenever they wish. Usually, there will not be enough workstations for everyone at the same time, and, as the demand by each user varies from time to time, a booking system is sometimes employed to schedule the use of the workstations to achieve the maximum utilisation of system facilities. Also, this method will ensure that everyone will have a fair share of system usage.

Operators or designers are generally needed to use the system for productive work. The number of operators required for an installation depends on the size of the system and the workload of the company. Their function is essentially to produce work using the CAD/CAM system. It is not easy to describe the attributes of a good CAD/CAM operator, but basically the right attitude is very important. Obviously, some experience with the manual method is advantageous, but there must be an enthusiasm to learn and use the latest technology. Good operators are often self-motivated, innovative, and keen to try out new things. It seems that operator proficiency has little correlation with education, work experience, or age.

Generally, potential operators should be chosen from within the company because these people are already familiar with the working procedures and are usually able to make a smooth transition to CAD/CAM. The selection of these operators should be carried out well before the equipment is delivered to allow time for training and system start-up.

Normally, turnkey system vendors will offer some sort of formal training course which may be included as part of the package arrangement when purchasing the system. This training can take place at the user's site or at the vendor's site, whichever is more convenient to the parties involved. As there are start-up problems, interruptions, and other distractions at the user's site, it seems advisable to provide the training at the vendor's site which has well-maintained systems already in top running order, a well-organised instruction environment and, most probably, the vendor's best instructors available on hand to answer any questions.

Training courses are offered for various levels of personnel and usually last for one or two weeks. They consist of structured class and practical work using the vendor facilities. Courses are often available for basic, intermediate and advanced operators, in addition to those for programmers and managers. Generally, all operators should attend the basic course to develop at least an elementary working knowledge of the system. The basic course covers entry-level skills such as how to operate the various workstation components, how to issue correctly function commands, and how to produce simple two-dimensional drawings. When basic operators have more than several months of working experience with the system, they are ready to start the more advanced operator courses which include NC tape generation, finite element model construction, three-dimensional draughting and so on. Application programmers should be trained to use specialised design or manufacture software packages. Typical applications are mapping, piping, mechanical design, numerical control and printed circuit, etc. Programmers are important to maintain the application software and to enhance it if the need arises. For the system managers, courses are available to help them plan, install and manage a CAD/CAM system. Course topics commonly include workstation layout and access, personnel selection and training, CAD/CAM career structure and productivity considerations.

The timing of personnel training is also an important factor. Operators should be trained at the same time as the system is being installed, and the training should finish just before the system is set up and running properly. Scheduling the training in this way will ensure that the momentum and knowledge acquired through the course will not be lost. During the initial period after training, new operators should be given some simple exercises that are similar to the actual design and drafting tasks. This provides an opportunity for them to 'play' and familiarise with the system. At this stage, the new operators should be encouraged to be more inventive and adventurous to try out various different functions of the system. When the operator is confident enough to start production work, a good first task is to develop a simple library of standardised parts from existing engineering drawings.

According to the results of some survey on CAD/CAM operator training, people of diverse backgrounds can learn to become proficient in operating CAD/CAM systems. Such people include engineers, draughtspersons, illustrators, technicians and clerks. The survey results also indicate that less than 10 per cent of people fail to develop even marginal skills. The reason for this is that they are not interested in learning the system and how to use it. Another 10 per cent somehow do not have any aptitude to become more than marginal operators. There are about 40 per cent who develop into good operators, and the rest, also about 40 per cent, go on to become expert operators.

Figure 19.2 shows a typical operator learning curve. It can be seen from this curve that a beginner operator passes through four phases of training during the first 30 months. The first phase is the basic training that lasts about 3 months, during which the operators get themselves acquainted with the equipment and develop symbol libraries, command menus and operating procedures. They are not likely to make any contributions to engineering or design productivity, unless they are extremely talented or previously trained. Gradually, operators should become reasonably familiar with the command language of the system between about 3 and 9 months. During this second period, the basic symbol libraries and operating procedures should be set up; a small amount of time-saving over previous techniques can be seen. In the third phase which is between 9 and 15 months, operators should come to understand the system fully, make short cuts to production and be able to produce some complex drawings with little difficulty. By this time, significant productivity increases should be evident. During the fourth phase of over 15 months after the third phase, further time savings can be achieved. These time savings are due to improvements in system utilisation, task prioritising, and staff organisation, and not so much from increased operator speed and efficiency. Another factor is that additional software and hardware features will have been incorporated into the system to allow extra tasks to be handled.

No matter how powerful a CAD/CAM system is, it will not be of much good to the company if the people who use it are not adequately motivated to cooperate and make it a success. After all, it is people who influence the design and manufacture of a finished product. Nowadays, most technical people such as engineers and designers are generally excited by computer graphics and look forward to using a CAD/CAM system. They have frequently come across the

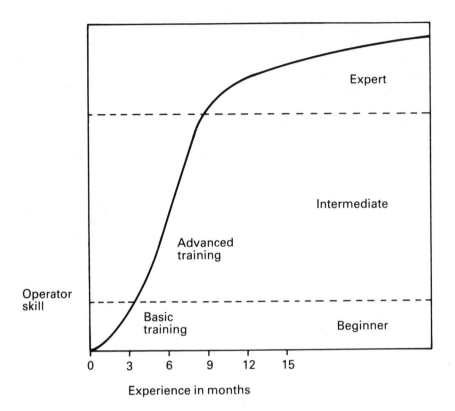

Fig. 19.2 – Typical operator learning curve.

coverage of CAD/CAM in the press and at trade exhibitions. Most have even seen the increasing number of installations in nearly all segments of industry and realise that it is just a matter of time before they get involved with CAD/CAM. Indeed, many professionals welcome the opportunity to learn more about it and sometimes try actively to collect information about the technology on their own. However, traditional attitudes die hard and it is important to make it clear to the people involved that CAD/CAM is not intended to challenge their skills but to supplement them.

The installation of a CAD/CAM system may create certain negative effects at a company. For example, potential CAD/CAM personnel may feel insecure and have doubts about their ability to learn the necessary skills. In some cases, they are not used to working with a CAD/CAM system and worry about causing damage to its hardware and software. Therefore, as soon as a company starts on the CAD/CAM route, it must invest in the confidence of its employees in order to overcome the initial psychological impact. These are mainly fears that can be easily allayed through an effective training program to educate and reassure the people. In general, most of these fears are dispelled as the employees discover

CAD/CAM to be friendly and forgiving. When necessary, any outstanding points may have to be clarified with the union if there is one at the company. It is vital that the majority of problems has been resolved in advance, and there are no real surprises for anyone.

REFERENCES

[1] Besant, C. B. *Computer-aided design and manufacture,* 2nd edn, Ellis Horwood, Chichester, 1983.
[2] Smith, W. A. *A Guide to CADCAM,* The Institution of Production Engineers, 1983.
[3] Groovers, M. P. and Zimmers Jr., E. W. *Computer-aided design and computer-aided manufacturing,* Prentice-Hall, 1984.
[4] *CAE systems and software annual,* Penton/IPC Publications, 1982.
[5] *CADCAM International,* Sept.–Nov. 1983, July and Oct. 1984 issues. EMAP Business & Computer Publications Ltd.
[6] *Engineering Materials and Design,* CAD Series, Feb.–Sept. 1982, Apr.–Sept. 1983 issues. Business Press International Ltd.

20

Implication of CAD/CAM for industry

We have seen in the previous chapters how the computer can be used to aid both the design and manufacturing process. The discussion so far has been biased towards the technical aspects of the applications of minicomputers and micro-computers in design and manufacture. The wider implications of CAD/CAM to industrial organisations will now be presented. This will include a discussion on the relationship between CAD/CAM and productivity, process planning, materials handling, inspection and quality control and selective assembly.

The computer used in the CAD/CAM context is the tool that makes it possible to collect, organise, analyse, transmit and store vast amounts of data and bring order and efficiency to manufacture. CAD/CAM can result in the removal of time consuming and mundane work from humans and is the route to a more productive, more creative and more easily managed manufacturing system.

There are two approaches to the introduction of CAD/CAM in a firm. One approach is to computerise the individual steps in the production process. Factories can then adopt these subsystems, in most cases, one at a time. The other approach is to tie all the subsystems together into a total integrated system. Of the two, the latter approach leads to the more difficult, longer and more crucial route. It should be remembered that CAD/CAM is a broad general concept and like all useful theories it is growing and in some cases changing.

The various parts of CAD/CAM fit into three general areas: design/draughting, planning/scheduling and fabrication/machining. There are of course links from these general areas in marketing, accounting and other ancillary activities within a firm. Fig. 20.1 gives a general picture, showing how the various activities are

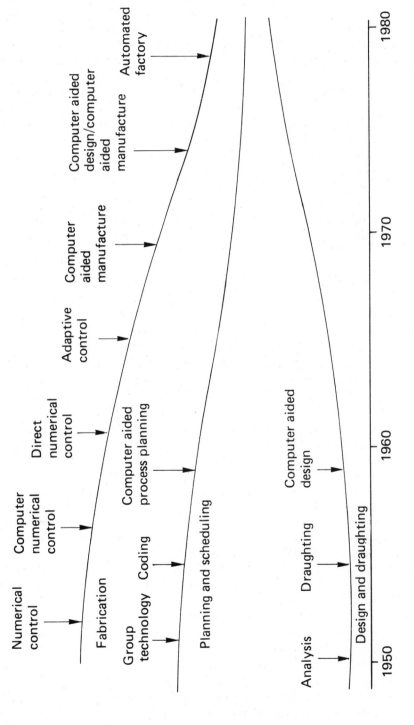

Fig. 20.1 — The growth of computer applications in the factory.

related. It also illustrates another important aspect of CAD/CAM, that is that CAD/CAM systems tend to blur and erase the company structures that usually separate the marketing, engineering and production departments. This change can occur without rearranging office walls or the layout of the factory. The sharing of essential information with other departments will be fast, more accurate and, to a large extent, automatic. There will be one common source of manufacturing information — the database — that can be kept current with relative ease an quickness; thus communication will be improved.

There are naturally many obstacles before CAD/CAM techniques can be introduced into an organisation. If these obstacles can be overcome then benefits to a firm can accrue since departments will function more closely with each other and the whole firm will react as a single entity. It is the ability to consider a product from the idea through to its ultimate production that is important, and this is where CAD/CAM can play a vital role. The goal of CAD/CAM is to make all manufacturing functions interdependent and interactive, the coordination being performed by a network of computers.

20.1 CAD/CAM AND PRODUCTIVITY

There are now many industrial experts who agree that technology is the most important factor contributing to productivity. Other factors such as government regulations, the attitudes of the work force and unions and variations in the money supply are also significant. Modern technology clearly has the best chance for improving productivity once it has the support and commitment of all those involved. It should be remembered that CAD/CAM is only one aspect of modern technology. Good research and development, modern and effective management, new materials and new techniques in production can outweigh CAD/CAM techniques in impact on productivity.

However, CAD/CAM can be singled out as a very promising area for improving productivity. Productivity is a ratio of labour, capital, raw material, energy time (input) to a real usable product (output). If this ratio is to be useful, both terms in the ratio must be easily measured. Furthermore, if these terms are used then they must be used consistently.

In the past, the tendency has been to examine small segments on each side of the ratio because they were easy to measure and compare. A good example is parts per hour for a given machine tool. While this ratio may be a sound basis for comparing machine tools it is not particularly useful for showing how a machine shop is making the best use of men, materials and machines. Many advances in machine tool design, such as electronic control, spindle speed and cutting tool materials, have resulted in improvements in productivity of individual machine tools but added little to the overall efficiency of the machine shop. In many cases the final result of these advances is the workpieces spend a longer time on pallets and less time on the machine. The number of finished items going out of the machine shop has been the same. However, CAD/CAM promises to have an impact on the total process of production. For this reason CAD/CAM and overall productivity will be closely linked. Another way of looking at the impact of

CAD/CAM is to consider the amount of time a workpiece spends waiting to be worked on. In most machine shops a part will only spend 5 per cent of the manufacturing cycle on the machine, the remaining time being spent on the shopfloor. This 95 per cent represents the time that the part is being moved from one work station to another, and for all this time the part is idle. Moreover, the relatively short time the part is on the machine can be broken down into setup and tooling versus actual metal removal or metal forming. These precentages are typically 70 and 30 respectively.

CAD/CAM, however, encompasses 100 per cent of the manufacturing process. What numerical control did for the productivity of the individual machine tool CAD/CAM can do for the entire machine shop or the entire production unit.

20.2 NC AND COMPUTER GRAPHICS

The start of the CAD/CAM process can be found in NC. Firstly, at an early stage, the computer played an active and essential role in NC. Secondly, the NC part-program was an obvious nucleus for instructing a file of manufacturing data — the database. Thirdly, NC provided highly flexible control of the individual machine tool.

NC programming is essentially a matter of encoding machine tool instructions derived from mathematics and performing the mathematical calculations quickly. Without the computer NC would not have become a practical tool. We have seen from previous chapters that the NC machine tool is tedious to program without some computer aid. The APT system was developed on a main frame computer so that the geometry of parts could be specified along with the relevant machining data in order to permit the speedy generation of the CL tape for the machine tool.

The next step was to examine how the users of computer graphics systems were generating shapes, particularly for mechanical engineering draughting applications. It was soon realised that designers were using computer graphics as an effective aid to generate complex geometries and that the data was in computer form. The next logical step was to pass this data which defined the geometry over to a program which could generate the manufacturing information. The joining of these two activities resulted in the necessity for a common data-base for design and manufacture and out of this marriage grew the CAD/CAM concept.

It is now realised that, as a form of communication, drawings and sketches are not very flexible and play a less important role in the overall design and manufacturing process. Storing information on paper is inefficient, handling many drawings is cumbersome, finding data from drawings is time consuming and as a means of transferring data from one process to another is outdated.

Computer graphics on the other hand is a revolutionary tool for the designer or draughtsman. Data and drawings can be recalled from files at high speed and displayed for viewing. New graphical data can be generated at high speed with the ability to simulate movement. For example, cutter paths can be generated and displayed in a simulation mode. Data for manufacturing processes can be simply and speedily transferred from one department to another.

Take, for example, a complex shape such as a turbine blade. It can be well defined by a computer graphics system, and in far more detail than is possible with a conventional drawing. The computerised data is then in a form that can be effectively used for generating the data for manufacture. Thus the transmission of information is akin to the telephone rather than the latter.

20.3 THE PLANNING OF MANUFACTURING PROCESSES

When the database for manufacturing information has been established, building an integrated CAD/CAM system is mainly concerned with the linking of various functions which have been computerised. Creating such a network may not be easy but the building blocks for such a system exist in some form in many organisations.

The first stage in the overall design and manufacture process is the product design. During this stage decisions are made on how components are to be made and assembled. Often detailed planning of manufacture will take place after the initial design process. Then the materials for the parts must be made available. These latter stages are ones of planning and outlining designs. Here, computers can play a significant role since the planning stage is often complex and time consuming.

Process planning refers to making an outline of the steps required to produce a finished part. Most modern factories employ sophisticated planning techniques in order to optimise production and many have introduced computers to aid this process. Computers enable the planning staff to assess the best method of production giving speedy access to machine tool types and characteristics, machine availability and so on. This new approach is a necessity today as work-pieces are much more complex and more precise. The materials are different and quality is often more demanding. Many kinds of machines, many different operations and many different workers may be involved in producing the finished part. Without a set plan to follow this production would be impossible.

All process plans must meet basic requirements. The plan must identify the part and give basic information about it, such as the material and the form in which it is provided. The plan must list all machines or processes in the order they are to be used. Finally the plan should specify the operation or requirements to be performed with each machine or process.

There are at least two ways of applying computers to this essential step of process planning. One is called variant process planning, the other generative process planning. The variant process planning method consists of producing new plans based on old ones, varying these to fit the new work in hand. The computer programs are based on information relating to past workpieces or similarly shaped parts that make up a collection of standard process models prestored in the database.

Generative process planning, relies on a much more powerful software program. A generative system develops a process plan based on the part geometry, the number of parts to be made, the material and information about available facilities in the factory. Using current information in the database, a generative system then determines the best machine, the optimum tooling and fixturing and finally calculating cost and time as it proceeds. Accurate and detailed coding of each part is essential. Generative process planning is more complicated than variant but is more flexible and potentially more useful.

Both methods of computer aided planning can be used to automatically plan the handling and procurement of raw materials as well the scheduling of production.

20.4 MATERIALS HANDLING AND CONTROL

Materials handling and inventory control are some areas where the use of computers is common and straightforward. They show the advantages of a computer as a high-speed filing and record-keeping system.

A process plan indicates that a certain supply of raw material must be ready for production to commence. If the request for material is generated and transmitted by computer, it will arrive at the stock control department immediately. There the request can be processed by other computer programs, such as inventory control, storage, raw material procurement, materials handling, parts tracking and progress chasing. These may have been independent systems at one time but under a CAD/CAM system they will be brought together so that they can interact.

It is now generally agreed that a computerised system of inventory control can save a firm a considerable amount of money. The computer, having access to lead times or ordering and production rates, can automatically re-order new parts of raw materials. Purchase orders with the correct information will automatically be generated. The system will also take into account any sudden fluctuations in production or lead times and take the appropriate action. As goods are received, the system of payment to the vendor is automatically generated through to the computer in the financial department.

Such a computerised system allows interrogation of the state or performance of a factory at any time so that action can be taken before the consequences become serious or difficult to correct.

20.5 INSPECTION AND QUALITY CONTROL

The computer has proved to be a valuable tool for inspection when linked to any automatic gauging system. A computer can be programmed to analyse the data from a gauging system so that parts within a certain range of measurements can be accepted and passed for assembly or further work. The computer can detect trends or shifts in dimensions, so that, if a part dimension is moving towards its upper or lower tolerance, then a correction can be made either automatically or manually, via a warning system. A man would have difficulty in gathering the necessary statistics for properly detecting that something was going wrong before a parts scrap situation arose. Computers can sample reliably and quickly so that the necessary control is achieved. The computer will either be programmed to sample every part or a certain number of parts at regular intervals in order to provide the necessary control.

20.6 SELECTIVE ASSEMBLY

There are many instances in manufacture where two parts must be made so that they fit together. Sometimes the problem is acute with very close tolerances on the mating parts such as in hydraulic pistons and cylinders. A solution to this problem is to manufacture one part of the mating unit to a lower tolerance than that required and to use the computer to control the gauging of each component, determine the statistical curve of distribution and then use that data to manufacture the mating part to the same statistic distribution curve. The computer then operates an automatic assembly machine that controls the measurement, qualification and selection of two mating components that have the required fit. The end result is a much lower manufacturing cost. This system is now used widely in industries such as the automotive industry for applications like power steering components.

20.7 CAD/CAM AND THE FUTURE

The application of computer technology to the individual aspects of design and manufacture has proved to be successful for many firms. However, bringing these applications together in a common network is not so simple. If the maximum advantage of CAD/CAM technology is to be achieved then this integration is essential. Integrated systems are being developed and they tend to show two main characteristics. One is a hierarchy of computers, the other is a two-way link between members of this hierarchy.

In CAD/CAM the computer system must be more than an information development and output device. It must be truly interactive and responsive. For example, if a gauge machine shows that parts being machined are moving out of tolerance, then the machine performing the manufacture must be automatically corrected from signals being sent by the gauging computer to the computer controlling the machine tool. This closed-loop self-correcting system is the first step to integrated computer aided manufacture.

So there could be many small computers at one level controlling machines and processes with all these computers being supervised by intermediate computers. While each low-level computer will serve a different machine or function, each intermediary computer will control departments such as inventory, machine control, tooling, planning or quality control. There can be a level of computer above the intermediate computer which will control interdepartmental activities.

The ultimate integrated CAD/CAM systems are not yet available because of the lack of experienced programmers caused by a huge expansion in the computer industry due to the falling of hardware costs. There is also a further problem in the creation of fully integrated CAD/CAM systems and this is due to a lack of standards.

Other problems are associated with the management and control of the database, the updating of this database and the security of the database. Finally, there can be problems associated with and their relationship to such systems as a CAD/CAM system.

The benefits of CAD/CAM systems are now such that they can readily be measured. For example, such systems are demonstrating a shorter design cycle and a reduced lead time for getting a product into production. The cost of development is also being reduced; as more use is made of simulation techniques fewer prototypes and a smaller amount of testing is needed; and better engineering and production analysis is improved by using computers.

On the manufacturing side CAD/CAM offers greater efficiency and better utlisation of resouces, reduction of inventory through closer control and better scheduling and an improvement in health and safety through the automation of hazardous tasks.

20.8 THE TRADE UNION VIEW

Perhaps the attributes of trade unions to the use of CAD and CAM techniques will be more significant than all the new developments. We have seen how unions and management disputes in the printing industry can result in the closure of a national newspaper over the proposed introduction of computerised typesetting and printing equipment. Many companies are finding it increasingly difficult to introduce CAD techniques in their drawing offices. Unions, such as the AUEW (TASS), have looked closely at the effects of the use of CAD techniques on their members and have come to the conclusion that such methods should not be introduced into a company without full consultation with the union.

The unions see many potential dangers in CAD and CAM and these dangers are well described by Cooley [1]. The unions see a continuing fragmentation of jobs, just as the designer's job in in the 1930s became broken down into a series of specialised jobs in the 1940s, with stressmen needed to perform calculations, metallurgists to select materials, tribologists to determine the form of lubrication, and the draughtsmen to perform the drawing.

The unions think that the employers will wish to exploit their members more and more as high capital cost equipment is used in design and manufacture. They see a higher degree of specialisation resulting in less job satisfaction since fewer will be able to see the panoramic picture of the complete job. It has been suggested that, as personnel become highly specialised, employers will wish to keep them in the same job until that job disappears, when the men will be 'scrapped' along with the machines.

It is claimed that talk in industry about dedicated machines and computers is commonplace, and this means that men will become a dedicated part of a system similar to a component in a machine. Even graduates will become more specialised, needing degrees in electronics, control or heavy engineering, rather than in electrical engineering. It is thought that CAD, in some cases, takes specialisation to a very high level so that personnel may only have a short working life. The unions further claim that students are being trained as industrial fodder rather than being trained to think.

Probably the most dangerous aspect seen by many is the high burn-up rate in staff which the intensive use of CAD facilities can bring about. For example, one American company quotes 95 per cent of a designer's time as being spent in searching for information, and only 5 per cent in the actual design decision-making. The introduction of computer graphics can eliminate much of the routine reference work and can intensify decision-making by over 1900 per cent.

Thus the stress put on the man can become very great. It is claimed that, because the job requires a greater intensity of work, only personnel within a certain age bracket will be recruited into those jobs, and that as staff grow older they may be subjected to career de-escalation with a consequent lowering of their status. Redundancies often show that it is the older men who are being eliminated.

It is the change in job structures and the de-skilling of traditional labour that most worries unions. For example, the setting of machines once performed by skilled men on the shopfloor is now moving to the drawing offices where, with the use of CAD equipment, control tapes are generated for NC machine tools.

The unions also point out that, apart from all the negative effects that CAD can bring to its members, the union itself can be in a strong position because its members are becoming increasingly important to the support of highly capitalised equipment, and that therefore employers are vulnerable to strikes from a few members of staff.

Some trade union members are aware of trends in automation in countries such as Japan and the USA. They realise that competitors from these countries are using computer controlled machinery to great effect, for instance the computer controlled robots used in the Japanese car and tractor plants. The increase in efficiency that accrues from the use of such machinery results in products that are far more competitive than similar products produced by more conventional techniques involving a high labour content. It is now being realised by some that new techniques in design and production must be accepted in the long term, otherwise jobs will disappear to overseas competitors.

20.9 THE EMPLOYERS' VIEW

The employers' view of the CAD picture is not quite clear as that presented by the unions. Undoubtedly there have been a number of companies which have used high-cost CAD equipment and which have gone to extreme lengths to maximise the return on the capital invested. Rating and work-study methods were introduced to increase productivity in some cases.

The latest trends in CAD and CAM are away from that equipment which has a high cost-to-user ratio. Minicomputers and, more recently, microprocessors are resulting in the development of CAD and CAM equipment which is sufficiently inexpensive for 'idle time' to be economically tolerated. In other words, people are recognising that a designer must have time to think, or otherwise he will be forced into errors by excessive pressure of work.

Many companies that introduce CAD systems into their drawing offices do so only after careful planning and consultation with their staff. Where automatic draughting systems have been placed in the drawing office, the draughtsmen have often welcomed their introduction, because much of the tedium associated with draughting is performed by the machine, so leaving more time for the designers to think and plan the work. A useful policy is to share one CAD work station among a number of draughtsmen or designers. Each man can then plan his work and go to the CAD system for short periods where sketches and ideas can be worked up into proper designs. Adopting this approach results in a considerable increase in productivity without subjecting the staff to any severe stress. In fact, job satisfaction usually increases, because the drudgery of the job is removed and the results of a day's work can be very impressive. Very often, designers will add software of their own to the CAD system, so as to perform additional tasks on the machine.

The use of CAD may not necessarily mean more specialisation but quite the reverse. There are considerable moves away from specialisation because of the inflexibilities that result. CAD helps this move away from the specialist because a designer with a good all-round training can call on specialist programs, such as finite element stress analysis programs, which can be built into the CAD system.

University establishments, for example, Imperial College, are moving away from producing highly specialised graduates for industry and are instead offering the three- and four-year 'Total Technology' engineering degrees. These courses are planned to give graduates a wide base of engineering subjects together with up to 25 per cent of course content in management, economics, sociology and languages. During their course, students spend time working in industry, often overseas as well as in their own sponsoring firms. The students are encouraged to read engineering rather than be taught and 'spoon-fed'. The object of the university is to turn out graduates who can think and so be useful throughout their entire working lives. They do not aim to produce graduates trained for a specific job which may only last for a few years.

Computing is playing an increasingly important role in educating engineers, because their skills can be widened by giving them a greater breadth of engineering knowledge.

The manner in which CAD and CAM are used in the future depends on every person concerned with engineering; it depends on each person taking a positive attitude to their introduction and having a real concern for the people who work in the engineering world, or indeed in any other field of human activity, because this problem will not stop at engineering but will affect society everywhere.

REFERENCES
[1] Cooley, M. *Computer-Aided Design – Its Nature and Implications.* AUEW (TASS), 1972.

Index

Accumulator, 27
Adaptive control, 354
Adaptive optimization, 359, 361
Annotation, 167
APT, 15, 215
  description of:–
  accumulator, 27
  geometry statements, 218
  motion statements, 220
  post-processor statements, 224
  symbols and words, 217
Arithmetic and logic unit, 28
ASCII, 43
Assembly language, 44
Automated guided vehicles, 323
  free-ranging, 328, 330
  painted-line guided, 328
  supervisory system, 338
  wire-guided, 324

Bandwidth minimisation
  of a stiffness matrix, 183
Batch processing, 47
Bayesian model, 362
Benchmark test, 379
Bill of materials, 370
Binary, 43
Bit, 27
BCD, 43
Boundary modelling, 160
Buttons, 99
Byte, 28

Computer system configuration, 48
Cost control, 366
CAD/CAM
  applications, 22
  benefits, 21
  concepts, 15
  evaluation, 386
  implementation, 376
  installation, 390
  justification, 378
  personnel, 392
  productivity, 400
  selection, 382
  software, 276
  specification, 382
  support, 392

Database, 21, 86
Digitiser, 54
Dimensioning, 168
Direct access storage, 32
Discontinuity plotting, 211, 212
Display processor, 125
Displays, 59
Distortion, 146
Distributed Intelligence Microcomputer Systems (DIMS), 303

EBCDIC, 43
Editing, 110
Eye coordinates system, 139

Factory management, 364
FANUC 3T system, 271
Files, 89, 90, 91, 92, 134
Finding data, 101, 102
Finite Element Analysis, 23
Flex operating system, 253
Flexible Manufacturing Systems (FMS), 323, 342
  applications, 343
  benefits, 343
  definition, 342
  economics of, 345
  machines in FMS, 344
  materials handling, 345
Floppy disk, 34, 247
FUJITSU system, 241

Generative process planning, 352
GENPLAN, 352
Geometric modelling, 152
Geometry, 160
GKS, 118
Graphic transformation, 114
Grids, 276
Group technology, 347

Hard copy unit, 72
Hexadecimal, 43
Hidden-line removal, 137
High-level language, 44

IGES, 119
Integration in design and manufacture, 15
Intel 8086 microprocessor, 246
Inspection, 403
Instruction register, 27
Inventory control, 365
Isometrics, 175
Isoparametric elements, 198

Joystick, 57, 144

Keyboard, 58

Linear transformations, 132

Machinability, 352
Machinability data system, 356
Machine language, 44
Machine tool control system, 263
Macroblocks, 196
Macros, 108
Magnetic disk, 33
Magnetic tape, 32
Management system, 367
Manufacturing resource planning (MRP II), 374
Marketing, 366
Material requirements planning (MRP), 365, 368
MAZATROL system, 271

MEDUSA, 230
Memory address register, 27
Menu, 96
Mesh generation, 181, 202
MICLASS system, 349
Microcomputers:–
  choosing a system, 245
  communications, 249
  definition, 245
  networks, 242
  primary memory, 246
  systems, 240
  use in CAD/CAM, 232, 240
Motorola 68B09 microprocessor, 241, 245
Motorola 68000 microprocessor, 241
Multiprogramming, 47

NC machining, 236
Network, 49
Nodal point numbering, 183

Octal, 43
On-line processing, 47
Operating system, 79
Opitz system, 348
Overlay, 81

Paper tape reader/punch, 37
Perspective transformation, 136, 140
Plasma panel displays, 66
Plotters, 67
  drum, 69
  electrostatic, 70
  flatbed, 68
  ink-jet, 72
Primitive modelling, 158
Printers, 39
Process optimization, 354
Process planning, 347
  parts classification and coding, 348
  planning, 350
  retrieval system, 350
Program counter, 27
Purchasing, 366

Quality control, 400

RAM, 247
Raster scan displays, 64
RD Projects Batchmatic controller, 233
Real-time, 48
Robots, 297
  assembly, 304
  axis control, 306
  controller, 300, 308
  control structure, 310
  coordinate transformation, 309
  different types, 298
  graphical simulation, 317
  grasp planning, 319
  mechanical structure, 306

programming of, 313, 314, 317, 318
sensors, 308
textual programming, 315
trajectory planner, 309
Rotational transformations, 131, 142

Sales, 366
Selective assembly, 404
Sequential access storage, 32
Shading, 137
Shop floor control, 365
Sketchpad, 14
Solid modelling, 155, 157
Status register, 28
Stiffness, 182
Stress, 182
Symbols, 106

Tablet, 55
Taylor-tool life equation, 358
Terminals, 38
Time-sharing, **47**
Tool-life curves, 360

Tool path generation, 237
Topology, 160
Tracker ball, 57
Transformation, 125
Translation, 116

UNIX, 246

Voice data entry system, 59
VLSI, 13

Winchester disk, 35
Windowing, 126
Wire frame modelling, 155
Workstation:–
    CAD/CAM cutter path derivation, 260
    CAD/CAM database, 257
    database definition, 254
    definition, 252
    display programs, 260

Zilog 8001 microprocessor, 246
Zoom, 146